1001

Solved
Engineering
Fundamentals

Problems

Michael R. Lindeburg, P.E.

PROFESSIONAL PUBLICATIONS, INC.
Belmont, CA 94002

In the ENGINEERING REVIEW MANUAL SERIES

Engineer-In-Training Review Manual
    Engineering Fundamentals Quick Reference Cards
    Mini-Exams for the E-I-T Exam
    1001 Solved Engineering Fundamentals Problems
    E-I-T Review: A Study Guide
Civil Engineering Reference Manual
    Civil Engineering Quick Reference Cards
    Civil Engineering Sample Examination
    Civil Engineering Review Course on Cassettes
    Seismic Design for the Civil P.E. Exam
    Timber Design for the Civil P.E. Exam
Structural Engineering Practice Problem Manual
Mechanical Engineering Reference Manual
    Mechanical Engineering Quick Reference Cards
    Mechanical Engineering Sample Examination
    Mechanical Engineering Review Course on Cassettes
    101 Solved Mechanical Engineering Problems
    Consolidated Gas Dynamics
Electrical Engineering Reference Manual
    Electrical Engineering Sample Examination
Chemical Engineering Reference Manual
    Chemical Engineering Practice Exam Set
Land Surveyor Reference Manual
Metallurgical Engineering Practice Problem Manual
Petroleum Engineering Practice Problem Manual
Expanded Interest Tables
Engineering Law, Design Liability, and Professional Ethics
Engineering Unit Conversions

In the ENGINEERING CAREER ADVANCEMENT SERIES

How to Become a Professional Engineer
The Expert Witness Handbook—A Guide for Engineers
Getting Started as a Consulting Engineer
Intellectual Property Protection—A Guide for Engineers
E-I-T/P.E. Course Coordinator's Handbook
Becoming a Professional Engineer

Distributed by: Professional Publications, Inc.
                1250 Fifth Avenue
                Department 77
                Belmont, CA 94002
                (415) 593-9119

1001 SOLVED ENGINEERING FUNDAMENTALS PROBLEMS

Printed in the United States of America

ISBN: 0-932276-90-3

Current printing of this edition (last number): 6 5 4 3

TABLE OF CONTENTS

PREFACE

1001 SOLVED ENGINEERING FUNDAMENTALS PROBLEMS was written for practicing engineers and students majoring in engineering who have a need to review basic engineering principles. Originally, this book was conceived of as an aid in preparing for the Engineer-in-Training (E-I-T), also known as Fundamentals of Engineering (F.E.), examination. However, its potential uses are broader than just examination review.

In the old days, when examinations for engineering licensing were in their infancy, most examination review books were compilations of problems with little, if any, supporting theory. These books placed a burden on the examinee to accumulate and study numerous textbooks and references, and contributed to the "shopping cart" syndrome (wherein examinees brought shopping carts of books to the exam).

That changed with the publication of my ENGINEER-IN-TRAINING REVIEW MANUAL. For once, examinees sitting for the E-I-T examination had a comprehensive source of review material.

However, some engineers seem more comfortable with a review method that involves working out the types of problems that might be expected on the exam. A massive collection of solved examination problems is perfect for these engineers. That is why I started working on 1001 SOLVED ENGINEERING FUNDAMENTALS PROBLEMS.

Unfortunately, the current examination is a "secure examination," and problems are not made public after the administration of the exam. Therefore, it was necessary to develop these problems from scratch, based on my knowledge of the examination.

I started by writing guidelines for the problem format, level of difficulty, etc., and then outlined the areas and types of problems I wanted in the book. Then, in order to achieve the highest degree of originality and objectivity, I turned these detailed instructions over to a team of engineering students at Stanford University. This team wrote the first drafts of most of the problems in this book.

Once the writing team was finished, it became my job to weed out some of the problems. I purposely ended up with many more problems than I could use. Some of them were too easy, some were too hard, and some were not relevant. It was a pleasant challenge choosing problems that would give this book "just the right flavor" for an engineering examination review.

I hope that you will find using this book valuable and rewarding.

Michael R. Lindeburg, P.E.
Belmont, CA
July, 1988

ACKNOWLEDGMENTS

The development of this book was a team project from the beginning. The ideas for the problems were contributed by more than 20 graduate and undergraduate students majoring in engineering at Stanford University. The diversity of problems in this book is a tribute to the success of that team approach.

My thanks to Janyce Mitchell and Scott Seu, both of whom were students at Stanford University at the time, for organizing the final compilation of problems, as well as for doing the initial editing, checking of solutions, and keyboarding. The dedication and effort you exhibited were nothing short of incredible.

A big thank you, also, to Mary Christensson and Sylvia Osias of Professional Publications for subsequent typesetting and making all of the changes and corrections I requested. I know both of you struggled with the inconsistent styles of engineers, the multiplicity of engineering disciplines in this one book, and the differences in conventions between English and SI solutions.

Thank you, Yves Martin and Cindy Arnold, of Professional Publications' art department, for bringing some sense of order and consistency to illustrations originally sketched by the many different contributors. I am pleased with the quality of the illustrations and page layouts.

And, it was Lynda Schembri of Professional Publications' accounting department who originated the cover design for this book. Thank you for taking the time out of your already-busy schedule to tackle something not in your job description.

Of course, every organism needs a head to guide it. This project relied on the organization and coordination skills of Lisa Rominger, Professional Publications' production department supervisor. I don't know how you did it, but you did. Thanks!

Other people, as well, contributed to this book in their own special ways. My omission of their names should not be taken as a lack of appreciation.

What a thrill it is to see a book like this develop. It wouldn't have been possible without all of you. Thank you, again!

Michael R. Lindeburg, P.E.
Belmont, CA
July 1988

1 MATHEMATICS

MATHEMATICS–1

The set A consists of elements $\{1, 3, 6\}$, and the set B consists of elements $\{1, 2, 6, 7\}$. Both sets come from the universe of $\{1, 2, 3, 4, 5, 6, 7, 8\}$. What is the intersection, $\bar{A} \cap B$?

(A) $\{2, 7\}$ (B) $\{2, 3, 7\}$ (C) $\{2, 4, 5, 7, 8\}$

     (D) $\{4, 5, 8\}$ (E) $\{1, 2, 4, 5, 6, 7, 8\}$

The set of "not A" consists of all universe elements not in set A: $\{2, 4, 5, 7, 8\}$.

The intersection of $\{2, 4, 5, 7, 8\}$ and $\{1, 2, 6, 7\}$ is the set of all elements appearing in both.

Thus $\bar{A} \cap B$ is $\{2, 7\}$

> Answer is (A)

MATHEMATICS–2

For a given function, it is found that $f(t) = f(-t)$. What type of symmetry does $f(t)$ have?

(A) odd symmetry
(B) even symmetry
(C) rotational symmetry
(D) quarter-wave symmetry
(E) no symmetry

When $f(t) = f(-t)$, the function is "mirrored" on either side of the vertical axis. This is known as even symmetry. Thus, $f(t)$ has even symmetry.

Answer is (B)

MATHEMATICS–3

What is the value of each interior angle of a regular pentagon?

(A) $\dfrac{\pi}{5}$ (B) $\dfrac{\pi}{3}$ (C) $\dfrac{2\pi}{5}$ (D) $\dfrac{\pi}{2}$ (E) $\dfrac{3\pi}{5}$

For a regular polygon, the value of each interior angle, θ, is

$$\theta = \frac{\pi(\text{number of sides} - 2)}{\text{number of sides}}$$

For a regular pentagon:

$$\theta = \frac{\pi(5 - 2)}{5}$$
$$= \frac{3\pi}{5}$$

Answer is (E)

MATHEMATICS–4

A cubical container that measures 2″ on a side is tightly packed with 8 marbles and is filled with water. All 8 marbles are in contact with the walls of the container and the adjacent marbles. All of the marbles are the same size. What is the volume of water in the container?

(A) 0.38 in^3 (B) 2.5 in^3 (C) 3.8 in^3 (D) 4.2 in^3 (E) 4.9 in^3

Because they are tightly packed, $r_{\text{marble}} = 0.5$ in.

$$V_{\text{water}} = V_{\text{box}} - 8V_{\text{marble}}$$
$$= 2^3 - 8\left[\frac{4}{3}\pi(0.5)^3\right]$$
$$= 3.81 \text{ in}^3$$

Answer is (C)

MATHEMATICS-5

Which number has four significant figures?

(A) 0.0014   (B) 0.01414   (C) 0.141   (D) 1.4140   (E) 1414.0

The number of significant digits for each choice is:

(A)	2
(B)	4
(C)	3
(D)	5
(E)	6

Thus only (B) has four significant figures.

Answer is (B)

MATHEMATICS-6

What is the solution of the equation $50x^2 + 5(x - 2)^2 = -1$, where x is a real-valued variable?

(A) -6.12 or -3.88 (B) -0.52 or 0.70 (C) 7.55

(D) $\dfrac{5 \pm \sqrt{-275}}{100}$ (E) no solution

For real valued x, the left-hand side of the equation must always be greater than or equal to zero, since all terms containing x are squared. Thus, there is no solution to this equation for real values of x.

Answer is (E)

MATHEMATICS-7

What are the roots of the cubic equation $x^3 - 8x - 3 = 0$?

(A) $x = -7.90, -3, -0.38$
(B) $x = -3, -2, 2$
(C) $x = -3, -0.38, 2$
(D) $x = -2.62, -0.38, 3$
(E) $x = 2.62, 3, 7.90$

By inspection, $(+3)$ is a root, and $(x-3)$ is a factor. Factor out $(x-3)$

$$\frac{x^3 - 8x - 3}{x - 3} = x^2 + 3x + 1$$

Use the quadratic equation to solve $x^2 + 3x + 1 = 0$.

$$x = 3, \frac{-3 \pm \sqrt{9 - 4}}{2}$$
$$= -2.62, \ -0.38, \ 3$$

Answer is (D)

MATHEMATICS–8

Naperian logarithms have a base closest to which number?

(A) 2.17   (B) 2.72   (C) 3.14   (D) 10   (E) 16

The base of Naperian logarithms is the number $e \approx 2.7183$. Of the choices given, 2.72 is the closest to e.

Answer is (B)

MATHEMATICS–9

What is the base-10 logarithm of $(1000)^3$?

(A) 3   (B) 6   (C) 9   (D) 27   (E) 3000

By definition:

$$\log_{10} 10^G = G$$
$$1000 = 10^3$$
$$(1000)^3 = (10^3)^3$$
$$= 10^9$$
$$\log_{10} 10^9 = 9$$

Answer is (C)

MATHEMATICS-10

What is the natural logarithm of e^{xy}?

(A) $\dfrac{1}{xy}$ (B) xy (C) $2.718xy$

(D) $\dfrac{2.718}{xy}$ (E) $\dfrac{xy}{2.718}$

By definition, the natural logarithm of a number is:

$$\ln e^G = G$$
$$\ln e^{xy} = xy$$

Answer is (B)

MATHEMATICS-11

What is the value of $(0.001)^{\frac{2}{3}}$?

(A) antilog $\left[\dfrac{3}{2}\log 0.001\right]$

(B) $\dfrac{2}{3}$ antilog $[\log 0.001]$

(C) antilog $\left[\log\left(\dfrac{0.001}{\frac{2}{3}}\right)\right]$

(D) $\dfrac{3}{2}$ antilog $[\log 0.001]$

(E) antilog $\left[\dfrac{2}{3}\log 0.001\right]$

$$\log x^a = a \log x$$
$$\log(0.001)^{\frac{2}{3}} = \frac{2}{3}\log 0.001$$
$$(0.001)^{\frac{2}{3}} = \text{antilog}\left[\frac{2}{3}\log 0.001\right]$$

Answer is (E)

MATHEMATICS–12

The salary of an employee's job has five levels, each one 5% greater than the one below it. Due to circumstances, the salary of the employee must be reduced from the top (fifth) level to the second level, which means a reduction of $122.00 per month. What is the employee's present salary per month?

(A) $440   (B) $570   (C) $680   (D) $890   (E) $1190

The salary levels can be seen as a geometric sequence. Let S_i be the salary at level i.

$$S_3 = 1.05 S_2$$
$$S_4 = 1.05 S_3$$
$$S_5 = 1.05 S_4$$
$$= (1.05)^3 S_2$$
$$S_5 - 122 = S_2$$
$$S_5 = (1.05)^3 (S_5 - 122)$$
$$= \$893$$

Thus the fifth level salary is approximately $890 per month.

Answer is (D)

MATHEMATICS–13

Which of the following statements regarding matrices is not true?

(A) $(\mathbf{A}^T)^T = \mathbf{A}$

(B) $\mathbf{A}(\mathbf{B} + \mathbf{C}) = \mathbf{AB} + \mathbf{AC}$

(C) $\begin{pmatrix} 2 & 5 \\ 1 & 0 \end{pmatrix} \begin{pmatrix} 1 \\ 2 \end{pmatrix} = \begin{pmatrix} 12 \\ 1 \end{pmatrix}$

(D) determinant of $\mathbf{A}$ = determinant of $\mathbf{B}$

(E) $(\mathbf{AB})^{-1} = \mathbf{A}^{-1}\mathbf{B}^{-1}$

The inverse of a product of two matrices is the product of the inverses, in reverse order.

$$(\mathbf{AB})^{-1} = \mathbf{B}^{-1}\mathbf{A}^{-1}$$

Answer is (E)

MATHEMATICS-14

What is the determinant of the following (2×2) matrix?

$$\begin{pmatrix} 5 & 9 \\ 7 & 6 \end{pmatrix}$$

(A) -33 (B) -27 (C) 3   (D) 27   (E) 93

The determinant, **D**, is calculated as follows:

$$\mathbf{D} = \begin{vmatrix} 5 & 9 \\ 7 & 6 \end{vmatrix}$$
$$= (5)(6) - (7)(9)$$
$$= -33$$

Answer is (A)

MATHEMATICS-15

What is the determinant of the following matrix?

$$\begin{pmatrix} 1 & 1 & 1 \\ 2 & -1 & 1 \\ 1 & 2 & -1 \end{pmatrix}$$

(A) 0   (B) 1   (C) 5   (D) 7   (E) 13

To find the determinant, expand by minors across the top row.

$$\mathbf{D} = 1\begin{vmatrix} -1 & 1 \\ 2 & -1 \end{vmatrix} - 1\begin{vmatrix} 2 & 1 \\ 1 & -1 \end{vmatrix} + 1\begin{vmatrix} 2 & -1 \\ 1 & 2 \end{vmatrix}$$
$$= [(-1)(-1) - (2)(1)] - [(2)(-1) - (1)(1)] + [(2)(2) - (1)(-1)]$$
$$= 7$$

Answer is (D)

MATHEMATICS–16

What is the inverse of the matrix $\mathbf{A}$?

$$\mathbf{A} = \begin{pmatrix} \cos\theta & -\sin\theta \\ \sin\theta & \cos\theta \end{pmatrix}$$

(A) $\begin{pmatrix} \cos\theta & -\sin\theta \\ \sin\theta & \cos\theta \end{pmatrix}$

(B) $\begin{pmatrix} -\cos\theta & \sin\theta \\ \sin\theta & \cos\theta \end{pmatrix}$

(C) $\begin{pmatrix} \cos\theta & \sin\theta \\ -\sin\theta & \cos\theta \end{pmatrix}$

(D) $\begin{pmatrix} \cos\theta\sin\theta & 0 \\ 0 & \sin\theta\cos\theta \end{pmatrix}$

(E) The inverse does not exist.

For a (2×2) matrix, $\mathbf{X}$,

$$\mathbf{X} = \begin{pmatrix} a & b \\ c & d \end{pmatrix}$$

The inverse, $\mathbf{X}^{-1}$, is:

$$\mathbf{X}^{-1} = \frac{1}{\mathbf{D}} \begin{pmatrix} d & -b \\ -c & a \end{pmatrix}$$

where $\mathbf{D}$ is the determinant of $\mathbf{X}$. For matrix $\mathbf{A}$:

$$\mathbf{D} = \cos^2\theta - (\sin\theta)(-\sin\theta)$$
$$= \cos^2\theta + \sin^2\theta$$
$$= 1$$
$$\mathbf{A}^{-1} = \begin{pmatrix} \cos\theta & \sin\theta \\ -\sin\theta & \cos\theta \end{pmatrix}$$

Answer is (C)

MATHEMATICS–17

What is the rank of the matrix $\mathbf{A}$?

$$\mathbf{A} = \begin{pmatrix} 1 & 1 & 0 & 1 \\ 3 & 1 & 1 & -1 \\ 0 & 1 & -1 & 1 \\ 2 & 0 & 1 & -2 \end{pmatrix}$$

(A) 0   (B) 1   (C) 2   (D) 3   (E) 4

The rank of a matrix is the number of independent vectors (rows). The rank can be found by row-reducing (diagonalizing) the matrix and counting the number of pivots in the row-reduced form of the matrix.

$$\text{Row } 2 = (-2)(\text{Row } 2) + (3)(\text{Row } 4)$$
$$\text{Row } 4 = \text{Row } 4 - \text{Row } 2$$
$$\text{Row } 3 = (2)(\text{Row } 3) + \text{Row } 2$$

The row-reduced form of $\mathbf{A}$ is:

$$\mathbf{A} = \begin{pmatrix} 1 & 1 & 0 & 1 \\ 0 & -2 & 1 & 4 \\ 0 & 0 & -1 & -2 \\ 0 & 0 & 0 & 0 \end{pmatrix}$$

The matrix cannot be further row-reduced. There are three pivots and, therefore, three independent rows. Thus, the rank of matrix $\mathbf{A}$ is 3.

Answer is (D)

MATHEMATICS–18

Determine the values of x_1 and x_2 which satisfy the following linear system:

$$\begin{pmatrix} 3 & 7 \\ 2 & 6 \end{pmatrix} \begin{pmatrix} x_1 \\ x_2 \end{pmatrix} = \begin{pmatrix} 2 \\ 4 \end{pmatrix}$$

(A) $\begin{pmatrix} 4 \\ -2 \end{pmatrix}$ (B) $\begin{pmatrix} 2 \\ -4 \end{pmatrix}$ (C) $\begin{pmatrix} -2 \\ 4 \end{pmatrix}$ (D) $\begin{pmatrix} -4 \\ 2 \end{pmatrix}$ (E) $\begin{pmatrix} 4 \\ 2 \end{pmatrix}$

The linear equations represented by this system are:

$$3x_1 + 7x_2 = 2$$
$$2x_1 + 6x_2 = 4$$

Use Cramer's rule to solve the system of equations.

$$x_1 = \frac{\begin{vmatrix} 2 & 7 \\ 4 & 6 \end{vmatrix}}{\begin{vmatrix} 3 & 7 \\ 2 & 6 \end{vmatrix}}$$

$$= \frac{-16}{4}$$

$$x_1 = -4$$

$$x_2 = \frac{\begin{vmatrix} 3 & 2 \\ 2 & 4 \end{vmatrix}}{\begin{vmatrix} 3 & 7 \\ 2 & 6 \end{vmatrix}}$$

$$= \frac{8}{4}$$

$$x_2 = 2$$

$$\begin{pmatrix} x_1 \\ x_2 \end{pmatrix} = \begin{pmatrix} -4 \\ 2 \end{pmatrix}$$

Answer is (D)

MATHEMATICS–19

What is the cosine of 120°?

(A) −0.500   (B) −0.450   (C) −0.866   (D) 0.500   (E) 0.866

An angle of 120° is in the second quandrant. Therefore, the cosine is negative.

$$\cos 120° = -\cos(180° - 120°)$$
$$= -\cos 60°$$
$$= -0.5$$

Answer is (A)

MATHEMATICS–20

What is the sine of 840°?

(A) −0.866   (B) −0.500   (C) 0.300   (D) 0.500   (E) 0.866

Two complete revolutions around the unit circle is 720°, and 840° − 720° = 120°. This angle is in the second quadrant where the sine is positive.

$$\sin 120° = \sin(180° − 120°)$$
$$= \sin 60°$$
$$\sin 840° = 0.866$$

Answer is (E)

MATHEMATICS–21

If the sine of angle A is given as K, what would be the tangent of angle A?

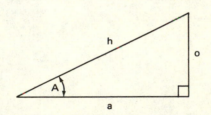

(A) $\dfrac{hK}{o}$ (B) $\dfrac{aK}{h}$ (C) $\dfrac{ha}{K}$ (D) $\dfrac{oK}{a}$ (E) $\dfrac{hK}{a}$

$$\sin A = \frac{o}{h}$$
$$= K$$
$$\tan A = \frac{o}{a}$$
$$= \left(\frac{h}{a}\right)\left(\frac{o}{h}\right)$$
$$= \frac{hK}{a}$$

Answer is (E)

MATHEMATICS–22

Which is true regarding the signs of the natural functions for angles between 90° and 180°?

(A) The tangent is positive.
(B) The cotangent is positive.
(C) The cosine is negative.
(D) The sine is negative.
(E) The secant is positive.

In the second quadrant the natural functions and their signs are as follows:

sin	positive
cos	negative
tan	negative
cot	negative
sec	negative
csc	positive

Answer is (C)

MATHEMATICS–23

What is the inverse natural function of the cosecant?

(A) secant   (B) sine   (C) cosine

(D) tangent   (E) cotangent

In a right triangle, the cosecant is calculated as the hypotenuse divided by the ordinate. The sine is defined as the ordinate divided by the hypotenuse. Thus, for any angle:

$$\sin \theta = \frac{1}{\csc \theta}$$

Answer is (B)

MATHEMATICS-24

What is the sum of the squares of the sine and cosine of an angle?

(A) 0   (B) 1   (C) $\sqrt{3}$ (D) 2   (E) 3

For any angle:

$$\cos^2 x + \sin^2 x = 1$$

Answer is (B)

MATHEMATICS-25

What is an equivalent expression for $\sin 2x$?

(A) $\dfrac{1}{2}\sin x \cos x$ (B) $2\sin x \cos\left(\dfrac{1}{2}x\right)$ (C) $-2\sin x$

(D) $-2\sin x \cos x$ (E) $\dfrac{2\sin x}{\sec x}$

The double angle formula for the sine function is:

$$\sin 2x = 2\sin x \cos x$$
$$= \frac{2\sin x}{\sec x}$$

Answer is (E)

MATHEMATICS-26

The series expansion for $\cos x$ contains which powers of x?

(A) 0, 2, 4, 6, 8, ...
(B) 1, 3, 5, 9, ...
(C) 1, 2, 3, 4, 5, ...
(D) 1/2, 3/2, 5/2, 7/2, ...
(E) −1, −3, −5, −9, ...

The Taylor expansion for $\cos x$ is as follows:

$$\cos x = 1 - \frac{x^2}{2!} + \frac{x^4}{4!} - \frac{x^6}{6!} + \cdots$$

Thus, only the positive even powers of x are contained in the expansion of $\cos x$.

Answer is (A)

MATHEMATICS-27

A transit set up 112.1 feet from the base of a vertical chimney reads 32° 30' with the cross hairs set on the top of the chimney. With the telescope level, the vertical rod at the base of the chimney is 5.1 feet. How tall is the chimney?

(A) 66.3 ft   (B) 71.4 ft   (C) 76.5 ft   (D) 170.9 ft   (E) 181.1 ft

To find the height, H, of the chimney, refer to the following figure.

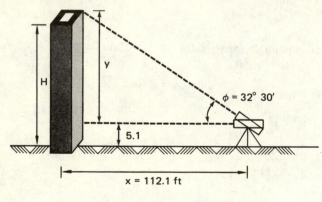

$$\tan \phi = \frac{y}{x}$$
$$y = (112.1)\tan 32.5°$$
$$= 71.4 \,\text{ft}$$
$$H = 5.1 + y$$
$$= 5.1 + 71.4$$
$$= 76.5 \,\text{ft}$$

Answer is (C)

MATHEMATICS–28

At approximately what time between the hours of 12:00 noon and 1:00 p.m. would the angle between the hour hand and the minute hand of a continuously driven clock be exactly 180°?

(A) 12:28 p.m.   (B) 12:30 p.m.   (C) 12:33 p.m.

           (D) 12:37 p.m.   (E) 12:41 p.m.

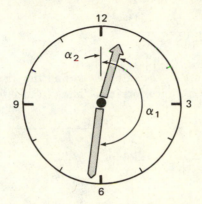

The change in the angle of the minute hand between 12:00 p.m. and 1:00 p.m., α_1, is

$$\alpha_1 = \frac{360°}{60}t$$
$$= (6t)°$$

and the change in the angle of the hour hand between 12:00 noon and 1:00 p.m., α_2, is

$$\alpha_2 = \frac{360°}{12(60)}t$$
$$= (0.5t)°$$

where t is in minutes past 12:00 noon. The angle between the two hands is $\alpha_1 - \alpha_2$.

$$\alpha_1 - \alpha_2 = 180°$$
$$6t - 0.5t = 180$$
$$5.5t = 180$$
$$t = 32.7 \, \text{min}$$

Thus, the time is approximately 12:33 p.m.

Answer is (C)

MATHEMATICS–29

In finding the distance, d, between two points, which equation is the appropriate one to use?

(A) $d = \sqrt{(x_1 - x_2)^2 + (y_2 - y_1)^2}$

(B) $d = \sqrt{(x_1 - y_1)^2 + (x_2 - y_2)^2}$

(C) $d = \sqrt{(x_1^2 - x_2^2) + (y_1^2 - y_2^2)}$

(D) $d = \sqrt{(x_2 - x_1)^2 + (y_2 - y_1)^2}$

(E) $d = \sqrt{(x_1 - x_2)^2 - (y_1 - y_2)^2}$

The correct formula to use in order to find the distance, d, between two points is the distance formula. The distance formula is defined as follows:

$$d = \sqrt{(x_2 - x_1)^2 + (y_2 - y_1)^2}$$

Answer is (D)

MATHEMATICS–30

The equation $y = a_1 + a_2 x$ is an algebraic expression for which of the following choices?

(A) a cosine expansion series
(B) projectile motion
(C) potential energy
(D) a circle in polar form
(E) a straight line

$y = mx + b$ is the slope-intercept form of the equation of a straight line. Thus, $y = a_1 + a_2 x$ describes a straight line.

Answer is (E)

MATHEMATICS–31

Find the slope of the line defined by $y - x = 5$.

(A) $5 + x$ (B) $-\dfrac{1}{2}$ (C) $\dfrac{1}{4}$ (D) 1 (E) 2

The slope-intercept form of the equation of a straight line is $y = mx + b$, where m is the slope and b is the y-intercept.

$$y - x = 5$$
$$y = x + 5$$
$$m = \text{the coefficient of } x$$
$$= 1$$

Answer is (D)

MATHEMATICS–32

Find the equation of a line with slope $= 2$ and y-intercept $= -3$.

(A) $y = -3x + 2$
(B) $y = 2x - 3$
(C) $y = \frac{2}{3}x + 1$
(D) $y = 2x + 3$
(E) $y = 3x - 2$

The slope-intercept form of the given equation is:

$$y = 2x - 3$$

Answer is (B)

MATHEMATICS–33

Find the equation of the line that passes through the points $(0,0)$ and $(2,-2)$.

(A) $y = x$ (B) $y = -2x + 2$ (C) $y = -2x$
       (D) $y = -x$ (E) $y = x - 2$

Since the line passes through the origin, the y-intercept is 0. Thus, the equation simplifies to $y = mx$. Substituting for the known points,

$$y = \frac{-2 - 0}{2 - 0} x$$
$$= -x$$

Answer is (D)

MATHEMATICS-34

What is the name for a vector that represents the sum of two vectors?

(A) scalar   (B) resultant   (C) tensor

           (D) moment   (E) tangent

By definition, the sum of two vectors is known as the resultant.

Answer is (B)

MATHEMATICS-35

What is the resultant, R, of the vectors F_1, F_2, and F_3?

$$F_1 = 4i + 7j + 6k$$
$$F_2 = 9i + 2j + 11k$$
$$F_3 = 5i - 3j - 8k$$

(A) $R = 18i + 6j + 9k$
(B) $R = -18i - 6j - 9k$
(C) $R = 18i + 12j + 25k$
(D) $R = 21i$
(E) $R = 442$

The resultant of vectors given in unit-vector form is the sum of the components.

$$R = (4 + 9 + 5)i + (7 + 2 - 3)j + (6 + 11 - 8)k$$
$$= 18i + 6j + 9k$$

Answer is (A)

MATHEMATICS-36

Simplify the expression $(\mathbf{A} \times \mathbf{B}) \cdot \mathbf{C}$, given:

$$\mathbf{A} = 3i + 2j$$
$$\mathbf{B} = 2i + 3j + k$$
$$\mathbf{C} = 5i + 2k$$

(A) 0   (B) 20   (C) $60i + 24k$

     (D) $5i + 2k$ (E) 180

First find $\mathbf{A} \times \mathbf{B}$.

$$\mathbf{A} \times \mathbf{B} = (3i + 2j) \times (2i + 3j + k)$$
$$= \begin{vmatrix} i & j & k \\ 3 & 2 & 0 \\ 2 & 3 & 1 \end{vmatrix}$$
$$= i(2 - 0) - j(3 - 0) + k(9 - 4)$$
$$= 2i - 3j - 5k$$
$$(\mathbf{A} \times \mathbf{B}) \cdot \mathbf{C} = (2i - 3j + 5k) \cdot (5i + 0j + 2k)$$
$$= (2)(5) + (-3)(0) + (5)(2)$$
$$= 20$$

Answer is (B)

MATHEMATICS-37

What type of curve is generated by a point which moves in uniform circular motion about an axis, while travelling with a constant speed, v, parallel to the axis?

(A) a cycloid   (B) an epicycloid   (C) a hypocycloid

     (D) a spiral of Archimedes   (E) a helix

A curve generated by the method described is called a helix and is illustrated in the following figure.

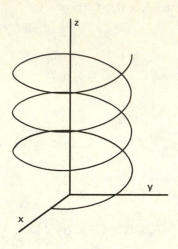

Answer is (E)

MATHEMATICS–38

What is a possible outcome of an experiment called?

(A) a sample space   (B) a random point   (C) an event
            (D) a finite set   (E) a discrete set

By definition, an event is a possible outcome of a trial or experiment.

Answer is (C)

MATHEMATICS–39

In probability theory, what is the set of all possible outcomes of an experiment called?

(A) a set of random events
(B) a fuzzy set
(C) a cumulative distribution
(D) a sample space
(E) a set of random variables

By definition, the sample space is the set of all possible outcomes of
an experiment.

Answer is (D)

MATHEMATICS–40

How can the values of a random variable defined over a sample space be
described?

(A) always continuous
(B) always numerical
(C) strictly nonzero
(D) defined only over a finite horizon
(E) always mutually exclusive of other sample spaces

The values of a random variable can be continuous or discrete over
a finite or infinite domain. The values in the sample space can be
shared by other sample spaces. However, the values of a random
variable must be numerical.

Answer is (B)

MATHEMATICS–41

If two random variables are independently distributed, what is their relationship?

(A) They are not identically distributed.
(B) They are uncorrelated.
(C) They are mutually exclusive.
(D) (A) or (B)
(E) (A) or (B) or (C)

By definition, two independently distributed random variable are uncorrelated. Any two random variables may or may not be identically
distributed. Independent events cannot be mutually exclusive.

Answer is (B)

MATHEMATICS–42

Which of the following properties of probability is not valid?

(A) The probability of an event is always positive and less than one.

(B) If E_0 is an event which cannot occur in the sample space, the probability of E_0 is zero.

(C) If events E_1 and E_2 are mutually exclusive, then the probability of both events occurring is zero.

(D) If events E_1 and E_2 are mutually exclusive, then $P(E_1 + E_2) = P(E_1) + P(E_2) - P(E_1 E_2)$.

(E) The expected value of a discrete distribution of random variables does not have to be one of the distribution's values.

The probability law given in choice (D) is valid for independent events, not mutually exclusive events. The correct rule for mutually exclusive events is:

$$P(E_1 + E_2) = P(E_1) + P(E_2)$$

Answer is (D)

MATHEMATICS–43

Which one of the following functions cannot be a probability density function for the variable x?

(A)

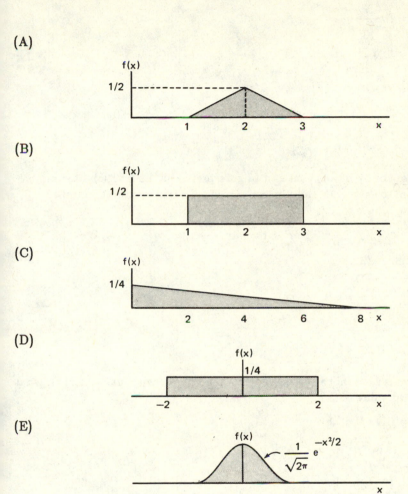

(B)

(C)

(D)

(E)

To be a density function, the area under the curve must equal 1. That is, the cumulative density function must sum to 1. The area under the curve for choice (A) is $\frac{1}{2}$. Therefore, it cannot be a density function.

Answer is (A)

MATHEMATICS–44

If n is the number of trials and m is the number of successes, what is the frequency based interpretation of the probability of event E?

(A) $P(E) = \lim\limits_{n \to \infty} \dfrac{n - m}{n}$

(B) $P(E) = \lim\limits_{n \to \infty} \dfrac{n}{m}$

(C) $P(E) = \lim\limits_{n \to \infty} \dfrac{m}{m - n}$

(D) $P(E) = \lim\limits_{n \to \infty} \dfrac{m}{n}$

(E) $P(E) = \lim\limits_{n \to \infty} \dfrac{n}{m - n}$

The probability of an event can be interpreted as the fraction of successful outcomes when the experiment is performed an infinite number of times. Thus:

$$P(E) = \lim\limits_{n \to \infty} \frac{m}{n}$$

Answer is (D)

MATHEMATICS–45

For a continuous random variable X with probability density function $f(x)$, what is the expected value of X?

(A) $E(X) = \int_0^\infty x f(x)\, dx$

(B) $E(X) = \int_{-\infty}^\infty x f(x)\, dx$

(C) $E(X) = \int_0^\infty f(x)\, dx$

(D) $E(X) = \int_0^\infty x\, dx$

(E) $E(X) = \int_0^\infty (x - \bar{x}) f(x)\, dx$

The expected value or average of X can be defined mathematically as follows:

$$E(X) = \int_{-\infty}^\infty x f(x)\, dx$$

Answer is (B)

MATHEMATICS–46

If $P(B) \neq 1$, and A and B are not independent events, what is $P(A|B)$?

(A) $P(A) \times P(B)$
(B) $P(B|A) \times P(A) \div P(B)$
(C) $P(A)$
(D) $P(A|B) \times P(B) \div P(A)$
(E) $P(A) \div P(B)$

The probability of event A occurring, given that the dependent event B has occurred, is predicted by the conditional probability law, commonly known as Bayes theorem.

$$P(B|A) = \frac{P(AB)}{P(A)}$$

$$\text{Similarly: } P(A|B) = \frac{P(AB)}{P(B)}$$

$$\text{Therefore: } P(A|B) = \frac{P(B|A) \times P(A)}{P(B)}$$

Answer is (B)

MATHEMATICS–47

If the discrete random variable X has a geometric distribution parameter P and smallest mass point O, what is the expected value of X?

(A) P (B) P^{-1} (C) P^{1-P} (D) $\dfrac{1-P}{P}$ (E) P^2

The geometric distribution is a special case of the negative binomial distribution. The mean is $\dfrac{1-P}{P}$, and the variance is $\dfrac{1-P}{P^2}$. Note: some authors define the geometric distribution with the smallest mass point being 1 (instead of 0). In that case, the mean is $\dfrac{1}{P}$ and the variance is the same as before.

Answer is (D)

MATHEMATICS–48

If the variable X has a Poisson distribution with parameter λ, what is the expected value of X?

(A) λ^2 (B) $\lambda(1 - \lambda)$ (C) λ^{-1}

           (D) $\dfrac{\lambda^{k-\lambda}}{k!}$ (E) λ

Most probability books contain derivations for the mean and variance of the Poisson distribution. Both the mean and variance are equal to λ.

> **Answer is (E)**

MATHEMATICS–49

If X is a binomial random variable with parameters n and p, what is the expected value of X?

(A) $n(1 - p)$ (B) $np(1 - p)$ (C) p^{-1}

           (D) np (E) n

For a binomial distribution, the mean is np, and the variance is $np(1 - p)$.

> **Answer is (D)**

MATHEMATICS–50

For a discrete random variable X with probability mass function $P(X)$, what is the expected value of X?

(A) $E(X) = \displaystyle\sum_{\text{all } x_i} x_i P(x_i)$

(B) $E(X) = \displaystyle\sum_{\text{all } x_i} x_i^2 P(x_i)$

(C) $E(X) = \displaystyle\sum_{\text{all } x_i} P(x_i)$

(D) $E(X) = \displaystyle\sum_{\text{all } x_i} (x_i - \bar{x}) P(x_i)$

(E) $E(X) = \displaystyle\sum_{\text{all } x_i} (P(x_i))^2$

The expected value of a discrete function is given by the following:

$$E(X) = x_1 P(x_1) + x_2 P(x_2) + \cdots$$
$$= \sum_{\text{all } x_i} x_i P(x_i)$$

Answer is (A)

MATHEMATICS–51

An item's cost distribution is given as a function of the probability. What is the expected cost?

cost	probability
1	0.07
2	0.23
3	0.46
4	0.17
5	0.04
6	0.03

(A) 2.5   (B) 2.9   (C) 3.0   (D) 3.1   (E) 3.4

The expected value is the sum of the products of the individual values and their respective probabilities.

$$E(\text{cost}) = 1(0.07) + 2(0.23) + 3(0.46) + 4(0.17) + 5(0.04) + 6(0.03)$$
$$= 2.97$$

Answer is (C)

MATHEMATICS-52

In a dice game, one fair die is used. The player wins $10 if he rolls either a 1 or a 6. He loses $5 if he turns up any other face. What is the expected winning for one roll of the die?

(A) $0   (B) $3.33   (C) $5.00   (D) $6.67   (E) $10.00

For a fair die, the probability of any face turning up is $\frac{1}{6}$. Therefore, the expected value is:

$$10 \times \left(2 \times \frac{1}{6}\right) - 5 \times \left(4 \times \frac{1}{6}\right) = 0$$

Answer is (A)

MATHEMATICS-53

An urn contains four black balls and six white balls. What is the probability of getting one black ball and one white ball in two consecutive draws from the urn?

(A) 0.04   (B) 0.24   (C) 0.27   (D) 0.48   (E) 0.53

$$P(\text{black and white}) = P(\text{black then white}) + P(\text{white then black})$$
$$= \left(\frac{4}{10}\right)\left(\frac{6}{9}\right) + \left(\frac{6}{10}\right)\left(\frac{4}{9}\right)$$
$$= 0.53$$

Answer is (E)

MATHEMATICS–54

The probability that both stages of a two-stage rocket will function correctly is 0.95. the reliability of the first stage is 0.98. What is the reliability of the second stage?

(A) 0.95   (B) 0.96   (C) 0.97   (D) 0.98   (E) 0.99

In a serial system:

$$R_t = R_1 R_2$$

$$R_2 = \frac{R_t}{R_1}$$

where R_n is the reliability of stage n and R_t is the total reliability of all stages. For the rocket:

$$R_2 = \frac{0.95}{0.98}$$

$$= 0.97$$

Answer is (C)

MATHEMATICS–55

What is the exponential form of the complex number $3 + 4i$?

(A) $e^{i53.1°}$ (B) $5e^{i53.1°}$ (C) $5e^{i126.9°}$

(D) $7e^{i53.1°}$ (E) $7e^{i126.9°}$

Any complex number, $a + bi$ can be converted to its equivalent exponential form as follows:

$$a + bi = \sqrt{a^2 + b^2}e^{i \arctan \frac{b}{a}}$$

Therefore,

$$3 + 4i = \sqrt{3^2 + 4^2}e^{i \arctan\left(\frac{4}{3}\right)}$$

$$\arctan\left(\frac{4}{3}\right) = 53.1°$$

$$3 + 4i = 5e^{i53.1°}$$

Answer is (B)

MATHEMATICS-56

What is the product of the complex numbers $3 + 4i$ and $7 - 2i$?

(A) $10 + 2i$

(B) $13 + 22i$

(C) $13 + 34i$

(D) $29 + 22i$

(E) $29 + 34i$

$$(3 + 4i)(7 - 2i) = 21 - 8i^2 + 28i - 6i$$
$$= 21 + 8 + 28i - 6i$$
$$= 29 + 22i$$

Answer is (D)

MATHEMATICS-57

What is the rectangular form of the complex number $7.2e^{i\frac{7\pi}{13}}$?

(A) $7.15 + 0.87i$

(B) $7.15 - 0.87i$

(C) $-0.87 + 7.15i$

(D) $-0.87 - 7.15i$

(E) $12.18e^i$

A complex number of the form ce^{id} can be converted to rectangular form as follows:

$$ce^{id} = c\cos d + (c\sin d)i$$
$$7.2e^{i\frac{7\pi}{13}} = 7.2\left[\cos\frac{7\pi}{13} + \left(\sin\frac{7\pi}{13}\right)i\right]$$
$$= -0.87 + 7.15i$$

Answer is (C)

MATHEMATICS-58

What is the product of the complex numbers $2 - 2i$ and $\sqrt{32}e^{i\frac{\pi}{4}}$?

(A) 16

(B) $16i$

(C) $16e^{i\frac{\pi}{4}}$

(D) $16(1-i)$

(E) $32(1+i)$

$$2 - 2i = \sqrt{2^2 + 2^2}\, e^{i\,\arctan \frac{-2}{2}}$$
$$= \sqrt{8}\, e^{-i\frac{\pi}{4}}$$

Therefore,

$$(2 - 2i)\left(\sqrt{32}\, e^{i\frac{\pi}{4}}\right) = \left(\sqrt{8}\, e^{-i\frac{\pi}{4}}\right)\left(\sqrt{32}\, e^{i\frac{\pi}{4}}\right)$$
$$= \left(\sqrt{8}\right)\left(\sqrt{32}\right) e^{i\left(\frac{\pi}{4} - \frac{\pi}{4}\right)}$$
$$= 16$$

Answer is (A)

MATHEMATICS–59

What is the rationalized value of the complex number $\dfrac{6 + 2.5i}{3 + 4i}$?

(A) $-0.32 + 0.66i$ (B) $0.32 - 0.66i$ (C) $1.12 - 0.66i$
          (D) $-1.75 + 1.03i$ (E) $1.75 - 1.03i$

In order to rationalize a complex number, multiply the numerator and denominator by the complex conjugate of the denominator and simplify.

$$\frac{6 + 2.5i}{3 + 4i} = \frac{(6 + 2.5i)(3 - 4i)}{(3 + 4i)(3 - 4i)}$$
$$= \frac{28 - 16.5i}{25}$$
$$= 1.12 - 0.66i$$

Answer is (C)

MATHEMATICS–60

What is the first derivative with respect to x of the function $g(x) = 4\sqrt{9}$?

(A) 0   (B) $\dfrac{4}{9}$ (C) 4   (D) $4(9)^{\frac{1}{2}}$ (E) $4\left(\dfrac{9}{2}\right)$

The derivative of a constant is zero. Therefore $g'(x) = 0$.

Answer is (A)

MATHEMATICS–61

If a is a simple constant, what is the derivative of $y = x^a$?

(A) ax (B) x^{a-1} (C) ax^{a-1}

(D) $(a-1)x$ (E) ax^{a+1}

$$y = x^a$$
$$y' = ax^{a-1}$$

Answer is (C)

MATHEMATICS–62

Find the derivative of $f(x) = \left[x^3 - (x-1)^3\right]^3$?

(A) $3x^2 - 3(x-1)^2$

(B) $3\left[x^3 - (x-1)^3\right]^2$

(C) $9\left[x^3 - (x-1)^3\right]\left[x^2 - (x-1)^2\right]$

(D) $9\left[x^3 - (x-1)^3\right]^2\left[x^2 - (x-1)^2\right]$

(E) $9x^2 - 9(x-1)^2$

$$f(x) = \left[x^3 - (x-1)^3\right]^3$$
$$f'(x) = 3\left[x^3 - (x-1)^3\right]\frac{2d}{dx}\left[x^3 - (x-1)^3\right]$$
$$= 3\left[x^3 - (x-1)^3\right]^2\left[3x^2 - 3(x-1)^2(1)\right]$$
$$= 9\left[x^3 - (x-1)^3\right]^2\left[x^2 - (x-1)^2\right]$$

Answer is (D)

MATHEMATICS–63

Differentiate $f(x) = \sqrt{2x^2 + 4x + 1}$.

(A) $2x + 2$

(B) $\frac{1}{2}\sqrt{2x^2 + 4x + 1}$

(C) $\dfrac{2x + 2}{\sqrt{2x^2 + 4x + 1}}$

(D) $\dfrac{4x + 4}{\sqrt{2x^2 + 4x + 1}}$

(E) $4(x + 1)$

$$f(x) = \sqrt{2x^2 + 4x + 1}$$
$$= \left(2x^2 + 4x + 1\right)^{\frac{1}{2}}$$
$$f'(x) = \frac{1}{2}\left(2x^2 + 4x + 1\right)^{-\frac{1}{2}} \frac{d}{dx}\left(2x^2 + 4x + 1\right)$$
$$= \frac{1}{2}\left(2x^2 + 4x + 1\right)^{-\frac{1}{2}}\left(4x + 4\right)$$
$$= \frac{2x + 2}{\sqrt{2x^2 + 4x + 1}}$$

Answer is (C)

MATHEMATICS–64

Find the second derivative of $y = \sqrt{x^2} + x^{-2}$.

(A) $1 - 2x^{-3}$

(B) $1 - 6x^{-4}$

(C) 3

(D) $\dfrac{6}{x^4}$

(E) $6x^4$

$$y = \sqrt{x^2} + x^{-2}$$
$$y' = \frac{x}{\sqrt{x^2}} - 2x^{-3}$$
$$= \pm 1 - 2x^{-3}$$
$$y'' = 6x^{-4}$$

Note: $\dfrac{x}{\sqrt{x^2}} = \pm 1$ because by definition, $\sqrt{x^2} = |x|$.

Answer is (D)

MATHEMATICS-65

Find $\dfrac{dy}{dt}$ given the following two simultaneous differential equations.

$$2\frac{dx}{dt} - 3\frac{dy}{dt} + x - y = k$$

$$3\frac{dx}{dt} + 2\frac{dy}{dt} - x \quad = \cos t$$

(A) $\dfrac{2}{13}\left(\cos t + \dfrac{5}{2}x - \dfrac{3}{2}y - \dfrac{3}{2}k\right)$

(B) $\dfrac{1}{3}\left(\sin t + \dfrac{1}{9}x - y^3 - \dfrac{3}{2}k\right)$

(C) $-\dfrac{1}{6}\left(\sin t + \dfrac{1}{9}x + y^2 - \dfrac{3}{2}k\right)$

(D) $\dfrac{2}{9}\left(\cos t + \dfrac{3}{2}x - \dfrac{5}{2}y - \dfrac{3}{2}k\right)$

(E) $\dfrac{3}{7}\left(\cos t + \dfrac{5}{2}x + \dfrac{3}{2}y + \dfrac{3}{2}k\right)$

Solve both equations for $\dfrac{dx}{dt}$.

$$\frac{dx}{dt} = \frac{1}{2}\left(k + y - x + 3\frac{dy}{dt}\right)$$

$$\frac{dx}{dt} = \frac{1}{3}\left(\cos t + x - 2\frac{dy}{dt}\right)$$

Combine and solve for $\dfrac{dy}{dt}$.

$$\frac{1}{2}\left(k + y - x + 3\frac{dy}{dt}\right) = \frac{1}{3}\left(\cos t + x - 2\frac{dy}{dt}\right)$$

$$9\frac{dy}{dt} + 4\frac{dy}{dt} = -3k - 3y + 3x + 2\cos t + 2x$$

$$13\frac{dy}{dt} = 2\cos t + 5x - 3y - 3k$$

$$\frac{dy}{dt} = \frac{2}{13}\left(\cos t + \frac{5}{2}x - \frac{3}{2}y - \frac{3}{2}k\right)$$

Answer is (A)

MATHEMATICS–66

If $y = \cos x$, what is $\dfrac{dy}{dx}$?

(A) $\sec x$ (B) $-\sec x$ (C) $\csc x$ (D) $\sin x$ (E) $-\sin x$

$$\frac{d}{dx}(\cos x) = -\sin x$$

Answer is (E)

MATHEMATICS–67

If the second derivative of the equation of a curve is proportional to the negative of the equation of the same curve, what is that curve?

(A) a hyperbola   (B) a square wave   (C) a sinusoid

(D) a cycloid   (E) an epicycloid

The only type of function that fits the description is a sinusoidal one.

$$\frac{d^2}{dx^2}(\sin x) = \frac{d}{dx}(\cos x)$$
$$= -\sin x$$

Answer is (C)

MATHEMATICS–68

Given $P = 2R^2 S^3 T^{\frac{1}{2}} + R^{\frac{1}{3}} S \sin 2T$, what is $\dfrac{\partial P}{\partial T}$?

(A) $R^2 S^3 T^{\frac{3}{2}} + 2R^{\frac{1}{3}} \cos 2T$

(B) $6RS^2 T^{-\frac{1}{2}} + \dfrac{2}{3} R^{-\frac{2}{3}} \cos 2T$

(C) $2R^2 S^3 T^{\frac{1}{2}} + R^{\frac{1}{3}} S \cos 2T$

(D) $R^2 S^3 T^{-\frac{1}{2}} + 2R^{\frac{1}{3}} S \cos 2T$

(E) $R^2 S^3 T^{-\frac{1}{2}} + R^{\frac{1}{3}} \cos 2T$

All variables other than T are treated as constants.

$$\frac{\partial P}{\partial T} = 2R^2 S^3 \left(\frac{1}{2} T^{-\frac{1}{2}}\right) + R^{\frac{1}{3}} S (\cos 2T)(2)$$

$$= R^2 S^3 T^{-\frac{1}{2}} + 2R^{\frac{1}{3}} S \cos 2T$$

Answer is (D)

MATHEMATICS–69

Which of the following describes the first derivative at point A of the function shown in the figure?

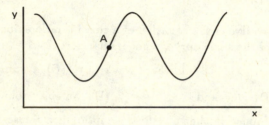

(A) positive only
(B) negative only
(C) zero
(D) positive or negative
(E) none of the above

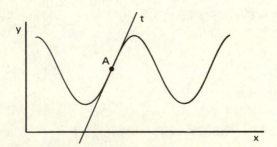

The first derivative corresponds to the slope of a tangent line at the point. The slope of this tangent line is positive. Therefore, the first derivative of the function at point A is also positive.

Answer is (A)

MATHEMATICS–70

Which of the following describes the second derivative at point A of the function shown?

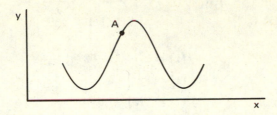

(A) positive only
(B) negative only
(C) zero
(D) positive or negative
(E) none of the above

> The second derivative corresponds to the concavity of the function. Since the curvature at this point is concave down, the second derivative is negative. The second derivative also indicates what is happening to the first derivative, the slope. Since the slope is decreasing at point A, the second derivative must be negative.

Answer is (B)

MATHEMATICS–71

What is the slope of the graph $y = -x^2$ at the point $(2, 3)$?

(A) -4 (B) -2 (C) 1 (D) 3 (E) 4

> The slope of a curve is given by the first derivative.

$$y(x) = -x^2$$
$$y'(x) = -2x$$

at the point $(2, 3)$:

$$y'(2) = -2(2)$$
$$= -4$$

Answer is (A)

PROFESSIONAL PUBLICATIONS, INC. • Belmont, CA

MATHEMATICS–72

Given the function $f(x) = x^3 - 5x + 2$, find the value of the first derivative at $x = 2$, $f'(2)$.

(A) 2

(B) $3x^2 - 5$

(C) 7

(D) 8

(E) 10

$$f(x) = x^3 - 5x + 2$$
$$f'(x) = 3x^2 - 5$$
$$f'(2) = 3(2)^2 - 5$$
$$= 7$$

Answer is (C)

MATHEMATICS–73

Find the slope of the tangent to a parabola, $y = x^2$, at a point on the curve where $x = \dfrac{1}{2}$.

(A) 0

(B) $\dfrac{1}{4}$

(C) $-\dfrac{1}{2}$

(D) $\dfrac{1}{2}$

(E) 1

$$y = x^2$$
$$y' = 2x$$
$$y'\left(\frac{1}{2}\right) = 2\left(\frac{1}{2}\right)$$
$$= 1$$

Answer is (E)

MATHEMATICS–74

What is the slope of the curve $y = x^2 - 4x$ as it passes through the origin?

(A) 0

(B) −3

(C) −4

(D) 4

(E) infinite

$$y = x^2 - 4x$$
$$\frac{dy}{dx} = 2x - 4$$
$$\left.\frac{dy}{dx}\right|_{x=0} = (2)(0) - 4$$
$$= -4$$

Answer is (C)

MATHEMATICS–75

Find the slope of the line tangent to the curve $y = x^3 - 2x + 1$ at the point $(1, 2)$.

(A) $\dfrac{1}{4}$ (B) $\dfrac{1}{3}$ (C) $\dfrac{1}{2}$ (D) 1   (E) $\dfrac{3}{2}$

$$y = x^3 - 2x + 1$$
$$y' = 3x^2 - 2$$
$$y'(1) = 3 - 2$$
$$= 1$$

Answer is (D)

MATHEMATICS–76

Determine the equation of the line tangent to the graph $y = 2x^2 + 1$, at the point $(1, 3)$.

(A) $y = 2x + 1$
(B) $y = 4x - 1$
(C) $y = 2x - 1$
(D) $y = 4x + 1$
(E) $y = \dfrac{1}{4}x - 1$

First, determine the slope of the graph at $x = 1$.

$$y = 2x^2 + 1$$
$$y' = 4x$$
$$y'(1) = 4$$

Since the tangent line intersects the graph at $(1,3)$, the equation of the tangent line is:

$$y = 4x + b$$
$$3 = 4(1) + b$$
$$b = -1$$
$$y = 4x - 1$$

Answer is (B)

MATHEMATICS-77

Given $y_1 = 4x + 3$ and $y_2 = x^2 + C$, find C such that y_2 is tangent to y_1.

(A) 2   (B) 4   (C) 5   (D) 6   (E) 7

The slope of $y_1 = 4x + 3$ is 4 everywhere. Therefore, y_2 must have a slope of 4 at the tangent point.

$$y_2' = 2x$$
$$4 = 2x$$
$$x = 2$$

Therefore, $x = 2$ at the tangent point. Now find $y_1 = y_2$ at the tangent point and substitute in to find C.

$$y_1 = 4(2) + 3$$
$$= 11$$
$$y_2 = 11$$
$$11 = (2)^2 + C$$
$$C = 7$$

Answer is (E)

MATHEMATICS-78

Given:

$$\frac{dy_1}{dx} = \frac{2}{13}\left(1 + \frac{5}{2}x - \frac{3}{2} - \frac{3}{4}k\right)$$

What is the value of k such that y_1 is perpendicular to the curve $y_2 = 2x$ at $(1, 1)$?

(A) 2   (B) 3   (C) 6   (D) 7   (E) 10

For two lines to be perpendicular, $m_1 m_2 = -1$ where m_n is the slope of line n.

$$\frac{dy_2}{dx} = 2$$

Therefore, at $(1, 1)$

$$\frac{dy_1}{dx} = -\frac{1}{2}$$

$$= \frac{2}{13}\left(1 + \frac{5}{2}x - \frac{3}{2} - \frac{3}{4}k\right)$$

$$-\frac{1}{2} = \frac{2}{13}\left(1 + \frac{5}{2}(1) - \frac{3}{2} - \frac{3}{4}k\right)$$

$$-\frac{13}{4} = 2 - \frac{3}{4}k$$

$$\frac{3}{4}k = \frac{21}{4}$$

$$k = 7$$

Answer is (D)

MATHEMATICS–79

The distance a body travels is a function of time and is given by $x(t) = 18t + 9t^2$. Find its velocity at $t = 2$.

(A) 20   (B) 24   (C) 36   (D) 54   (E) 60

Velocity is the first time derivative of the position function.

$$x(t) = 18t + 9t^2$$
$$v(t) = x'(t)$$
$$= 18 + 18t$$
$$v(2) = 18 + 18(2)$$
$$= 54$$

Answer is (D)

MATHEMATICS-80

A particle moves according to the following functions of time:

$$x(t) = 3\sin t$$
$$y(t) = 4\cos t$$

What is the resultant velocity at $t = \pi$?

(A) 0   (B) 3   (C) 4   (D) 9   (E) 25

$$v = \sqrt{\left(\frac{dx}{dt}\right)^2 + \left(\frac{dy}{dt}\right)^2}$$

$$\frac{dx}{dt} = 3\cos t$$

$$\frac{dy}{dt} = -4\sin t$$

$$v(t) = \sqrt{9\cos^2 t + 16\sin^2 t}$$

$$v(\pi) = \sqrt{9(-1)^2 + 16(0)^2}$$

$$= 3$$

Answer is (B)

MATHEMATICS-81

Water is pouring into a swimming pool. After t hours, there are $t + \sqrt{t}$ gallons in the pool. At what rate is the water pouring into the pool when $t = 9$ hours?

(A) $\frac{1}{6}$ gph   (B) $\frac{1}{2}$ gph   (C) 1 gph

(D) $\frac{7}{6}$ gph   (E) $\frac{4}{3}$ gph

let   V = volume of water in the tank

$$V = t + \sqrt{t}$$

$$V' = 1 + \left(\frac{1}{2}\right)\left(\frac{1}{\sqrt{t}}\right)$$

$$V'(9) = 1 + \left(\frac{1}{2}\right)\left(\frac{1}{3}\right)$$

$$= \frac{7}{6} \text{ gph}$$

Answer is (D)

MATHEMATICS-82

If x increases uniformly at the rate of 0.001 feet per second, at what rate is the expression $(1 + x)^3$ increasing when x becomes 9 feet?

(A) 0.001 cfs   (B) 0.003 cfs   (C) 0.3 cfs

          (D) 1.003 cfs   (E) 300 cfs

$$\frac{dx}{dt} = 0.001$$

$$f(x) = (1 + x)^3$$

$$\frac{df}{dx} = 3(1 + x)^2$$

$$\frac{df}{dt} = \frac{df}{dx}\frac{dx}{dt}$$

$$= (0.003)(1 + x)^2$$

$$\left.\frac{df}{dt}\right|_{x=9} = (0.003)(1 + 9)^3$$

$$= 0.3 \text{ cfs}$$

Answer is (C)

MATHEMATICS-83

A spherical balloon is being filled with air at a rate of 1 cubic foot per second. Compute the time rate of change of the surface area of the balloon at the instant when its volume is 113.1 cubic feet.

(A) $0.67 \text{ ft}^2/\text{sec}$ (B) $1.73 \text{ ft}^2/\text{sec}$ (C) $3.0 \text{ ft}^2/\text{sec}$

          (D) $3.7 \text{ ft}^2/\text{sec}$ (E) $\dfrac{4\pi}{3} \text{ ft}^2/\text{sec}$

$$V = \frac{4}{3}\pi r^3$$

$$A = 4\pi r^2$$

$$\frac{dV}{dt} = 4\pi r^2 \frac{dr}{dt}$$

$$\frac{dr}{dt} = \left(\frac{1}{4\pi r^2}\right)\frac{dV}{dt}$$

$$\frac{dA}{dt} = \left(\frac{dA}{dr}\right)\frac{dr}{dt}$$

$$= (8\pi r)\frac{dr}{dt}$$

$$= \left(\frac{8\pi r}{4\pi r^2}\right)\frac{dV}{dt}$$

$$= \left(\frac{2}{r}\right)\frac{dV}{dt}$$

Solve for the radius of the sphere when the volume is 113.1 ft^3, and substitute into the equation for the rate of change of the surface area.

$$r = \left(\frac{3V}{4\pi}\right)^{\frac{1}{3}}$$

$$= \left(\frac{(3)(113.1)}{4\pi}\right)^{\frac{1}{3}}$$

$$r = 3\text{ ft}$$

$$\frac{dA}{dt} = \left(\frac{2}{3}\right)(1)$$

$$= 0.67\text{ ft}^2/\text{sec}$$

Answer is (A)

MATHEMATICS-84

Consider a strictly convex function of one variable, x, with a lower bound and an upper bound on x. For what value(s) of x will the function be minimized?

(A) at the upper bound of x
(B)   at the lower bound of x
(C) strictly between the upper and lower bounds of x
(D) (A) or (B)
(E) (A) or (B) or (C)

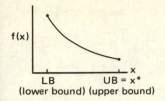 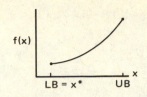 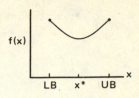

The examples above demonstrate that for a concave up function, the minimum could occur at the lower bound, the upper bound, or somewhere in between. Thus, choices (A), (B), or (C) could be correct.

Answer is (E)

MATHEMATICS–85

Consider a strictly concave function in one variable, x, with a lower bound and an upper bound on x. For what value(s) of x will the function be minimized?

(A) at the upper bound of x
(B) at the lower bound of x
(C) strictly between the upper and lower bounds of x
(D) (A) or (B)
(E) (A) or (B) or (C)

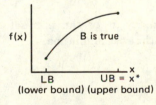 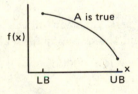 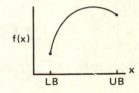

The figures above demonstrate that for a concave down curve, the minimum could occur at either the lower or the upper bound. Therefore, choice (D) is correct.

Answer is (D)

MATHEMATICS–86

What is the maximum of the function $y = -x^3 + 3x$, for $x \geq -1$?

(A) -2 (B) -1 (C) 0 (D) 1 (E) 2

The maximum occurs where $y' = 0$ and $y'' < 0$ or at an endpoint.

$$y = -x^3 + 3x$$
$$y' = -3x^2 + 3$$
$$y'' = -6x$$

When $\quad y' = 0$

$$0 = -3x^2 + 3$$
$$x^2 = 1$$
$$x = \pm 1 \quad (-1 \text{ is also an endpoint})$$
$$y(-1) = -(-1)^3 + 3(-1)$$
$$= -2$$
$$y(1) = -(1)^3 + 3(1)$$
$$= 2$$

Therefore,

$$y_{max} = 2$$

Answer is (E)

MATHEMATICS–87

The cost C, of a product is a function of the quantity x, of the product: $C(x) = x^2 - 4000x + 50$. Find the quantity for which the cost is minimum.

(A) 1000   (B) 1500   (C) 2000   (D) 3000   (E) 3500

$$C = x^2 - 4000x + 50$$
$$C' = 2x - 4000$$
$$C'' = 2$$

When

$$C' = 0$$
$$2x - 4000 = 0$$
$$x = 2000$$
$$C'' > 0$$

Thus, cost is a minimum when $x = 2000$.

Answer is (C)

MATHEMATICS–88

Compute the following limit: $\lim\limits_{x\to\infty} \dfrac{x+2}{x-2}$.

(A) 0   (B) 1   (C) 2

       (D) ∞ (E) indefinite

Divide both the numerator and denominator by x, and allow x to approach infinity.

$$\lim_{x\to\infty} \frac{x+2}{x-2} = \lim_{x\to\infty} \frac{1+\frac{2}{x}}{1-\frac{2}{x}}$$
$$= \frac{1+0}{1-0}$$
$$= 1$$

Answer is (B)

MATHEMATICS–89

Simplify the expression $\lim\limits_{x\to 4} \dfrac{x^2-16}{x-4}$.

(A) 0   (B) 8   (C) 12   (D) 16   (E) ∞

Factor the numerator and simplify the fraction before taking the limit.

$$\lim_{x\to 4} \frac{x^2-16}{x-4} = \lim_{x\to 4} \frac{(x-4)(x+4)}{x-4}$$
$$= \lim_{x\to 4} (x+4)$$
$$= 8$$

Answer is (B)

MATHEMATICS–90

Compute the following limit: $\lim\limits_{x\to 0} \dfrac{1-\cos x}{x^2}$.

(A) 0   (B) $\dfrac{1}{4}$ (C) $\dfrac{1}{2}$ (D) 1   (E) ∞

Since both the numerator and denominator approach zero, use L'Hospital's rule. (L'Hospital's rule states that the derivative of the numerator divided by the derivative of the denominator has the same limit as the original fraction, provided that both the numerator and denominator of the original fraction approach zero.)

$$\lim_{x \to 0} \frac{1 - \cos x}{x^2} = \lim_{x \to 0} \frac{\sin x}{2x}$$

Since the numerator and denominator both approach zero, apply L'Hospital's rule again.

$$\lim_{x \to 0} \frac{1 - \cos x}{x^2} = \lim_{x \to 0} \frac{\cos x}{2}$$
$$= \frac{\cos 0}{2}$$
$$= \frac{1}{2}$$

Answer is (C)

MATHEMATICS-91

The existence of the two equations $y' = f(x)$, and $y = \phi(x)$ implies that which of the following equations is true?

(A) $\phi(x) = \int f(x)dx + C$
(B) $\phi(x) = f(x)$
(C) $\phi'(x) = \int f(x)dx + C$
(D) $\phi'(x) = y$
(E) $\phi(x) = \int ydy + C$

$$y = \phi(x)$$
$$\frac{d\phi}{dx} = y'$$
$$\phi(x) = \int y'\,dx + C$$

Since

$$y' = f(x)$$
$$\phi(x) = \int f(x)dx + C$$

Answer is (A)

MATHEMATICS–92

Fill in the blank in the following statement:

The integral of a function between certain limits divided by the difference in abscissas between those limits gives the _____ of the function.

(A) average   (B) middle   (C) intercept

   (D) asymptote   (E) limit

By definition:

$$\frac{1}{b-a}\int_a^b f(x) = \text{the average value of the function}$$

Answer is (A)

MATHEMATICS–93

Find the area under the curve $y = \dfrac{1}{x}$ between the limits $y = 2$ and $y = 10$.

(A) 1.61   (B) 2.39   (C) 3.71   (D) 3.97   (E) 5.16

The area under the curve $f(x)$ between x_1 and x_2, A, is given by

$$A = \int_{x_1}^{x_2} f(x)dx.$$ The x limits corresponding to the y limits are

$x = \dfrac{1}{2}$ and $x = \dfrac{1}{10}$.

$$A = \int_{\frac{1}{10}}^{\frac{1}{2}} \frac{1}{x}dx$$

$$= \ln x \Big|_{\frac{1}{10}}^{\frac{1}{2}}$$

$$= 1.61$$

Answer is (A)

MATHEMATICS–94

Find the area of the shaded region between $y = 6x - 1$ and $y = \dfrac{1}{4}x + 3$, bounded by $x = 0$ and the intersection point.

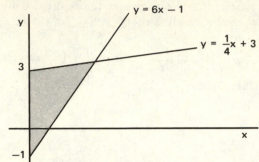

(A) $\dfrac{32}{529}$

(B) $\dfrac{16}{23}$

(C) $\dfrac{32}{23}$

(D) $\dfrac{1440}{529}$

(E) $\dfrac{64}{23}$

The area between curve 1 and curve 2 is equal to the area under curve 1 minus the area under curve 2. The intersection point of the two curves is found by equating both functions.

$$6x - 1 = \frac{1}{4}x + 3$$

$$\frac{23}{4} = 4$$

$$x = \frac{16}{23}$$

The area, A, is:

$$A = \int_0^{\frac{16}{23}} \left(\frac{1}{4}x + 3 - 6x + 1 \right) dx$$

$$= \int_0^{\frac{16}{23}} \left(-\frac{23}{4}x + 4 \right) dx$$

$$= \left[-\frac{23}{8}x^2 + 4x \right]_0^{\frac{16}{23}}$$

$$= -\frac{23}{8} \left(\frac{16}{23} \right)^2 + 4 \left(\frac{16}{23} \right)$$

$$= -\frac{32}{23} + \frac{64}{23}$$

$$= \frac{32}{23}$$

Answer is (C)

MATHEMATICS–95

If it is known that $y = 1$ when $x = 1$, what is the constant of integration for the following integral?

$$y(x) = \int \left(e^{2x} - 2x\right) dx$$

(A) $C = 2 - e^2$

(B) $C = 3 - e^2$

(C) $C = 4 - e^2$

(D) $C = \dfrac{1}{2}\left(3 - e^2\right)$

(E) $C = \dfrac{1}{2}\left(4 - e^2\right)$

$$y(x) = \int e^{2x} dx - 2 \int x\, dx$$

$$= \frac{1}{2}e^{2x} - x^2 + C$$

$$= \frac{1}{2}\left(e^{2x} - 2x^2\right) + C$$

However, $y(1) = 1$.

$$1 = \frac{1}{2}\left(e^{2(1)} - 2(1)^2\right) + C$$

$$= \frac{1}{2}\left(e^2 - 2\right) + C$$

$$C = 1 + 1 - \frac{1}{2}e^2$$

$$= \frac{1}{2}\left(4 - e^2\right)$$

Answer is (E)

MATHEMATICS–96

It is known that $y(x)$ passes through the points (0, 2) and (1, 4). Solve for $y(x)$ if the second derivative is:

$$\frac{d^2y}{dx^2} = 1$$

(A) $y = \left(x^2 + 3x\right) + 2$

(B) $y = \frac{1}{2}\left(x^2 + 3x\right) + 2$

(C) $y = \frac{1}{2}\left(x^2 - 3x\right) - 2$

(D) $y = \frac{1}{2}\left(x^2 + 3x\right) - 2$

(E) $y = \frac{1}{2}\left(x^2 - 3x\right) + 2$

Integrate twice to get the general form of the equation.

$$\frac{d^2y}{dx^2} = 1$$

$$\frac{dy}{dx} = \int 1dx$$

$$= x + C_1$$

$$y = \int (x + C_1)$$

$$= \frac{1}{2}x^2 + C_1 x + C_2$$

Now solve for C_1 and C_2 using the given conditions.

$$2 = \frac{1}{2}(0) + C_1(0) + C_2$$

$$C_2 = 2$$

$$4 = \frac{1}{2}(1)^2 + C_1(1) + 2$$

$$C_1 = \frac{3}{2}$$

$$y = \frac{1}{2}x^2 + \frac{3}{2}x + 2$$

Answer is (B)

MATHEMATICS–97

What is a solution of the first order difference equation $y(k+1) = y(k) + 5$?

(A) $y(k) = 4 - \dfrac{5}{k}$

(B) $y(k) = C - k$, where C is a constant

(C) $y(k) = 5^k + \dfrac{1 - 5^k}{-4}$

(D) $y(k) = 20 + 5k$

(E) The solution is nonexistent for real-valued y.

Assume the solution has the form:

$$y(k) = 20 + 5k$$

Substitute the assumed solution into the difference equation.

$$\begin{aligned} y(k+1) &= 20 + 5(k+1) \\ &= 20 + 5k + 5 \\ &= y(k) + 5 \end{aligned}$$

Answer is (D)

MATHEMATICS–98

What is the solution of the linear difference equation $y(k+1) = 15y(k)$?

(A) $y(k) = \dfrac{15}{1 + 15k}$

(B) $y(k) = \dfrac{15k}{16}$

(C) $y(k) = C + 15^k$, where C is a constant

(D) $y(k) = 15^k$

(E) (C) or (D)

Assume the solution has the form:

$$y(k) = 15^k$$

Substitute into the difference equation.

$$y(k + 1) = 15^{k+1}$$
$$= 15 \left(15^k\right)$$
$$= 15y(k)$$

Note: if $y(k) = C + 15^k$, then $y(k + 1) = C + 15^{k+1} \neq 15y(k)$.

Answer is (D)

MATHEMATICS-99

What is the solution of the linear difference equation $(k+1)y(k+1) - ky(k) = 1$?

(A) $y(k) = 12 - \dfrac{1}{k}$

(B) $y(k) = 1 - \dfrac{12}{k}$

(C) $y(k) = 12 + 3k$

(D) $y(k) = 3 + \dfrac{1}{k}$

(E) $y(k) = 12^k + C$, where C is a constant

Assume the solution has the form:

$$y(k) = 1 - \frac{12}{k}$$

Substitute the solution into the difference equation.

$$(k + 1) \times y(k + 1) - k \times y(k) = 1$$
$$(k + 1)\left(\left(1 - \frac{12}{k + 1}\right) - k\left(1 - \frac{12}{k}\right)\right) = 1$$
$$(k + 1)\left(\frac{k + 1 - 12}{k + 1}\right) - k\left(\frac{k - 12}{k}\right) = 1$$
$$k + 1 - 12 - k + 12 = 1$$
$$1 = 1$$

Thus, $y(k) = 1 - \dfrac{12}{k}$ solves the difference equation.

Answer is (B)

MATHEMATICS–100

Which of the following is a differential equation of the first order?

(A) $(y'')^3 + 2y' = -3$

(B) $\dfrac{\partial Q}{\partial x} - \dfrac{\partial Q}{\partial y} = 0$

(C) $\dfrac{dy}{dx} + \dfrac{9 - x}{x} = y^3$

(D) $\left(\dfrac{dy}{dx}\right)^2 = -y + x$

(E) $x - \dfrac{d^3 y}{dx^3} = y$

A first order differential equation contains only first derivatives and does not have partial derivatives. The only choice that fulfills this requirement is (C).

Answer is (C)

MATHEMATICS–101

How can the differential equation $a\frac{d^2 x}{dt^2} + B(t)\frac{dx}{dt} + C = D(t)$ best be described?

(A) linear, homogeneous, and first order
(B) homogeneous and first order
(C) linear, second order, and nonhomogeneous
(D) linear, homogeneous, and second order
(E) second order and nonhomogeneous

The differential equation has a second derivative, so it is of second order. The forcing function is nonzero, so the equation is nonhomogeneous. All of the terms on the left-hand side only have coefficients that are either constant or a function of the independent variable. Therefore the equation is also linear.

Answer is (C)

MATHEMATICS–102

The differential equation given is correctly described by which one of the following choices?

$$a\frac{d^2y}{dx^2} + bxy\frac{dy}{dx} = f(x)$$

(A) first order
(B) linear, second order, homogeneous
(C) nonlinear, second order, homogeneous
(D) linear, second order, nonhomogeneous
(E) nonlinear, second order, nonhomogeneous

Since there is a second derivative, the differential equation is of second order. Since the coefficient of one of the terms contains the dependent variable, y, the equation is nonlinear. Since the forcing function, $f(x)$, is implied to be nonzero, the differential equation is also nonhomogeneous.

Answer is (E)

MATHEMATICS–103

Determine the solution of the following differential equation:

$$y' + 5y = 0$$

(A) $y = 5x + C$ (B) $y = Ce^{-5x}$ (C) $y = Ce^{5x}$
    (D) (A) or (B)   (E) (B) or (C)

This is a first order linear equation with characteristic equation $r + 5 = 0$. Therefore, the form of the solution is

$$y = Ce^{-5x}$$

where the constant, C, could be determined from additional information.

Answer is (B)

MATHEMATICS–104

What is the general solution of the differential equation $\dfrac{d^2y}{dx^2} + 4y = 0$?

(A) $y = \sin x + 2\tan x + C$
(B) $y = e^x - 2e^{-x} + C$
(C) $y = 2x^2 - x + C$
(D) $y = \sin 2x - x\cos 2x + C$
(E) $y = \sin 2x + \cos 2x + C$

Examination of the differential equation shows that a multiple of the function and its second derivative must sum to zero. Sines and cosines have the property that their second derivatives are the negatives of the original natural function.

Thus, if $y = \sin 2x + \cos 2x$, then:

$$y' = 2\cos 2x - 2\sin 2x$$
$$y'' = -4\sin 2x - 4\cos 2x$$
$$y'' + 4y = -4\sin 2x - 4\cos 2x + 4(\sin 2x + \cos 2x)$$
$$= 0$$

Thus, the function in choice (E) solves the differential equation.

Answer is (E)

MATHEMATICS–105

The differential equation $\dfrac{dx}{dt} + 4x = 0$ has the initial condition $x(0) = 12$. What is the value of $x(2)$?

(A) 3.35×10^{-4} (B) 4.03×10^{-3} (C) 3

(D) 6   (E) 22

This is a first-order, linear, homogeneous differential equation with characteristic equation $r + 4x = 0$.

$$\dot{x} + 4x = 0$$
$$x = x_0 e^{-4t}$$
$$x(0) = x_0 e^{-4(0)}$$
$$= 12$$
$$x_0 = 12$$
$$x = 12e^{-4t}$$
$$x(2) = 12e^{-4(2)}$$
$$= 12e^{-8}$$
$$= 4.03 \times 10^{-3}$$

Answer is (B)

MATHEMATICS-106

A curve passes through the point $(1, 1)$. Determine the absolute value of the slope of the curve at $x = 25$ if the differential equation of the curve is the exact equation $y^2 dx + 2xy dy = 0$.

(A) $\dfrac{1}{250}$ (B) $\dfrac{1}{125}$ (C) $\dfrac{1}{50\sqrt{5}}$ (D) $\dfrac{1}{\sqrt{125}}$ (E) $\dfrac{1}{5}$

$$y^2 dx + 2xy dy = 0$$
$$2xy dy = -y^2 dx$$
$$2\frac{dy}{y} = \frac{-dx}{x}$$

Integrating both sides,

$$2 \ln y = -\ln x + \ln C$$
$$\ln y^2 + \ln x = \ln C$$
$$\ln xy^2 = \ln C$$
$$xy^2 = C$$

Use the fact that the curve passes through the point $(1, 1)$ to solve for C, then determine the slope at $x = 25$.

$$1(1)^2 = C$$
$$C = 1$$
$$xy^2 = 1$$
$$y = \pm\sqrt{\frac{1}{x}}$$
$$y(25) = \pm\sqrt{\frac{1}{25}}$$
$$= \pm\frac{1}{5}$$
$$y^2\,dx + 2xy\,dy = 0$$
$$\frac{dy}{dx} = -\frac{y}{2x}$$
$$\left.\frac{dy}{dx}\right|_{x=25} = -\frac{\pm\frac{1}{5}}{2(25)}$$
$$= \pm\frac{1}{250}$$

Answer is (A)

MATHEMATICS-107

Determine the constant of integration for the separable differential equation $x\,dx + 6y^5\,dy = 0$. It is known that when $x = 0$, $y = 2$.

(A) 4   (B) 8   (C) 16   (D) 32   (E) 64

Since this differential equation is already separated, integrate to find the solution.

$$\int x\,dx + \int 6y^5\,dy = \int 0$$
$$\frac{1}{2}x^2 + y^6 = C$$

Use the initial conditions to solve for C.

$$\frac{1}{2}(0)^2 + (2)^6 = C$$
$$C = 64$$

Answer is (E)

MATHEMATICS–108

What is the Laplace transform of e^{-6t}?

(A) $\dfrac{1}{s+6}$

(B) $\dfrac{1}{s-6}$

(C) $e^{(-6+s)}$

(D) $e^{(6+s)}$

(E) $\dfrac{e^{(s+6)}}{s-a}$

The Laplace transform of a function, $\mathcal{L}(f)$, can be calculated for the definition of a transform. However, if a table of transforms is available, it is easier to refer to the table.

$$\begin{aligned}
\mathcal{L}(e^{-6t}) &= \int_0^\infty e^{-(s+6)t}\,dt \\
&= -\frac{e^{-(s+6)t}}{s+6}\Bigg|_0^\infty \\
&= \frac{1}{s+6}
\end{aligned}$$

Answer is (A)

2   ECONOMIC ANALYSIS

ECONOMICS–1

How is the capital recovery factor $(A/P, i, n)$ related to the uniform series sinking fund factor $(A/F, i, n)$? i is the effective annual rate of return, and n is in years.

(A) $(A/P, i, n) = (A/F, i, n) + i$

(B) $(A/P, i, n) = (A/F, i, n) - i$

(C) $(A/P, i, n) = \dfrac{(A/F, i, n)}{i}$

(D) $(A/P, i, n) = \dfrac{(A/F, i, n) + i}{n}$

(E) $(A/P, i, n) = \dfrac{(A/F, i, n) - i}{n}$

By definition:

$$(A/P, i, n) = (A/F, i, n) + i$$

Answer is (A)

ECONOMICS–2

What is an annuity?

(A) the future worth of a present amount
(B) an annual repayment of a loan
(C) a series of uniform amounts over a period of time
(D) a lump sum at the end of the year
(E) the present worth of a future amount

Answer is (C)

ECONOMICS–3

Which of the following expressions is incorrect?

(A) the future worth of a present amount, $(F/P, i, n) = \dfrac{1}{(P/F, i, n)}$

(B) the future worth of an annuity, $(F/A, i, n) = \dfrac{1}{(A/F, i, n)}$

(C) the present worth of an annuity, $(P/A, i, n) = \dfrac{1}{(A/P, i, n)}$

(D) $(A/F, i, n) \times (P/A, i, n) = (P/F, i, n)$

(E) $(A/F, i, n) - i = (A/P, i, n)$

$$(A/F, i, n) + i = (A/P, i, n)$$

Therefore, (E) is false.

Answer is (E)

ECONOMICS–4

When using net present worth calculations to compare two projects, which of the following could invalidate the calculation?

(A) differences in the magnitudes of the projects
(B) evaluating over different time periods
(C) mutually exclusive projects
(D) nonconventional cash flows
(E) use of the same discount rate for each period

(A), (C), (D), and (E) are all problems with internal rate of return calculations that net present worth handles nicely. However, the net present worth of two projects must be calculated for the same time period.

Answer is (B)

ECONOMICS–5

What is the present worth of a $100 annuity over a ten-year period, if the interest rate is 8%?

(A) $450   (B) $532   (C) $671   (D) $700   (E) $850

$$P = 100 \times (P/A, 8\%, 10)$$
$$= 100 \times 6.71$$
$$= \$671$$

The factor of $(P/A, 8\%, 10) = 6.71$ is obtained from the compound interest tables for an interest rate of 8% and a time period of 10 years.

Answer is (C)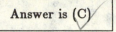

ECONOMICS-6

How much money must you invest today in order to withdraw $1000 per year for 10 years if the interest rate is 12%?

(A) $4800   (B) $5650   (C) $5808   (D) $6145   (E) $10,000

Use the compound interest tables for an interest rate of 12% and a time period of 10 years.

$$P = A(P/A, i, n)$$
$$P = \text{present worth}$$
$$A = \text{annual withdrawal}$$
$$i = 12\%$$
$$n = 10 \text{ years}$$
$$P = 1000 \times 5.650$$
$$= \$5650$$

Answer is (B)

ECONOMICS-7

A machine is under consideration for investment. The cost of the machine is $25,000. Each year it operates, the machine will generate a savings of $15,000. Given an effective annual interest rate of 18%, what is the discounted payback period, in years, on the investment in the machine?

(A) 1.67 years   (B) 1.75 years   (C) 2.16 years
          (D) 3.17 years   (E) 3.67 years

$$P = 25{,}000$$
$$A = 15{,}000$$
$$25{,}000 = 15{,}000(P/A, 18\%, n)$$
$$\frac{5}{3} = (P/A, 18\%, n)$$
$$= \frac{(1.18^n - 1)}{0.18(1.18)^n}$$
$$0.3(1.18)^n = 1.18^n - 1$$
$$0.7(1.18)^n = 1$$
$$n = \frac{\ln\left(\frac{10}{7}\right)}{\ln(1.18)}$$
$$= 2.16 \text{ years}$$

Answer is (C)

ECONOMICS–8

What is the present worth of a $100 annuity starting at the end of the third year and continuing to the end of the fourth year, if the annual interest rate is 8%?

(A) $122   (B) $153   (C) $160   (D) $162   (E) $165

$$P = 100\left[(P/A, 8\%, 4) - (P/A, 8\%, 2)\right]$$
$$= 100 \times (3.31 - 1.78)$$
$$= \$153$$

Answer is (B)

ECONOMICS–9

Consider a project which involves the investment of $100,000 now and $100,000 at the end of one year. Revenues of $150,000 will be generated at the end of years 1 and 2. What is the net present value of this project if the effective annual interest rate is 10%?

(A) − $150,000   (B) − $50,910   (C) $0

(D) $43,270   (E) $69,420

PROFESSIONAL PUBLICATIONS, INC. • Belmont, CA

$$P = -100,000 + 50,000(P/F, 10\%, 1) + 150,000(P/F, 10\%, 2)$$
$$= -100,000 + 50,000(0.9091) + 150,000(0.8264)$$
$$= 69,415$$
$$\approx \$69,420$$

Answer is (E)

ECONOMICS–10

At an annual rate of return of 8%, what is the future worth of $100 at the end of four years?

(A) $130   (B) $132   (C) $135

(D) $136   (E) $140

$$F = 100(F/P, 8\%, 4)$$
$$= 100 \times 1.3605$$
$$= 136.05$$
$$\approx \$136$$

Answer is (D)

ECONOMICS–11

A person invests $450 to be collected in eight years. Given that the interest rate on the investment is 14.5% per year, compounded annually, what sum, in dollars, will be collected eight years hence?

(A) $450   (B) $972   (C) $1014

(D) $1329   (E) $9240

$$F = 450(F/P, 14.5\%, 8)$$
$$= 450 \times 2.954$$
$$= \$1329$$

Answer is (D)

ECONOMICS–12

An investment of x dollars is made at the end of each year for three years, at an interest rate of 9% per year compounded annually. What will the dollar value of the total investment be upon the deposit of the third payment?

(A) $0.772x$ (B) $1.295x$ (C) $2.278x$

         (D) $3x$ (E) $3.278x$

$$F = A(F/A, i, n)$$
$$= A\left[\frac{(1+i)^n - 1}{i}\right]$$
$$= x\left[\frac{(1.09)^3 - 1}{0.09}\right]$$
$$= x\,(3.278)$$
$$= 3.278x \text{ dollars}$$

Answer is (E)

ECONOMICS–13

If \$500 is invested at the end of each year for six years, at an effective annual interest rate of 7%, what is the total dollar amount available upon the deposit of the sixth payment?

(A) \$3000   (B) \$3210   (C) \$3577

         (D) \$4260   (E) \$4502

$$F = 500(F/A, 7\%, 6)$$
$$= 500 \times 7.153$$
$$= \$3577$$

Answer is (C)

ECONOMICS–14

Assuming i = annual rate of return, n = number of years, F = future worth, and P = present worth, what is the future worth of a present amount $P(F/P, i, n)$?

(A) $P(1+i)^n$ (B) $P(1+i)^{n-1}$ (C) $P(1+i)^{-n}$

         (D) $P(1+n)^i$ (E) $P(1+n)^{-i}$

This situation corresponds to a single payment compound amount. Therefore:

$$P = (1+i)^n$$

Answer is (A)

ECONOMICS-15

You deposit $1000 into a 9% account today. At the end of two years, you will deposit another $3000. In five years, you plan a $4000 purchase. How much is left in the account one year after the purchase?

(A) $925
(B) $1424
(C) $1552
(D) $1691
(E) Nothing, the account will be closed.

year	cash flow (dollars)
0	1000
1	0
2	3000
3	0
4	0
5	−4000
6	0

$F = 1000(F/P, 9\%, 6) + 3000(F/P, 9\%, 4) - 4000(F/P, 9\%, 1)$

$\quad = 1000(1.6671) + 3000(1.4116) - 4000(1.0900)$

$\quad = \$1552$

Answer is (C)

ECONOMICS-16

You need $4000 per year for four years to go to college. Your father invested $5000 in a 7% account for your education when you were born. If you withdraw the $4000 at the end of your 17th, 18th, 19th and 21st years, how much money will be left in the account at the end of your 21st year?

(A) $1700   (B) $2500   (C) $3400
         (D) $4000   (E) $6152

$$F = 5000(F/P, 7\%, 21)$$
$$- 4000(F/P, 7\%, 4) - 4000(F/P, 7\%, 3)$$
$$- 4000(F/P, 7\%, 2) - 4000(F/P, 7\%, 1)$$

$$= 5000(4.1406) - 4000(1.3108)$$
$$- 4000(1.2250) - 4000(1.1449)$$
$$- 4000(1.0700)$$

$$= \$1700$$

Answer is (A)

ECONOMICS–17

The following schedule of funds is available to form a sinking fund.

current year (n)	5000
$n + 1$	4000
$n + 2$	3000
$n + 3$	2000

At the end of the fourth year, equipment costing $25,000 will have to be purchased as a replacement for old equipment. Money is valued at 20% by the company. At the time of purchase, how much money will be needed to supplement the sinking fund?

(A) $820   (B) $1000   (C) $2000
          (D) $8200   (E) $24,000

First, find the future worth of the available funds.

$$F = P_1(F/P, 20\%, 4)$$
$$+ P_2(F/P, 20\%, 3) + P_3(F/P, 20\%, 2)$$
$$+ P_4(F/P, 20\%, 1)$$
$$= 5000(2.074) + 4000(1.728)$$
$$+ 3000(1.44) + 2000(1.20)$$
$$= \$24,000$$

The additional funds necessary, A, are:

$$A = 25,000 - 24,000$$
$$= \$1000$$

Answer is (B)

ECONOMICS–18

In year zero, you invest \$10,000 in a 15% security for five years. During that time, the average annual inflation is 6%. How much, in terms of year zero dollars, will be in the account at maturity ?

(A) \$6,653   (B) \$13,382   (C) \$15,030

          (D) \$15,386   (E) \$20,113

First, find F, the future worth without accounting for inflation.

$$F = 10,000(F/P, 15\%, 5)$$
$$= 10,000 \times 2.0114$$
$$= \$20,114$$

Now, figure in inflation and express F in terms of real dollars (F_{real})

$$F_{real} = 20,114(P/F, 6\%, 5)$$
$$= 20,114 \times 0.7473$$
$$= \$15,030$$

Answer is (C)

ECONOMICS–19

A firm borrows \$2000 for six years at 8%. At the end of six years, it renews the loan for the amount due plus \$2000 more for two years at 8%. What is the lump sum due?

(A) \$5280   (B) \$5754   (C) \$6035

          (D) \$6135   (E) \$6215

$$F = [P_1(F/P_1, 8\%, 6) + 2000](F/P, 8\%, 2)$$
$$= [2000(1.587) + 2000](1.166)$$
$$= \$6035$$

Answer is (C)

ECONOMICS–20

A company invests $10,000 today to be repaid in five years in one lump sum at 12% compounded annually. If the rate of inflation is 3% compounded annually, how much profit, in present day dollars, is realized over the five years?

(A) $3202   (B) $5202   (C) $5626

              (D) $7623   (E) $8626

First, find the future worth of the investment without accounting for inflation.

$$F = 10,000(F/P,\ 12\%,\ 5)$$
$$= 10,000 \times 1.7623$$
$$= \$17,623$$

Next, find the present worth accounting for inflation, F_{real}.

$$F_{real} = 17,623(P/F,\ 3\%,\ 5)$$
$$= 17,623 \times 0.8626$$
$$= \$15,202$$
$$\text{profit} = 15,202 - 10,000$$
$$= \$5202$$

Answer is (B)

ECONOMICS–21

What must two investments with the same present worth and unequal lives have?

(A) identical salvage values
(B) different salvage values
(C) identical equivalent uniform annual cash flows
(D) different equivalent uniform annual cash flows
(E) identical tax consequences

Answer is (D)

ECONOMICS–22

The following cash flow diagram represents an investment of 400 dollars and a revenue of x dollars at the end of years one and two. Given a discount rate of 15% compounded annually, what must x be for this set of cash flows to have a net present worth of approximately zero?

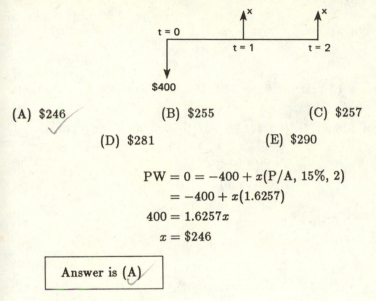

(A) $246

(B) $255

(C) $257

(D) $281

(E) $290

$$PW = 0 = -400 + x(P/A, 15\%, 2)$$
$$= -400 + x(1.6257)$$
$$400 = 1.6257x$$
$$x = \$246$$

Answer is (A)

ECONOMICS–23

An electric replacement pump is being considered for purchase. It is capable of providing 200 horsepower. The pertinent data are as follows:

cost	$3200
electric efficiency	0.85
maintenance cost per year	$50
life expectancy	14 years

The pump is used for 400 hours per year and the cost of electricity is $0.04 per kilowatt-hour (1 horsepower = 0.746 kilowatt). Assuming the pump will have no salvage value, and using straight line depreciation, what will be the monthly cost?

(A) $234.60

(B) $238.20

(C) $647.60

(D) $667.60

(E) $687.60

$$\text{monthly cost} = \frac{\left[\left(400\frac{hr}{yr}\right)\left(0.04\frac{\$}{kW\cdot h}\right)\left(200\,hp\right)\left(0.746\frac{kW}{hp}\right)\left(\frac{1}{0.85}\right)\right] + \$50}{12\frac{months}{yr}}$$

$$= \$238.20$$

Answer is (B)

ECONOMICS–24

What annuity over a 10 year period at 8% interest is equivalent to a present worth of $100?

(A) $12   (B) $13.80   (C) $14.10   (D) $14.90   (E) $15

$$A = 100(A/P, 8\%, 10)$$
$$= 100(0.149)$$
$$= \$14.90$$

Answer is (D)

ECONOMICS–25

Mr. Richardson borrowed $15,000 two years ago. The terms of the loan are 10% interest for 10 years with uniform payments. He just made his second annual payment. How much principal does he still owe?

(A) $10,117   (B) $11,700   (C) $12,000   (D) $13,024   (E) $13,117

The annual payments, A, are:

$$A = 15,000(A/P, 10\%, 10)$$
$$= 15,000(0.1627)$$
$$= \$2441 \text{ (in year zero dollars)}$$

year	amount owed		interest owed		payment		balance
1	15,000	+	1500	−	2441	=	$14,059
2	14,059	+	1406	−	2441	=	$13,024

Thus, Mr. Richardson still owes $13,024 on the principal.

Answer is (D)

ECONOMICS–26

Given that the discount rate is 15%, what is the equivalent uniform annual cash flow of the following stream of cash flows?

year 0	−$100,000
year 1	−$200,000
year 2	−$50,000
year 3	−$75,000

(A) −$90,260   (B) −$106,250   (C) −$124,200
          (D) −$126,500   (E) −$158,100

$$EUAC = (A/P, 15\%, 3)\big[-100,000 - 200,000(P/F, 15\%, 1)$$
$$-50,000(P/F, 15\%, 2) - 75,000(P/F, 15\%, 3)\big]$$
$$= 0.4380\big[-100,000 - 200,000(0.8696)$$
$$-50,000(0.7561) - 75,000(0.6575)\big]$$
$$= (0.4380)(-361,000)$$
$$= -\$158,100$$

Answer is (E)

ECONOMICS–27

A company must relocate one of its factories in three years. Equipment for the loading dock is being considered for purchase. The original cost is $20,000, the salvage value after three years is $8000. The company's rate of return (i) on money is 10%. Determine the capital recovery rate (CR) per year.

(A) $4805 per year   (B) $4946 per year   (C) $5115 per year
          (D) $5625 per year   (E) $5846 per year

$$CR = P(A/P, 10\%, 3) - F(A/F, 10\%, 3)$$
$$= 20,000(A/P) - 8000(A/F)$$
$$= 20,000(0.4021) - 8000(0.3021)$$
$$= \$5625 \text{ per year}$$

Answer is (D)

ECONOMICS–28

In five years, $18,000 will be needed to pay for a building renovation. In order to generate this sum, a sinking fund consisting of three annual payments is established now. For tax purposes, no further payments will be made after three years. What payments are necessary if money is worth 15% per annum?

(A) $2670   (B) $2870   (C) $3919   (D) $5100   (E) $5184

The present worth of $18,000 at the end of the third year is:

$$P_3 = F_5(P/F, 15\%, 2)$$
$$= 18,000(0.7561)$$
$$= \$13,610$$

Thus, the sinking fund must generate $13,610 in three years. The payments that are necessary, A, are:

$$A = 13,610(A/F, 15\%, 3)$$
$$= 13,610(0.2880)$$
$$= \$3919$$

Answer is (C)

ECONOMICS–29

Mr. Johnson borrows $100,000 at 10% effective annual interest. He must pay back the loan over 30 years with uniform monthly payments due on the first day of each month. What does Mr. Johnson pay each month?

(A) $839   (B) $846   (C) $870   (D) $878   (E) $884

An effective annual interst rate of 10% is equivalent to an effective monthly rate of:

$$(1+i)^{12} - 1 = 0.1$$
$$i = (1.1)^{\frac{1}{2}} - 1$$
$$= 0.7974\% \text{ per month}$$

The number of months, n, that Mr. Johnson has to pay off his loan is:

$$n = \left(30\,\text{yrs}\right)\left(12\,\frac{\text{months}}{\text{yr}}\right)$$

$$= 360 \text{ months}$$

$$\text{end of month payment} = 100,000(A/P, 0.7974\%, 360)$$

$$= 100,000(0.0084591)$$

$$= \$846$$

$$\text{beginning of month payment} = 846(P/F, 0.7974\%, 360)$$

$$= 846(0.9921)$$

$$= \$839$$

Answer is (A)

ECONOMICS–30

What is the formula for a straight-line depreciation rate?

(A) $\dfrac{100\% - \%\ \text{net salvage value}}{\text{estimated service life}}$

(B) $\dfrac{\%\ \text{net salvage value}}{\text{estimated service life}}$

(C) $\dfrac{100\%\ \text{net salvage value}}{\text{estimated service life}}$

(D) $\dfrac{\text{average net salvage value}}{\text{estimated service life}}$

(E) $\dfrac{\text{average gross salvage value}}{\text{estimated service life}}$

The straight-line depreciation rate $= \dfrac{100\% - \%\ \text{net salvage value}}{\text{estimated service life}}$.

Answer is (A)

ECONOMICS-31

What is the book value of equipment purchased three years ago for $15,000 if it is depreciated using the sum of years digits (SOYD) method? The expected life is five years.

(A) $3000   (B) $4000   (C) $6000   (D) $9000   (E) $10,000

In the SOYD method, the digits corresponding to n, the number of years of estimated life, are added.

$$T = \text{ total SOYD}$$
$$= \frac{n(n+1)}{2}$$
$$= \frac{5(5+1)}{2}$$
$$= 15$$

The depreciation charge for the first year is:

$$D_1 = \left(\frac{n}{T}\right) P$$
$$= \left(\frac{5}{15}\right) 15,000$$
$$= 5000$$

The depreciation charge for year two is:

$$D_2 = \left(\frac{n-1}{T}\right) P$$
$$= \left(\frac{4}{15}\right) 15,000$$
$$= 4000$$

For year three, the depreciation charge is:

$$D_3 = \left(\frac{n-2}{T}\right) P$$
$$= \left(\frac{3}{15}\right) 15,0000$$
$$= 3000$$

The total depreciation is:

$$D_{tot} = D_1 + D_2 + D_3$$
$$= 5000 + 4000 + 3000$$
$$= 12,000$$
$$\text{book value} = P - D_{tot}$$
$$= 15,000 - 12,000$$
$$= \$3000$$

Answer is (A)

ECONOMICS–32

The purchase of a motor for \$6000 and a generator for \$4000 will allow a company to produce its own energy. The configuration can be assembled for \$500. The service will operate for 1600 hours per year for 10 years. The maintenance cost is \$300 per year and the cost to operate is \$0.85 per hour for fuel and related costs. Using straight-line depreciation, what is the annual cost for the operation? There is \$400 in salvage value for the system at the end of 10 years.

(A) \$2480 per year   (B) \$2630 per year   (C) \$2670 per year

(D) \$2710   (E) \$3480

$$\text{total initial cost} = 6000 + 4000 + 500$$
$$= \$10,500$$
$$\text{salvage value} = \$400$$
$$\text{straight-line depreciation per year} = \frac{10,500 - 400}{10}$$
$$= \$1010$$
$$\text{maintenance per year} = \$300$$
$$\text{operation cost per year} = (0.85)(1600)$$
$$= \$1360$$
$$\text{annual cost} = 1010 + 300 + 1360$$
$$= \$2670$$

Answer is (C)

ECONOMICS-33

Company A purchases $200,000 of equipment in year zero. It decides to use straight-line depreciation over the expected 20 year life of the equipment. The interest rate is 14%. If its overall tax rate is 40%, what is the present worth of the depreciation tax shield?

(A) $3500

(B) $4000

(C) $26,500

(D) $39,700

(E) $80,000

$$\text{straight-line depreciation} = \frac{200,000}{20}$$
$$= \$10,000 \text{ per year write off}$$
$$PW = 10,000(P/A, 14\%, 20)(\text{tax rate})$$
$$= 10,000(6.623)(0.40)$$
$$= \$26,500$$

Answer is (C)

ECONOMICS-34

Which of the following is true regarding the minimum attractive rate of return used in judging proposed investments?

(A) It is the same for every organization.

(B) It is larger than the interest rate used to discount expected cash flow from investments.

(C) It is frequently a policy decision made by an organization's management

(D) It is not relevant in engineering economy studies.

(E) It is much smaller than the interest rate used to discount expected cash flows from investments.

Answer is (C)

ECONOMICS–35

What is the effective annual interest rate on a loan if the nominal interest rate is 12% per year compounded quarterly?

(A) 11.75%   (B) 12%   (C) 12.25%   (D) 12.55%   (E) 12.75%

$$\text{nominal interest rate} = r$$
$$\text{number of times compounded per year} = m$$
$$\text{effective interest rate} = i = \left(1 + \frac{r}{m}\right)^m - 1$$
$$= \left(1 + \frac{0.12}{4}\right)^4 - 1$$
$$= 0.1255$$
$$= 12.55\%$$

Answer is (D)

$$= \left(1 + \frac{0.14}{4}\right)^4 - 1$$

ECONOMICS–36

A person pays interest on a loan semiannually at a nominal annual interest rate of 16%. What is the effective annual interest rate?

(A) 15.5%   (B) 15.65%   (C) 16%   (D) 16.5%   (E) 16.64%

$$r = \text{nominal interest rate}$$
$$= 16\%$$
$$m = \text{number of times compounded}$$
$$= 2$$
$$i = \text{effective rate}$$
$$= \left(1 + \frac{r}{m}\right)^m - 1$$
$$= \left(1 + \frac{0.16}{2}\right)^2 - 1$$
$$= 16.64\%$$

Answer is (E)

ECONOMICS-37

Which of the following statements is not correct?

(A) A nominal rate of 12% per annum compounded quarterly is the same as $(12\%) \div 4 = 3\%$ per quarter.

(B) One dollar compounded quarterly at 3% for n years has a worth of $(1.03)^{4n}$ dollars.

(C) Compounding quarterly at a nominal rate of 12% per year is equivalent to compounding annually at a rate of 12.55%

(D) 12.55% annual compounding is the effective rate for a situation in which the principal is actually compounded quarterly at an annual rate of 12%

(E) Effective rate of return in choices (A) through (D) is the difference between 12.55% and 12%.

> **Answer is (E)**

ECONOMICS-38

A bank is advertising 9.5% accounts that yield 9.84% annually. How often is the interest compounded?

(A) daily   (B) monthly   (C) bi-monthly

(D) quarterly   (E) semiannually

$$r = \text{nominal interest rate} = 0.095$$

$$i = \text{effective interest rate} = 0.0984$$

$$m = \text{number of times interest is compounded per year}$$

$$i = \left(1 + \frac{r}{m}\right)^{m} - 1$$

$$0.0984 = \left(1 + \frac{0.095}{m}\right)^{m} - 1$$

Solve for m algebraically, or by trial-and-error using the five choices. The solution is $m = 4$.

> **Answer is (D)**

ECONOMICS-39

A firm is considering renting a trailer at \$300 per month. The unit is needed for five years. The leasing company offers a lump sum payment of \$24,000 at the end of five years as an alternative payment plan, but is willing to discount this figure. The firm places a value of 10% (effective annual rate) on invested capital. How large should the discount be in order to be acceptable as an equivalent?

(A) \$750   (B) \$820   (C) \$980

         (D) \$1030   (E) none of the above

For a 10% effective rate per year, the effective monthly rate, i, is:

$$0.1 = (1+i)^{12} - 1$$

$$i = 0.007974 \text{ per month}$$

At \$300 per month for five years:

$$F = 300(F/A, 0.7974\%, 60)$$
$$= 300(76.561)$$
$$= \$22,970$$
$$\text{discount} = 24,000 - F$$
$$= 24,000 - 22,970$$
$$= \$1030$$

Answer is (B)

ECONOMICS-40

Consider a deposit of \$1000, to be paid back in one year by \$975. What are the conditions on the rate of interest, $i\%$ per year compounded annually, such that the net present worth of the investment is positive? Assume $i \geq 0$.

(A) $0 \leq i < 50\%$
(B) $0 \leq i < 90\%$
(C) $12.5\% \leq i < 100\%$
(D) $0 \leq i \leq 16.7\%$
(E) There are no conditions on i that will make this possible.

set $P = 0$

$$= -1000 + 975(P/F, i, 1)$$

$$0 = -1000 + 975(P/F, i, 1)$$

$$= -1000 + 975 \left(\frac{1}{1+i} \right)$$

$$1 + i = 0.975$$

$$i = -0.025$$

Therefore, there is no solution such that $i \geq 0$, and $P \geq 0$.

$$\boxed{\text{Answer is (E)}}$$

ECONOMICS–41

Consider a deposit of $600, to be paid back in one year by $700. What are the conditions on the rate of interest, $i\%$ per year compounded annually, such that the net present worth of the investment is positive? Assume $i \geq 0$.

(A) $12.5\% \leq i < 14.3\%$
(B) $0 \leq i < 14.3\%$
(C) $0 \leq i < 16.7\%$
(D) $16.7 \leq i \leq 100\%$
(E) There are no conditions on i that will make this possible.

set $P > 0$

$$P > -600 + 700(P/F, i, 1)$$

$$0 > -600 + 700 \left(\frac{1}{1+i} \right)$$

$$\frac{7}{6} > 1 + i$$

$$i < 16.7\%$$

Thus, for $P > 0$, $0 \leq i < 16.7\%$.

$$\boxed{\text{Answer is (C)}}$$

ECONOMICS–42

A company has $100,000 to spend on various projects listed below. Using these projects only, what should this company consider its minimum attractive rate of return to be?

project	investment required	expected return
A	$10,000	14%
B	$25,000	10%
C	$50,000	12%
D	$40,000	16%
E	$25,000	11%
F	$30,000	10%
G	$20,000	12%

(A) 10%   (B) 11%   (C) 12%   (D) 14%   (E) 16%

The highest return projects should be chosen until all of the company's money is spent.

project	investment	return	cumulative investment
D	$40,000	16%	$40,000
A	$10,000	14%	$50,000
C	$50,000	12%	$100,000

The minimum attractive rate of return is 12%, the return on project C.

Answer is (C)

ECONOMICS–43

What is the internal rate of return on the following cash flow?

t_0	spend $100,000
t_1	spend $50,000
t_2	receive $100,000
t_3	receive $103,000

(A) 15.0%   (B) 17.5%   (C) 18.2%   (D) 20.0%   (E) 21.7%

The IRR is the interest rate which makes the present worth of the cash flow zero. It must be found by trial and error.

$$0 = -100,000 - 50,000(P/F, i\%, 1)$$
$$+ 100,000(P/F, i\%, 2) + 103,000(P/F, i\%, 3)$$

For an interest rate of 15%, the equation for the present worth of the cash flow yields:

$$\text{present worth} = -100,000 - 50,000(0.8696)$$
$$+ 100,000(0.7561) + 103,000(0.6575)$$
$$= -100,000 - 43,480 + 75,610 + 67,772$$
$$= \$148$$

For the size of the payments in the cash flow, $148 can be considered to be zero. Thus, the IRR is 15%.

Answer is (A)

ECONOMICS–44

Which of the following situations has a conventional cash flow so that an internal rate of return can be safely calculated and used?

(A) You purchase a house and pay the bank in monthly installments.

(B) You lease a car and pay by the month.

(C) Your company undertakes a mining project in which the land must be re-claimed at the end of the project.

(D) You invest in a safe dividend stock and receive dividends each year.

(E) Your company invests heavily in a new product that will generate profits for two years. To keep profits high for 10 years, the company plans to reinvest heavily after the two years.

The situation in choice (D) has a negative cash flow, one sign change, then positive cash flow. Thus, it is the only situation that has a conventional cash flow so that an IRR can be safely calculated and used.

Answer is (D)

ECONOMICS–45

A project has the cash flow shown in the figure. Theoretically, how many internal rates of return can be calculated for it?

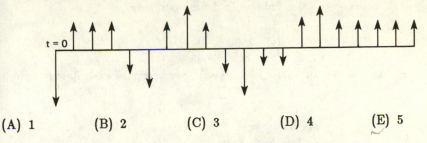

(A) 1   (B) 2   (C) 3   (D) 4   (E) 5

There are five places in the cash flow where there are sign changes. Thus, theoretically, five internal rates of return could be calculated for it.

Answer is (E)

ECONOMICS–46

An investment of $350,000 is made, to be followed by payments of $200,000 each year for three years. What is the annual rate of return on investment for this project?

(A) 15%   (B) 32.7%   (C) 41.7%   (D) 57.1%   (E) 71.4%

$$A = 200$$
$$= 350(A/P, i, 3)$$
$$(A/P, i, 3) = \frac{200}{350}$$
$$= 0.57143$$

interpolating from the tables,

i	$(A/P, i, 3)$
30%	0.55063
32.66%	0.57143
35%	0.58966

Thus, the IRR is approximately 32.7%.

Answer is (B)

ECONOMICS–47

A steel drum manufacturer incurs a yearly fixed operating cost of $200,000. Each drum manufactured costs $160 to produce and sells for $200. What is the manufacture's break-even sales volume in drums per year?

(A) 1000   (B) 1250   (C) 2500   (D) 5000   (E) 10,000

Given that x is the number of drums sold per year, the cash flow per year is:

operating cost	−$200,000
manufacturing cost	−$160x
sales	$200x

In order to break even, total cash flow must be zero.

$$-200,000 - 160x + 200x = 0$$
$$40x = 200,000$$
$$x = 5000$$

Answer is (D)

ECONOMICS–48

XYZ Corporation manufactures bookcases that it sells for $65 each. It costs XYZ $35,000 per year to operate its plant. This sum includes rent, depreciation charges on equipment, and salary payments. If the additional cost to produce one bookcase is $50, how many cases must be sold each year for XYZ to avoid taking a loss?

(A) 539 cases   (B) 750 cases   (C) 2333 cases

(D) 2334 cases   (E) 5390 cases

If Q = quantity sold
     a = incremental cost
     f = fixed cost
     p = fixed revenue

PROFESSIONAL PUBLICATIONS, INC. • Belmont, CA

Then,
$$Q = \frac{f}{p-a}$$
$$= \frac{35,000}{65-50}$$
$$= \frac{35,000}{15}$$
$$= 2333\tfrac{1}{3} \text{ cases/yr}$$

Therefore, XYZ must sell 2334 cases to avoid taking a loss.

Answer is (D)

ECONOMICS–49

A manufacturing firm maintains one product assembly line to produce signal generators. Weekly demand for the generators is 35 units, and the line operates for seven hours per day, five days per week. What is the maximum production time per unit, in hours, required of the line in order to meet demand?

(A) 0.75 hour   (B) 1 hour   (C) 2.25 hours
        (D) 5 hours   (E) 7 hours

$t = $ maximum production time per unit

$$\frac{1 \text{ week}}{35 \text{ units}} = t \left(\frac{1 \text{ day}}{7 \text{ hours}}\right) \left(\frac{1 \text{ week}}{5 \text{ days}}\right)$$
$$t = 1 \text{ hour per unit}$$

Answer is (B)

ECONOMICS–50

In determining the cost involved in fabricating subassembly B within a company, the following data have been gathered:

Total Cost of Manufacturing Subassembly B

item	cost
direct material	$0.30 per unit
direct labor	$0.50 per unit
tooling set up	$300 per set up

It is decided to subcontract the manufacturing of assembly B to an outside company. For an order of 100 units, which of the following unit price bids is unacceptable to the company?

(A) $3.50 per unit
(B) $3.65 per unit
(C) $3.75 per unit
(D) $4.10 per unit
(E) They are all reasonable bids.

For 100 units:

$$\text{cost per unit} = \frac{300}{100} + 0.30 + 0.50$$

$$\text{cost per unit} = \$3.80 \text{ per unit}$$

Thus, a bid of $4.10 is unacceptable.

Answer is (D)

ECONOMICS–51

The Economic Order Quantity (EOQ) is defined as the order quantity which minimizes the inventory cost per unit time. Which of the following is not an assumption of the basic EOQ model with no shortages?

(A) The demand rate is uniform and constant.
(B) There is a positive cost on each unit inventoried.
(C) The entire reorder quantity is delivered instantaneously.
(D) There is an upper bound on the quantity ordered.
(E) Reordering is done when the inventory is zero.

Recall:

$$EOQ = \sqrt{\frac{2aK}{h}}$$

where $a =$ the constant depletion rate (items per unit time)

$K =$ the fixed cost per order in dollars

$h =$ the inventory storage cost (dollars per item per unit time)

Thus, there is no upper bound on the quantity ordered.

Answer is (D)

ECONOMICS-52

Which of the following events will cause the optimal lot size, given by the classic EOQ model with no shortages, to increase?

(A) a decrease in inventory carrying cost
(B) a decrease in demand
(C) an increase in demand
(D) (A) or (C)
(E) none of the above

$$EOQ = \sqrt{\frac{2aK}{h}}$$

where EOQ = the order quantity

a = the constant depletion rate (items per unit time)

K = the fixed cost per order in dollars

h = the inventory storage cost (dollars per item per unit time)

Thus, a decrease in inventory carrying cost (h), or an increase in demand (a) will cause the optimal lot size to increase.

Answer is (D)

ECONOMICS-53

A manufacturer of sports equipment produces tennis rackets for which there is a demand of 200 per month. The production set up cost for each lot of rackets is $300. In addition, the inventory carrying cost for each racket is $24 per year. Using the EOQ model, which is the best production batch size for the rackets?

(A) 71 units   (B) 173 units   (C) 245 units
          (D) 346 units   (E) 353 units

$$EOQ = \sqrt{\frac{2aK}{h}}$$

$$a = 200 \; \frac{\text{rackets}}{\text{month}}$$

$$K = \$300$$

$$h = 24 \; \frac{\text{dollars}}{\text{yr}} \left(\frac{1 \; \text{yr}}{12 \; \text{months}} \right)$$

$$= 2\frac{\text{dollars}}{\text{rackets/month}}$$

$$\text{Therefore, EOQ} = \sqrt{\frac{2(200)(300)}{2}}$$
$$= 244.9$$
$$= 245 \text{ units}$$

Answer is (C)

ECONOMICS–54

For what range of discount rates is project 2 the most attractive project?

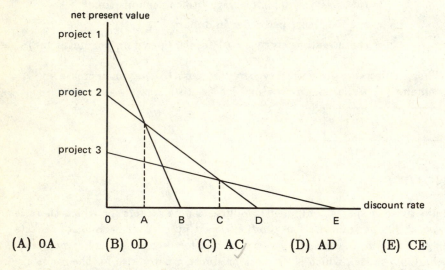

(A) 0A   (B) 0D   (C) AC   (D) AD   (E) CE

Project 2 has the highest net present worth over this range.

Answer is (C)

ECONOMICS–55

The internal rate of return of a project that involves an initial investment with subsequent positive cash flows is 18%. Five companies are considering the project. Given the following respective minimum attractive rates of return (MARR), which company will be most likely to accept the project?

company	MARR
A	12%
B	15%
C	17.5%
D	18%
E	19%

(A) company A (B) company B (C) company C

(D) company D (E) company E

The project is good for companies A, B, and C. However, it is the best for company A.

Answer is (A)

ECONOMICS–56

Two mutually exclusive projects are being considered. Project A requires an investment of $1,000,000 at year zero. Project A will pay $200,000 per year forever. Project B also requires an investment of $1,000,000 at year zero. However, it pays $1,500,000 the next year, and nothing after that. The internal rate of return (IRR) on project A is 20%. The IRR for project B is 50%. Which is the better project? The borrowing rate is 5%.

(A) Project A, because it has a lower IRR.
(B) Project B, because it has a higher IRR.
(C) The two projects are equivalent.
(D) Project A, because its net present worth is higher.
(E) Project B, because its net present worth is higher.

The use of IRR breaks down for mutually exclusive projects. There-fore, consider the net present worth (P) of each project. The net present worth of project A is:

$$P_A = -1,000,000 + 200,000(P/A, 5\%, n \longrightarrow \infty)$$
$$= -1,000,000 + 200,000 \lim_{n \to \infty} \left(\frac{1.05^n - 1}{0.05(1.05)^n} \right)$$
$$= -1,000,000 + 200,000(20)$$
$$= \$3,000,000$$

The net present worth of project B is:

$$P_B = -1,000,000 + 1,500,000(P/F, 5\%, 1)$$
$$= -1,000,000 + 1,500,000(0.9524)$$
$$= \$428,600$$

Since the net present worth of project A is higher, the company should choose project A.

Answer is (D)

ECONOMICS-57

Which plan is the least expensive way to purchase plant maintenance equipment? The discount rate is 11%.

plan A: $50,000 down, equal payments of
$25,115.12 for 20 years
plan B: nothing down, equal payments of
$31,393.91 for 20 years
plan C: $100,00 down, equal payments of
$21,975.74 for 20 years

(A) plan A   (B) plan B   (C) plan C

(D) plan A or B   (E) plan A or B or C

plan A
$$P = 50,000 + 25,115.12(P/A, 11\%, 20)$$
$$= \$250,000$$

plan B
$$P = 31,393.91(P/A, 11\%, 20)$$
$$= \$250,000$$

plan C
$$P = 100,000 + 21975.74(P/A, 11\%, 20)$$
$$= \$275,000$$

Thus, plan C is the most expensive and plans A and B are equivalent.

Answer is (D)

ECONOMICS-58

The volatility, β, of a stock is found to be 1.5 times the stock market average. If the risk primium for buying stocks averages 8.3% and the present treasury bill rate (assumed to be risk-free) is 7%, what is the expected return (ER) on the stock?

(A) 12.45%   (B) 15.3%   (C) 18.8%   (D) 19.45%   (E) 22.95%

$$ER = \text{risk free rate} + \beta(\text{market premium})$$
$$= 7\% + 1.5(8.3\%)$$
$$= 19.45\%$$

Answer is (D)

ECONOMICS-59

What is a borrower of a particular loan almost always required to do durin repayment?

(A) pay exactly the same amount of interest each payment
(B) repay the loan over an agreed-upon amount of time
(C) pay exactly the same amount of principal each payment
(D) (A) and (C)
(E) (B) and (C)

Answer is (B)

ECONOMICS–60

What is work-in-process classified as?

(A) an asset   (B) a liability   (C) an expense
          (D) a revenue   (E) owner's equity

Work in process is included in the working fund investments. The working fund investments is an asset not subject to depreciation.

Answer is (A)

ECONOMICS–61

What is the indirect product cost (IPC) spending variance?

(A) the difference between actual IPC and IPC absorbed

(B) the difference between actual IPC and IPC volume adjusted budget

(C) the IPC volume adjusted budget (fixed + volume(variable IPC rate))

(D) the IPC volume adjusted budget minus the total IPC absorbed

(E) [fixed + volume(variable IPC rate)]

$$- \left[\frac{\text{fixed + expected volume(variable IPC rate)}}{\text{expected volume}} + \text{actual volume} \right]$$

Answer is (B)

ECONOMICS–62

Firm A uses full absorption costing while firm B uses variable product costing. How will the financial statements of these companies differ?

(A) Firm A has a higher cost of goods sold and, therefore, a smaller profit.

(B) Firm A has a higher cost of goods sold, higher inventory value, and higher retained earnings.

(C) Firm A has a smaller cost of goods sold and a larger profit.

(D) Firm A has a smaller cost of goods sold, no change in inventory value, and no change in retained earnings.

(E) There are no differences.

Full absorption costing includes all direct and indirect, fixed and variable production costs. Variable product costing leaves fixed costs for the expense accounts. Therefore, the cost of goods sold is less under variable costing. Inventory value (an asset) is higher under full absorption. Since assets equal liabilities plus owner's equity, the owner's equity (retained earnings) must increase as the assets have.

Answer is (B)

ECONOMICS–63

How is the material purchase price variance defined?

(A) (quantity purchased × actual price)−(quantity purchased × standard price)
(B) (quantity issued − standard quantity)(standard price)
(C) (actual price − standard price)(quantity used)
(D) (quantity purchased − quantity used)(actual price)
(E) (quantity purchased − standard quantity)(actual price)

The definition of material purchase price variance is given in choice (A).

Note: choice (B) is the material usage variance.

Answer is (A)

ECONOMICS–64

Which of the following does not affect owner's equity?

(A) dividends paid
(B) license to start business
(C) invested capital
(D) expense to get license of start business
(E) revenues

The license to start business is a company asset, not a part of owner's equity.

Answer is (B)

ECONOMICS–65

Higrow Company is planning to grow 30% during the next fiscal year. What has to increase if Higrow is to achieve their goal?

(A) the ratio of sales to total assets
(B) the ratio of total assets to equity
(C) equity
(D) gross margin
(E) any combination of (A), (B), and (C)

$$\text{sales} = \frac{\text{sales}}{\text{total assets}} \times \frac{\text{total assets}}{\text{equity}} \times \text{equity}$$

Sales can grow only if at least one of the three terms on the right hand side of the equation grows. Gross margin is simply a measure of how much a product costs to make.

Answer is (E)

ECONOMICS–66

Tops Corporation's gross margin is 45% of sales. Operating expenses such as sales and administration are 15% of sales. Tops is in a 40% tax bracket. What percent of sales is their profit after taxes?

(A) 0%   (B) 5%   (C) 18%   (D) 24%   (E) 30%

$$\text{gross margin} = 45\%$$
$$\text{sales and administrative costs} = 15\%$$
$$\text{before tax profit} = 45 - 15$$
$$= 30\%$$
$$\text{after tax profit} = (1 - 0.40)(30\%)$$
$$= 18\%$$

Answer is (C)

ECONOMICS–67

Z Corporation is applying for a short term loan. In reviewing Z Corporation's financial records, the banker finds a current ratio of 2.0, an acid test ratio of 0.5, and an accounts receivable period of 70 days. What should the banker do?

(A) extend the loan, as Z Corporation will have no trouble repaying it
(B) be concerned that Z Corporation will be unable to meet the payments
(C) suggest that Z Corporation lower its inventories
(D) suggest that Z Corporation be more aggressive in collecting on its invoices
(E) (B), (C), and (D)

Z Corporation has invested heavily in inventory and accounts receivable. If it could change its accounts receivable collection period to 30 to 60 days and invest less in inventory, the company would probably not need the loan.

Answer is (E)

ECONOMICS–68

Companies A and B are identical except for their inventory accounting systems. Company A uses the last-in, first-out convention while company B uses the first-in, first-out convention. How will their financial statements differ in an inflationary environment?

(A) Company A's profits will be higher and the book value of their inventory will be higher than for company B.

(B) Company A's profits will be higher and the book value of their inventory will be lower than for company B.

(C) Company B's profits and inventory book value will be higher than for company A.

(D) Company B's profits will be higher than A's, but inventory book value will be lower.

(E) The statements will be identical.

Last-in, first-out (LIFO) puts a higher value on the inventory that went into the cost of goods sold. Thus, the gross margin is lowered and profits are lowered. The remaining inventory is still valued at old prices, so its value is also low.

Answer is (C)

ECONOMICS–69

What is the acid test ratio?

(A) the ratio of owner's equity to total current liabilities
(B) the ratio of all assets to total liabilities
(C) the ratio of current assets (exclusive of inventory) to total current liabilities
(D) the ratio of gross margin to operating, sales, and administrative expenses
(E) the ratio of profit after taxes to equity

Answer is (C)

ECONOMICS–70

The balance sheet of Allied Company is as follows:

assets		liabilities	
cash	$10,000	payables	$17,000
receivables	12,000	notes due	6,000
inventory	7,000	long term debt	3,000
capital equipment	20,000	owner's equity	23,000
total:	$49,000	total:	$49,000

What is its acid test ratio?

(A) 0.385   (B) 0.592   (C) 0.846   (D) 1.115   (E) 1.706

$$\text{acid test ratio} = \frac{\text{cash} + \text{accounts receivable}}{\text{total liabilities}}$$

$$= \frac{10,000 + 12,000}{17,000 + 6000 + 3000}$$

$$= \frac{22}{26}$$

$$= 0.846$$

Answer is (C)

3 SYSTEMS OF UNITS

UNITS–1

An iron block weighs 5 newtons and has a volume of 200 cubic centimeters. What is the density of the block?

(A) 800 kg/m³

(B) 988 kg/m³

(C) 1255 kg/m³

(D) 2550 kg/m³

(E) 3450 kg/m³

$$\rho = \frac{m}{V}$$

$$W = mg$$

$$m = \frac{W}{g}$$

$$= \frac{5\,\mathrm{N}}{9.81\,\frac{\mathrm{m}}{\mathrm{s}^2}}$$

$$= 0.51\,\mathrm{kg}$$

$$\rho = \frac{0.51\,\mathrm{kg}}{200\,\mathrm{cm}^3}$$

$$= 0.0025\,\frac{\mathrm{kg}}{\mathrm{cm}^3}\left(\frac{100\,\mathrm{cm}}{1\,\mathrm{m}}\right)^3$$

$$= 2550\,\mathrm{kg/m}^3$$

Answer is (D)

UNITS-2

If the density of a gas is 0.003 slugs per cubic foot, what is the specific weight of the gas?

(A) 9.04 N/m³   (B) 15.2 N/m³   (C) 76.3 N/m³
          (D) 98.2 N/m³   (E) 111.1 N/m³

The specific weight, γ, is defined as follows:

$$\gamma = \rho g$$

$$= 0.003 \frac{\text{slug}}{\text{ft}^3} \left(32.2 \frac{\text{ft}}{\text{sec}^2}\right) \left(14.59 \frac{\text{kg}}{\text{slug}}\right) \left(\frac{1 \text{ ft}}{0.3048 \text{ m}}\right)^2$$

$$= 15.2 \frac{\text{kg}}{\text{s}^2 \cdot \text{m}^2}$$

$$= 15.2 \text{ N/m}^3$$

Answer is (B)

UNITS-3

The specific gravity of mercury relative to water is 13.55. What is the specific weight of mercury? (The specific weight of water is 62.4 lbf per cubic foot.)

(A) 82.2 kN/m³   (B) 102.3 kN/m³   (C) 132.9 kN/m³
          (D) 150.9 kN/m³   (E) 240.1 kN/m³

$$\text{specific weight} = \gamma$$
$$\text{specific gravity} = \rho$$
$$\gamma = \rho g$$
$$\frac{\rho_{mercury}}{\rho_{water}} = 13.55$$
$$\frac{\gamma_{mercury}}{\gamma_{water}} = 13.55$$
$$\gamma_{mercury} = 13.55 \gamma_{water}$$

$$= 13.55 \left(62.4 \frac{\text{lbf}}{\text{ft}^3}\right) \left(4.45 \frac{\text{N}}{\text{lbf}}\right) \left(\frac{1 \text{ ft}}{0.3048 \text{ m}}\right)^3$$

$$= 132.9 \text{ kN/m}^3$$

Answer is (C)

UNITS-4

If the specific weight of a liquid is 58.5 lbf per cubic foot, what is the specific volume of the liquid?

(A) 0.5321 cm^3/g   (B) 0.6748 cm^3/g   (C) 0.9504 cm^3/g
        (D) 1.0675 cm^3/g   (E) 1.5502 cm^3/g

$$\text{specific weight} = \gamma$$

$$\text{specific volume} = \frac{1}{\rho}$$

$$\gamma = 58.5 \, \frac{\text{lbf}}{\text{ft}^3}$$

$$= 58.5 \, \frac{\text{lbf}}{\text{ft}^3} \left(4.449 \, \frac{\text{N}}{\text{lbf}} \right) \left(\frac{1 \, \text{ft}}{0.3048 \, \text{m}} \right)^3$$

$$= 9189.6 \, \frac{\text{N}}{\text{m}^3}$$

$$= \rho g$$

$$\rho = \frac{\gamma}{g}$$

$$= 936.8 \, \frac{\text{kg}}{\text{m}^3}$$

$$v = \frac{1}{\rho}$$

$$= 0.0010675 \, \frac{\text{m}^3}{\text{kg}} \left(100 \, \frac{\text{cm}}{\text{m}} \right)^3 \left(\frac{1 \, \text{kg}}{1000 \, \text{g}} \right)$$

$$= 1.0675 \, \text{cm}^3/\text{g}$$

Answer is (D)

UNITS-5

Which of the following are not units of pressure?

(A) Pa   (B) N/m^2 (C) kg/m·s^2
        (D) bars   (E) kg/m^2

All of the above are units of pressure (force over area) except for (E), which has units of mass over length squared.

Answer is (E)

UNITS-6

A cylinder weighs 150 lbf. Its cross-sectional area is 40 square inches. When the cylinder stands vertically on one end, what pressure does the cylinder exert on the floor?

(A) 14.1 kPa   (B) 25.8 kPa   (C) 63.2 kPa

(D) 89.7 kPa   (E) 123 kPa

$$p = \frac{150\,\text{lbf}}{40\,\text{in}^2}$$

$$= 3.75\,\text{psi} \left(4.448\,\frac{\text{N}}{\text{lbf}}\right) \left(\frac{1\,\text{in}}{2.54\,\text{cm}}\right)^2 \left(\frac{1\,\text{cm}}{100\,\text{m}}\right)^2$$

$$= 25.8\,\text{kPa}$$

Answer is (B)

UNITS-7

What pressure is a column of water 100 centimeters high equivalent to?

(A) 9810 dyne/cm^2 (B) 9810 N/m^2 (C) 0.1 bar

(D) 0.1 atm   (E) 98,100 N/m^2

$$p = \rho g h$$

$$= 1000\,\frac{\text{kg}}{\text{m}^3} \left(9.81\,\frac{\text{m}}{\text{s}^2}\right) \left(0.1\,\text{m}\right)$$

$$= 9810\,\frac{\text{kg}}{\text{m}\cdot\text{s}^2}$$

$$= 9810\,\text{N/m}^2$$

Answer is (B)

UNITS-8

Determine the stress on a circular rod 10″ in diameter when a force of 10 newtons acts on one end.

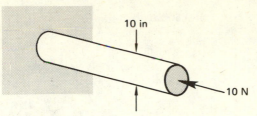

(A) 0.05 kPa   (B) 0.10 kPa   (C) 0.15 kPa
     (D) 0.20 kPa   (E) 0.25 kPa

$$\text{stress} = \sigma$$
$$= \frac{F}{A}$$
$$= \frac{10\,\text{N}}{A}$$
$$A = \pi r^2$$
$$= \pi \left[\left(5\,\text{in}\right) \left(\frac{2.54\,\text{cm}}{1\,\text{in}}\right) \left(\frac{1\,\text{m}}{100\,\text{cm}}\right) \right]^2$$
$$= 0.0507\,\text{m}^2$$
$$\sigma = \frac{10\,\text{N}}{0.0507\,\text{m}^2}$$
$$= 200\,\text{Pa}$$
$$= 0.20\,\text{kPa}$$

Answer is (D)

UNITS-9

Water is flowing in a pipe with a radius of 10″ at a velocity of 5 m/s. At the temperature in the pipe, the density and viscosity of the water are as follows:

$$\rho = 997.9\ \text{kg/m}^3$$
$$\mu = 1.131\ \text{Pa·s}$$

What is the Reynold's number for this situation?

(A) 44.1   (B) 88.2   (C) 1140   (D) 2241   (E) 3100

$$\text{Re} \equiv \frac{\rho v D}{\mu}$$

$$D = 2\left(10\,\text{in}\right)\left(0.0254\,\frac{\text{m}}{\text{in}}\right)$$

$$= 0.508\,\text{m}$$

$$v = 5\,\frac{\text{m}}{\text{s}}$$

$$\mu = 1.131\,\text{Pa}\cdot\text{s}\left(\frac{\frac{\text{N}}{\text{m}^2}}{\text{Pa}}\right)\left(\frac{\text{kg}\cdot\frac{\text{m}}{\text{s}^2}}{\text{N}}\right)$$

$$= 1.131\,\frac{\text{kg}}{\text{m}\cdot\text{s}}$$

$$\text{Re} = \frac{\left(998\,\frac{\text{kg}}{\text{m}^3}\right)\left(5\,\frac{\text{m}}{\text{s}}\right)\left(0.508\,\text{m}\right)}{1.131\,\frac{\text{kg}}{\text{m}\cdot\text{s}}}$$

$$= 2241$$

(Reynold's number is dimensionless)

Answer is (D)

UNITS–10

What is the flow rate through a pipe 4″ in diameter carrying water at a velocity of 11 ft/sec?

(A) 590 cm³/s   (B) 726 cm³/s   (C) 993 cm³/s

(D) 19,200 cm³/s   (E) 27,200 cm³/s

$$Q = vA$$

$$A = \pi r^2$$

$$= \pi (2\,\text{in})^2$$

$$= 12.5\,\text{in}^2 \left(\frac{1\,\text{ft}}{12\,\text{in}}\right)^2$$

$$= 0.0872\,\text{ft}^2$$

$$Q = \left(11\,\frac{\text{ft}}{\text{sec}}\right)\left(0.0872\,\text{ft}^2\right)$$

$$= \left(0.959\,\frac{\text{ft}^3}{\text{sec}}\right)\left(\frac{12\,\text{in}}{1\,\text{ft}}\right)^3 \left(\frac{2.54\,\text{cm}}{1\,\text{in}}\right)^3$$

$$= 27,184.1\,\frac{\text{cm}^3}{\text{s}}$$

$$\approx 27,200\,\text{cm}^3/\text{s}$$

Answer is (E)

UNITS–11

How long must a current of 5.0 amperes pass through a 10 ohm resistor until a charge of 1200 coulombs passes through?

(A) 1 min   (B) 2 min   (C) 3 min   (D) 4 min   (E) 5 min

$$\text{current} \equiv I$$

$$= \frac{\text{charge}}{\text{time}}$$

$$t = \frac{1200\,\text{C}}{5.0\,\text{A}}$$

$$= 240\,\text{sec}$$

$$= 4\ \text{min}$$

Answer is (D)

UNITS–12

A car moving at 70 km/hr has a mass of 1700 kg. What force is necessary to decelerate it at a rate of 40 cm/s²?

(A) 0.680 N   (B) 42.5 N   (C) 680 N   (D) 4250 N   (E) 68 kN

Use Newton's second law.

$$\mathbf{F} = m\mathbf{a}$$
$$\mathbf{a} = 40\,\frac{\text{cm}}{\text{s}^2}\left(\frac{1\,\text{m}}{100\,\text{cm}}\right)$$
$$= 0.4\,\frac{\text{m}}{\text{s}^2}$$
$$\mathbf{F} = 1700\,\text{kg}\left(0.4\,\frac{\text{m}}{\text{s}^2}\right)$$
$$= 680\,\text{N}$$

Answer is (C)

UNITS–13

One hundred milliliters of water in a plastic bag of negligible mass is to be catapulted upwards with an initial acceleration of 20.0 m/s². What force is necessary to do this? Assume that gravity is 9.81 m/s² and the density of water is 1 g/cm³.

(A) 2.00 N   (B) 2.98 N   (C) 15.0 N   (D) 2.00 kN   (E) 2.98 kN

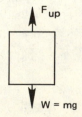

Use Newton's second law.

$$\mathbf{F} - mg = m\mathbf{a}$$
$$\mathbf{F} = m(g + \mathbf{a})$$
$$m = \rho V$$
$$= \left(1 \frac{g}{cm^3}\right)\left(100\,cm^3\right)\left(\frac{1\,kg}{100\,g}\right)$$
$$= 0.10\,kg$$
$$\mathbf{F} = \left(0.1\,kg\right)\left(9.81\,\frac{m}{s^2} + 20\,\frac{m}{s^2}\right)$$
$$= 2.981\,\frac{kg \cdot m}{s^2}$$
$$= 2.981\,N$$

Answer is (B)

UNITS–14

For the cantilever and applied force shown, calculate the resisting torque at the wall.

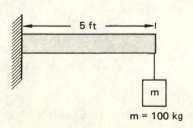

5 ft

m

m = 100 kg

(A) 5000 N·m

(B) 15,000 N·m

(C) 49,000 N·m

(D) 167,000 N·m

(E) 500,000 N·m

For the cantilever to be in static equilibrium, the sum of the moments taken at the wall must be zero.

$$\sum \text{moments} = 0$$

$$T_{resist} - T_{applied} = 0$$

$$T_{resist} = T_{applied}$$

$$= \ell \times F_{applied}$$

$$\ell = 5\,\text{ft}\left(\frac{0.3048\,\text{m}}{\text{ft}}\right)$$

$$= 1.524\,\text{m}$$

$$\mathbf{F} = m\mathbf{a}$$

$$= 100\,\text{kg}\left(9.81\,\frac{\text{m}}{\text{s}^2}\right)$$

$$= 9810\,\text{N}$$

$$T_{resist} = (1.524\,\text{m})(9810\,\text{N})$$

$$= 14{,}950\ \text{N·m}$$

$$\approx 15{,}000\ \text{N·m}$$

Answer is (B)

UNITS–15

Which of the following is not a unit of work?

(A) N·m   (B) erg   (C) kg · m^2/s^2
                  (D) dyne   (E) W·s

The units of work are force times distance or power multiplied by time. Therefore, all of the choices are units of work except for the dyne. A dyne is a unit of force.

Answer is (D)

UNITS–16

Which of the following is the definition of a joule?

(A) a unit of power
(B) a newton·meter
(C) a kg·m/s^2
(D) a rate of change of energy
(E) all of the above

A joule is a unit of energy and is defined as a newton·meter. None of the other choices are units of energy.

Answer is (B)

UNITS–17

A boy pulls a sled with a mass of 20 kg horizontally over a surface with a coefficient of friction of 0.20. It takes him 10 minutes to pull the sled 100 yards. What is his average power output over these 10 minutes?

(A) 4 W   (B) 6 W   (C) 8 W   (D) 10 W   (E) 12 W

$$\overline{P} = \frac{dW}{dt}$$
$$= \frac{\mathbf{F} \cdot d\mathbf{x}}{dt}$$

The force that the boy must pull with, $\mathbf{F}_b$, must be large enough to overcome the frictional force.

$$\mathbf{F}_b = \mathbf{F}_f$$
$$= \left(20\,\text{kg}\right)\left(9.81\,\frac{\text{m}}{\text{s}^2}\right)\left(0.20\right)$$
$$= 39.24\,\text{N}$$
$$dW = \mathbf{F}_b \cdot d\mathbf{x}$$
$$= \left(39.24\,\text{N}\right)\left(100\,\text{yd}\right)\left(\frac{3\,\text{ft}}{\text{yd}}\right)\left(\frac{0.3048\,\text{m}}{\text{ft}}\right)$$
$$W = 3588\,\text{J}$$
$$\overline{P} = \frac{3588\,\text{J}}{10\,\text{min}}\left(\frac{1\,\text{min}}{60\,\text{s}}\right)$$
$$= 5.98$$
$$\approx 6\,\text{W}$$

Answer is (B)

UNITS–18

A force of 200 lbf acts on a block at an angle of 28° with respect to horizontal. The block is pushed 2 feet horizontally. What is the work done by this force?

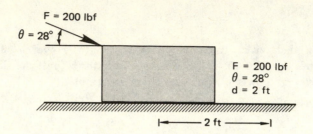

(A) 215 J   (B) 320 J   (C) 480 J

    (D) 540 J   (E) 680 J

The work done by the force is:

$$W = \mathbf{F} \cdot d\mathbf{x}$$
$$= Fx \cos \theta$$
$$= (200 \text{ lbf})(2 \text{ ft}) \cos 28°$$
$$= \left(353.18 \text{ ft-lbf}\right)\left(4.45 \frac{N}{\text{lbf}}\right)\left(0.3048 \frac{m}{\text{ft}}\right)$$
$$= 480 \text{ J}$$

Answer is (C)

UNITS–19

Two particles collide, stick together, and continue their motion together. Each particle has a mass of 10 g, and their respective velocities before the collision were 10 m/s and 100 m/s. What is the energy of the system after the collision?

(A) 21.8 J   (B) 30.2 J   (C) 48.2 J   (D) 77.9 J   (E) 100 J

$$E_f = \frac{1}{2} M_{tot} v_f^2$$

$$M_{tot} = m_1 + m_2$$

v_f = the velocity of the two masses after the collision

Use the principle of conservation of momentum.

$$m_1 v_1 + m_2 v_2 = M_{tot} v_f$$

$$\left(10\,\text{g}\right)\left(10\,\frac{\text{m}}{\text{s}} + 100\,\frac{\text{m}}{\text{s}}\right) = \left(20\,\text{g}\right)v_f$$

$$v_f = 55\,\frac{\text{m}}{\text{s}}$$

$$E_f = \frac{1}{2}\left(0.02\,\text{kg}\right)\left(55\,\frac{\text{m}}{\text{s}}\right)^2$$

$$= 30.25\,\frac{\text{kg}\cdot\text{m}^2}{\text{s}^2}$$

$$= 30.25\,\text{J}$$

Answer is (B)

UNITS–20

Two protons, each of charge 1.6×10^{-19} coulomb, are 3.4 micrometers apart. What is the change in the potential energy of the protons if they are brought 63 nanometers closer together?

(A) 6.4×10^{-29} J   (B) 7.16×10^{-29} J   (C) 1.28×10^{-24} J

(D) 3.21×10^{-24} J   (E) 5.80×10^{-20} J

The potential energy of a system of two charges is given by the following:

$$U = k\frac{q_1 q_2}{r}$$

$$k = 8.99 \times 10^9 \frac{N \cdot m^2}{C^2}$$

q_1, q_2 = charges

r = the distance between the charges

$$\Delta U = U_f - U_i$$

$$= k\frac{q_1 q_2}{r_f} - k\frac{q_1 q_2}{r_i}$$

$$= kq_1 q_2 \left(\frac{1}{r_f} - \frac{1}{r_i}\right)$$

$$= \left(8.99 \times 10^9 \frac{N \cdot m^2}{C^2}\right) \left(1.60 \times 10^{-19} C\right)^2$$

$$\times \left(\frac{1}{3.4 \times 10^{-6} m} - \frac{1}{63 \times 10^{-9} m}\right)$$

$$= 1.28 \times 10^{-24} J$$

Answer is (C)

UNITS-21

According to the Bohr model of the atom, the energy of the atom when the electron is at the first Bohr radius is given by:

$$E_0 = \frac{mk^2 e^4}{2\hbar^2}$$

where

m = mass of the electron = $9.1095 \times 10^{-31} kg$

k = Coulomb's constant = $8.99 \times 10^9 \frac{N \cdot m^2}{C^2}$

e = electron charge = $1.602 \times 10^{-19} C$

$\hbar = \dfrac{h}{2\pi}$

h = Planck's constant = $6.626 \times 10^{-34} J \cdot s$

Using these values in metric units, calculate the value of E_0.

(A) 2.18×10^{-36} W   (B) 2.18×10^{-36} J   (C) 2.18×10^{-18} J
   (D) 2.18×10^{-18} W   (E) 2.18×10^{-18} N

$$E_0 = \frac{\left(9.1095 \times 10^{-31}\,\text{kg}\right)\left(8.99 \times 10^9\,\frac{\text{N·m}^2}{\text{C}^2}\right)^2\left(1.602 \times 10^{-19}\,\text{C}\right)^4}{2\left(\frac{6.626 \times 10^{-34}\,\text{J·s}}{2\pi}\right)^2}$$

$$= 2.18 \times 10^{-18}\,\text{J}$$

Answer is (C)

UNITS-22

A copper bar is 90 centimeters long at 86 °F. What is the increase in its length when the bar is heated to 95 °F? The linear expansion coefficient for copper, α, is $1.7 \times 10^{-5}\,1/°\text{C}$.

(A) 2.12×10^{-5} m   (B) 3.22×10^{-5} m   (C) 5.25×10^{-5} m
    (D) 7.65×10^{-5} m   (E) 8.74×10^{-5} m

The change in length of the bar is given by the following:

$$\Delta L = \alpha L_0 \Delta T$$

Convert the temperatures from °F to °C.

$$°\text{C} = \frac{5}{9}(°\text{F} - 32)$$

$$T_1 = \frac{5}{9}(95 - 32) = 35°\text{C}$$

$$T_2 = \frac{5}{9}(86 - 32) = 30°\text{C}$$

$$\Delta T = T_2 - T_1$$

$$= 35 - 30$$

$$= 5°\text{C}$$

$$\Delta L = \left(1.77 \times 10^{-5}\,\frac{1}{°\text{C}}\right)\left(0.9\,\text{m}\right)\left(5\,°\text{C}\right)$$

$$= 7.65 \times 10^{-5}\,\text{m}$$

Answer is (D)

UNITS–23

A slab of iron with temperature, $T_{i1} = 48\,°C$ is used to heat a flat glass plate that has an initial temperature of $T_{g1} = 18\,°C$. Assuming no heat is lost to the environment, and the masses are $m_i = 0.40\,kg$ for the slab and $m_g = 310\,g$ for the plate, what is the amount of heat transferred when the two have reached equal temperatures? Assume $c_i = 0.11\,kcal/kg·\,°C$ for iron, and $c_g = 0.20\,kcal/kg·°C$ for glass.

(A) 860 cal   (B) 32 kcal   (C) 53 kcal
          (D) 320 kcal   (E) 860 kcal

The heat transferred by an object is given by the following equation:

$$Q = mc\Delta T$$

where

Q = heat transferred

m = mass of the object

ΔT = change in temperature of the object

c = specific heat capacity of the object

Since no heat has been lost, $Q_i = Q_g$.

$$m_i c_i \Delta T_i = m_g c_g \Delta T_g$$

The final temperature of each object is the same. Thus, the above equation becomes:

$$-m_i c_i\,(T_2 - T_{i1}) = m_g c_g\,(T_2 - T_{g1})$$

Solve the equation for T_2, then substitute in to find Q.

$$T_2 = \frac{m_i c_i T_{1i} + m_g c_g T_{1g}}{m_i c_i + m_g c_g}$$

$$= \frac{\left(0.49\,kg\right)\left(110\,\frac{cal}{kg\,·°\,C}\right)\left(48\,°C\right) + \left(0.310\,kg\right)\left(200\,\frac{cal}{kg·°C}\right)\left(18\,°C\right)}{\left(0.49\,kg\right)\left(110\,\frac{cal}{kg·°C}\right) + \left(0.310\,kg\right)\left(200\,\frac{cal}{kg·°C}\right)}$$

$$= 32.0\,°C$$

$$Q_i = m_i c_i\,(32\,°C - 48\,°C)$$

$$= \left(0.490\,kg\right)\left(0.11 \times 10^3\,\frac{cal}{kg\,·°\,C}\right)\left(-16\,°C\right)$$

$$= -860\,cal$$

Thus, 860 calories were transferred from the iron tray to the glass plate.

Answer is (A)

UNITS-24

What is the average thermal conductivity for the composite material shown?

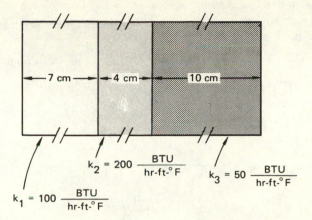

(A) 75 W/m · K   (B) 100 W/m · K   (C) 125 W/m · K
          (D) 150 W/m · K   (E) 175 W/m · K

The electrical analog to this configuration is resistors connected in parallel. Thus, the following equation is true for the conductivities of the material.

$$\frac{\ell_{tot}}{\bar{k}A} = \frac{1}{A}\left(\frac{\ell_1}{k_1} + \frac{\ell_2}{k_2} + \frac{\ell_3}{k_3}\right)$$

$$\frac{\ell_{tot}}{\bar{k}} = \left(\frac{0.07\,\text{m}}{100\,\frac{\text{BTU}}{\text{hr-ft-}^\circ\text{F}}} + \frac{0.04\,\text{m}}{200\,\frac{\text{BTU}}{\text{hr-ft-}^\circ\text{F}}} + \frac{0.10\,\text{m}}{50\,\frac{\text{BTU}}{\text{hr-ft-}^\circ\text{F}}}\right)\left(1.7305\,\frac{\frac{\text{BTU}}{\text{hr-ft-}^\circ\text{F}}}{\frac{\text{W}}{\text{m·K}}}\right)$$

$$= \frac{0.07\,\text{m}}{173.05\,\frac{\text{W}}{\text{m·K}}} + \frac{0.04\,\text{m}}{346.1\,\frac{\text{W}}{\text{m·K}}} + \frac{0.10\,\text{m}}{86.52\,\frac{\text{W}}{\text{m·K}}}$$

$$= 0.0017\,\frac{\text{m}^2\cdot\text{K}}{\text{W}}$$

$$\ell_{tot} = 0.21\,\text{m}$$

$$\bar{k} = \frac{0.21\,\text{m}}{0.0017\,\frac{\text{m}^2\text{·K}}{\text{W}}}$$

$$= 125\ \text{W/m} \cdot \text{K}$$

Answer is (C)

UNITS–25

Calculate the energy transfer rate across a 6″ wall of firebrick with a temperature difference across the wall of 50 °C. The thermal conductivity of firebrick is 0.65 BTU/hr-ft-°F at the temperature of interest.

(A) 112 W/m²   (B) 285 W/m²   (C) 369 W/m²

         (D) 429 W/m²   (E) 897 W/m²

$$\Delta T = 50\,^\circ\text{C} \times \frac{9}{5} = 90\,^\circ\text{F}$$

$$\dot{Q} = -k\frac{dT}{dx}$$

$$= -k\frac{\Delta T}{\Delta x} \quad \text{(for linear profile)}$$

$$\dot{Q} = \frac{\left(0.65\,\frac{\text{BTU}}{\text{hr-ft-}^\circ\text{F}}\right)(90\,^\circ\text{F})}{(6\,\text{in})\left(\frac{1\,\text{ft}}{12\,\text{in}}\right)} = 117\,\text{BTU/hr-ft}^2$$

$$Q = \left(117\frac{\text{BTU}}{\text{hr-ft}^2}\right)\left(0.2931\,\frac{\text{W-hr}}{\text{BTU}}\right)\left(3.281\,\frac{\text{ft}}{\text{m}}\right)^2$$

$$= 369\ \text{W/m}^2$$

Answer is (C)

UNITS–26

A house has brick walls 15 millimeters thick. On a cold winter day, the temperatures of the inner and outer surfaces of the walls are measured and found to be 20°C and −12 °C, respectively. If there is 120 m² of exterior wall space, and the thermal conductivity of brick is 0.711 J/m·s·° C, how much heat is lost through the walls per hour?

(A) 182 J   (B) 12.5 kJ   (C) 655 kJ   (D) 2150 kJ   (E) 655 MJ

$$\dot{Q} = \frac{kA\Delta T}{x}$$

where k = thermal conductivity

$$= 0.711 \frac{J}{m \cdot s \cdot °C}$$

A = area of the surface

$$= 120\,m^2$$

ΔT = temperature difference

$$= 20\,°C - (-12\,°C)$$

$$= 32\,°C$$

x = the thickness of the wall

$$= 15\,mm$$

$$= 0.015\,m$$

$$\dot{Q} = \frac{\left(0.711 \frac{J}{m \cdot s \cdot °C}\right)\left(120\,m^2\right)\left(3600 \frac{s}{hr}\right)\left(32\,°C\right)}{0.015\,m}$$

$$= 655 \times 10^6 \text{ J/hr}$$

$$= 655 \text{ MJ/hr}$$

Thus, the heat transferred per hour is 655 MJ.

Answer is (E)

UNITS–27

Air has a specific heat of $1\,kJ/kg \cdot K$. If 2 BTU of energy is added to 100 g of air, what is the change in air temperature?

(A) 10.0 °C (B) 21.1 °C (C) 44.1 °C

(D) 88.2 °C (E) none of the above

$$Q = mc_p \Delta T$$

$$\Delta T = \frac{Q}{mc_p}$$

$$Q = 2 \, \text{BTU} \left(1.055 \, \frac{\text{kJ}}{\text{BTU}} \right)$$

$$= 2.11 \, \text{kJ}$$

$$m = 100 \, \text{g} \left(\frac{1000 \, \text{g}}{\text{kg}} \right)$$

$$= 0.10 \, \text{kg}$$

$$c_p = 1 \, \frac{\text{kJ}}{\text{kg} \cdot \text{K}}$$

$$\Delta T = \frac{2.11 \, \text{kJ}}{\left(0.10 \, \text{kg} \right) \left(1 \, \frac{\text{kJ}}{\text{kg} \cdot \text{K}} \right)}$$

$$= 21.1 \, \text{K}$$

$$= 21.1 \, ^\circ\text{C}$$

Note: the temperature differences of 1 K and 1 °C are equivalent.

Answer is (B)

UNITS–28

Air has a specific heat (c_p) of $1 \, \text{kJ/kg} \cdot \text{K}$. If 100 g of air are heated with a 1500 W heater, which of the following occurs?

I.   The air heats up at a rate of 15 K/s.
II.   The air reaches a final temperature of 1500 K.
III.   The air undergoes a nonisentropic process.

(A) I only   (B) I and II   (C) I, II, and III

(D) II and III   (E) I and III

$$Q = mc_p\Delta T$$

$$\dot{Q} = mc_p\Delta\dot{T}$$

$$\Delta\dot{T} = \frac{\dot{Q}}{mc_p}$$

$$= \frac{1500\,\frac{\text{J}}{\text{s}}}{\left(0.1\,\text{kg}\right)\left(1\,\frac{\text{kJ}}{\text{kg}\cdot\text{K}}\right)}$$

$$= 15\,\text{K/s}$$

Thus, I is correct.

It is not possible to predict the final temperature of the air without knowing the length of time it is being heated. Therefore, II is false.

The addition of heat is a nonisentropic process (the entropy of the air changes). Therefore, III is correct.

$\boxed{\text{Answer is (E)}}$

UNITS–29

The change in enthalpy of an incompressible liquid with constant specific heat is given by:

$$h_2 - h_1 = c(T_2 - T_1) + v(p_2 - p_1)$$

where $\quad T_n$ = temperature at a state n

$\qquad p_n$ = pressure at state n

$\qquad v$ = specific volume of the liquid

Water, with $c_p = 4.18$ kJ/kg·K and $v = 1.00 \times 10^{-3}$ m^3/kg has the following final states:

$$\text{state I:} \quad T_1 = 19°\text{C} \quad p_1 = 1.013 \times 10^5\,\text{Pa}$$
$$\text{state II:} \quad T_2 = 30°\text{C} \quad p_2 = 0.113\,\text{MPa}$$

What is the change in enthalpy from state I to state II?

(A) 46.0 kJ/kg   (B) 46.0 kN/kg   (C) 46.0 kPa/kg

(D) 56.0 kJ/kg   (E) 56.0 kPa/kg

Before plugging into the equation, express the two pressures in consistent units.

$$p_1 = 101,300\,\text{Pa}$$
$$p_2 = 111,300\,\text{Pa}$$

Temperature differences need not be in K, since a temperature difference of $1\,°C$ equals a temperature difference of 1 K. The specific volume, v, and specific heat, c_p, are already in consistent units.

$$h_2 - h_1 = \left(4180\,\frac{\text{J}}{\text{kg}\cdot\text{K}}\right)\left(30\,°\text{C} - 19\,°\text{C}\right)$$
$$+ \left(111,300\,\text{Pa} - 101,300\,\text{Pa}\right)\left(0.001\,\frac{\text{m}^3}{\text{kg}}\right)$$
$$= 46,000\,\frac{\text{J}}{\text{kg}}$$
$$= 46.0\,\text{kJ/kg}$$

Answer is (A)

UNITS-30

In a constant temperature, closed system process, 100 BTU of heat is transferred to the working fluid at 100 °F. What is the change in entropy of the working fluid?

(A) 0.18 kJ/K   (B) 0.25 kJ/K   (C) 0.34 kJ/K
      (D) 0.57 kJ/K   (E) 1.0 kJ/K

For a closed system at a constant temperature:

$$S_2 - S_1 = \frac{dQ}{T}$$
$$= \frac{100\,\text{BTU}}{560\,°\text{R}}$$
$$= 0.179\,\frac{\text{BTU}}{°\text{R}}\left(\frac{1.06\,\text{kJ}}{\text{BTU}}\right)\left(\frac{1.8\,°\text{R}}{\text{K}}\right)$$
$$= 0.34\,\text{kJ/K}$$

Answer is (C)

UNITS-31

In the figure shown, pressure is plotted using $10,000 \text{ N/m}^2$ per division, volume is plotted using 1 liter per division, and the area enclosed by the cycle is 3 divisions squared. What is the work done during the cycle?

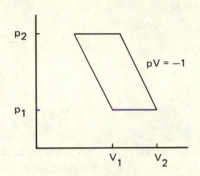

(A) 0.03 W·s   (B) 3×10^7 ergs   (C) 30 J

(D) 30,000 N·m   (E) 30,000 W·s

$$W = \mathbf{F} \cdot \mathbf{x}$$
$$= p \cdot V$$
$$= \text{area enclosed by the cycle}$$
$$= 3 \left(10,000 \, \frac{\text{N}}{\text{m}^2} \right) \left(1 \, l \right)$$
$$= \left(30,000 \, \frac{\text{N}}{\text{m}^2} \right) \left(0.001 \, \text{m}^3 \right)$$
$$= 30 \text{ N·m}$$
$$= 30 \text{ J}$$

Answer is (C)

UNITS-32

If a $\frac{1}{3}$ horsepower pump runs for 20 minutes, what is the energy used?

(A) 0.06 ergs   (B) 0.25 kW   (C) 0.30 MJ

(D) 0.11 kW·h   (E) 6.67 kW·h

$$W = \left(\frac{1}{3}\,hp\right)\left(745.7\,\frac{W}{hp}\right)\left(20\,min\right)\left(\frac{1\,hr}{60\,min}\right)$$

$$= 0.0828\ kW\cdot h \left(3600\,\frac{s}{h}\right)$$

$$= 0.298\ MJ$$

$$W \approx 0.30\ MJ$$

Answer is (C)

UNITS-33

A machine is capable of accelerating a 1 kg mass at 1 m/s^2 for 1 minute. The machine runs at 60 rpm. What is the power output of the machine?

(A) 1 erg   (B) 1 cal   (C) 1 J   (D) 1 W   (E) 1 kW

$$P = \frac{W}{t}$$

$$W = \mathbf{F}\cdot\mathbf{x}$$

$$\mathbf{F} = m\mathbf{a}$$

$$t = \text{period of revolution}$$

$$P = \frac{max}{t}$$

$$m = 1\,kg$$

$$a = 1\,\frac{m}{s^2}$$

$$f = 60\ \text{rpm}\left(\frac{1\,min}{60\,s}\right)$$

$$= 1\ hz$$

$$t = \frac{1}{f}$$

$$= 1\ s$$

$$P = \frac{\left(1\,kg\right)\left(1\,\frac{m}{s^2}\right)\left(1\,m\right)}{1\,s}$$

$$= 1\ W$$

Answer is (D)

UNITS-34

A power of 6 kW is supplied to the motor of a crane. The motor has an efficiency of 90%. With what constant speed does the crane lift an 800 lbf weight?

(A) 0.09 cm/s   (B) 0.32 cm/s   (C) 0.98 cm/s

          (D) 1.52 cm/s   (E) 2.10 cm/s

$$P = \frac{dW}{dt}$$
$$= \frac{\mathbf{F} \cdot \mathbf{x}}{t}$$
$$= \mathbf{F} \cdot \mathbf{v}$$
$$\mathbf{v} = \frac{P}{F}$$
$$\text{input power} = P_i$$
$$= 6\,\text{kW}$$
$$\text{efficiency} = \eta$$
$$= 0.90$$
$$\text{useful power} = P_r$$
$$= \eta P_i$$
$$= (0.90)(6000\,W)$$
$$= 5400\,W$$
$$\mathbf{v} = \frac{P_r}{F}$$
$$= \frac{P_r}{mg}$$
$$= \left(\frac{5400\,\text{W}}{800\,\text{lbf}}\right)\left(\frac{1\,\text{lbf}}{4.45\,\text{N}}\right)$$
$$= 1.52\,\frac{\text{W}}{\text{N}}$$
$$= 1.52\,\text{m/s}$$

Answer is (D)

UNITS–35

Given the following heat exchanger with specified inlet and outlet enthalpies, what is the energy requirement for the heating coil?

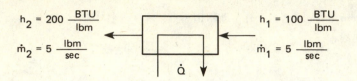

$$h_2 = 200 \ \frac{BTU}{lbm}$$

$$\dot{m}_2 = 5 \ \frac{lbm}{sec}$$

$$h_1 = 100 \ \frac{BTU}{lbm}$$

$$\dot{m}_1 = 5 \ \frac{lbm}{sec}$$

$$\dot{Q}$$

(A) 500 kW   (B) 528 kW   (C) 561 kW   (D) 601 kW   (E) 648 kW

Consider a control volume around the exchanger.

$$Q = \dot{m}(h_2 - h_1)$$
$$= \left(5 \frac{lbm}{s}\right)\left(100 \frac{BTU}{lbm}\right)$$
$$= \left(500 \frac{BTU}{s}\right)\left(1.055 \frac{kJ}{BTU}\right)$$
$$= 527.5 \frac{kJ}{s}$$
$$= 528 \ kW$$

Answer is (B)

UNITS–36

An engine has an efficiency of 26%. It uses 2 gallons of gasoline per hour. Gasoline has a heating value of 20,500 BTU/lbm and a specific gravity of 0.8. What is the power output of the engine?

(A) 0.33 kW   (B) 20.8 kW   (C) 26.0 kW

(D) 41.7 kW   (E) 80.2 kW

First, find $\dot{Q}$, the input power of the engine, in kilowatts.

$$\dot{Q} = \left(\frac{2\,\text{gal}}{\text{hr}}\right)\left(\frac{1\,\text{ft}^3}{7.48\,\text{gal}}\right)\left[\left(0.8\right)\left(62.4\,\frac{\text{lbm}}{\text{gal}}\right)\right]$$
$$\times \left(20,500\,\frac{\text{BTU}}{\text{lbm}}\right)\left(1.054\,\frac{\text{kJ}}{\text{BTU}}\right)$$
$$= \left(288,000\,\frac{\text{kJ}}{\text{hr}}\right)\left(\frac{1\,\text{hr}}{3600\,\text{s}}\right)$$
$$= 80\,\text{kW}$$

Next, find P, the power output of the engine.

$$P = \eta \dot{Q}$$
$$= 0.26(80\,\text{kW})$$
$$= 20.8\,\text{kW}$$

Answer is (B)

UNITS-37

Two liters of an ideal gas, at a temperature of $T_1 = 25\,°\text{C}$ and a pressure of $p_1 = 0.101\,\text{MPa}$, are in a 10 cm diameter cylinder with a piston at one end. The piston is depressed, so that the cylinder is shortened by 10 centimeters. The temperature increases by 2 °C. What is the change in pressure?

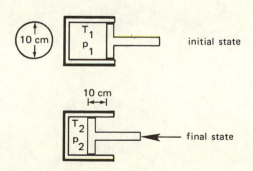

(A) 0.156 MPa   (B) 0.167 MPa   (C) 0.251 MPa

(D) 0.327 MPa   (E) 0.430 MPa

Apply the ideal gas law to the gas in the cylinder.

$$\frac{p_1 V_1}{T_1} = \frac{p_2 V_2}{T_2}$$

$$p_2 = \frac{p_1 V_1 T_2}{T_1 V_2}$$

$$T_1 = 25\,^{\circ}\text{C}$$

$$= 298\,\text{K}$$

$$T_2 = 27\,^{\circ}\text{C}$$

$$= 300\,\text{K}$$

$$\Delta V = \Delta l (\pi r^2)$$

$$= (-10\,\text{cm})\pi (5\,\text{cm})^2$$

$$= -785\,\text{cm}^3$$

$$(\Delta V < 0 \text{ because the piston is depressed})$$

$$V_2 = V_1 + \Delta V$$

$$= 2000\,\text{cm}^3 - 785\,\text{cm}^3$$

$$= 1215\,\text{cm}^3$$

$$p_2 = \frac{(0.101\,\text{MPa})(2000\,\text{cm}^3)(300\,\text{K})}{(298\,\text{K})(1215\,\text{cm}^3)}$$

$$= 0.167\,\text{MPa}$$

Answer is (B)

UNITS–38

The average power output of a cylinder in a combustion engine is given by:

$$\overline{P} = pLAN$$

where p = average pressure on the piston during the stroke
L = length of the piston stroke
A = area of the piston head
N = number of strokes per second

An 8-cylinder engine has the following specifications at optimum speed:

$$p = 283 \text{ kPa}$$
$$L = 14 \text{ cm}$$
$$d = \text{diameter of piston head}$$
$$= 12 \text{ cm}$$
$$N = 1500 \text{ strokes/min}$$

What is the average power output of this engine?

(A) 89.5 N/s   (B) 89.5 kW   (C) 89.5×10^3 J · m/s

(D) 89.5 kJ   (E) 89.5 kPa

$\overline{P} = pLAN$ for one cylinder

$\overline{P}_{tot} = 8\,pLAN$ for eight cylinders

$$A = \pi r^2 = \pi \left(\frac{d}{2}\right)^2$$

$$= \pi (6 \text{ cm})^2 = 113 \text{ cm}^2 \left(\frac{1 \text{ m}}{100 \text{ cm}}\right)^2$$

$$= 0.0113 \text{ m}^2$$

$$\overline{P}_{tot} = 8\left(283{,}000 \text{ Pa}\right)\left(0.14 \text{ m}\right)\left(0.0113 \text{ m}^2\right)\left(1500\,\frac{\text{strokes}}{\text{min}}\right)\left(\frac{1 \text{ min}}{60 \text{ s}}\right)$$

$$= 89{,}500 \text{ Pa·m}^3/\text{s}$$

$$= 89{,}500\,\frac{\text{J}}{\text{s}}$$

$$= 89.5 \text{ kW}$$

Answer is (B)

UNITS–39

What is the power required to transfer 97,000 coulombs of charge through a potential rise of 50 volts in one hour?

(A) 0.5 kW (B) 0.9 kW (C) 1.3 kW

(D) 2.8 kW (E) 3.4 kW

$$W = qV$$
$$q = \text{charge}$$
$$V = \text{potential rise}$$
$$P = \frac{W}{t}$$
$$= \frac{qV}{t}$$
$$= \frac{(97,000\,\text{C})(50\,\text{V})}{3600\,\text{s}}$$
$$= \frac{4.85 \times 10^6\,\text{J}}{3600\,\text{s}}$$
$$= 1.34\,\text{kW}$$

Answer is (C)

UNITS–40

A current of 7 amperes passes through a 12 ohm resistor. What is the power dissipated in the resistor?

(A) 84 W (B) 0.59 hp (C) 0.79 hp (D) 7 hp (E) 84 hp

$$P = I^2 R$$
$$I = \text{current}$$
$$R = \text{resistance}$$
$$P = (7\,\text{A})^2 (12\,\Omega)$$
$$= 588\,\text{W} \left(\frac{1\,\text{hp}}{745.7\,\text{W}} \right)$$
$$= 0.79\,\text{hp}$$

Answer is (C)

UNITS–41

If the average energy in a nuclear reaction is 200 MeV/fission, what is the power output of a reactor if there are 2.34×10^{19} fissions per second?

(A) 550 W (B) 120 kW (C) 30 MW
(D) 750 MW (E) 35 GW

The power output of the reactor is simply the energy per fission times the number of fissions per second.

$$P = \left(200 \, \frac{\text{MeV}}{\text{fission}}\right) \left(2.34 \times 10^{19} \, \frac{\text{fissions}}{\text{s}}\right) \left(\frac{1 \times 10^6 \, \text{eV}}{\text{MeV}}\right)$$
$$= \left(4.69 \times 10^{27} \, \frac{\text{eV}}{\text{s}}\right) \left(1.602 \times 10^{-19} \, \frac{\text{J}}{\text{eV}}\right)$$
$$= 750 \times 10^6 \, \text{W}$$
$$= 750 \, \text{MW}$$

Answer is (D)

4

FLUID STATICS AND DYNAMICS

FLUIDS-1

Which statement is true for a fluid?

(A) It cannot sustain a shear force.
(B) It cannot sustain a shear force at rest.
(C) It is a liquid only.
(D) It has a very regular molecular structure.
(E) It can strain.

A fluid is defined as a substance which deforms continuously under the application of a shear force. This means that it cannot sustain a shear force at rest. Therefore, (B) is true.

Answer is (B)

FLUIDS-2

Which of the following is not a basic component of motion of a fluid element?

(A) translation   (B) rotation   (C) angular distortion

(D) volume distortion   (E) twist

The motion of a fluid element may be divided into three categories: translation, rotation, and distortion. Distortion can be further subdivided into angular and volume distortion. The only choice that is not a basic component of fluid element motion is twist.

Answer is (E)

FLUIDS-3

Which of the following must be satisfied by the flow of any fluid, real or ideal?

I.   Newton's second law of motion
II.   the continuity equation
III.   the requirement of a uniform velocity distribution
IV.   Newton's law of viscosity
V.   the principle of conservation of energy

(A) I, II, and III   (B) I, II, and IV   (C) I, II, and V
            (D) I, II, III, and IV   (E) I, II, IV, and V

Newton's second law, the continuity equation, and the principle of conservation of energy always apply for any fluid.

Answer is (C)

FLUIDS-4

What is the definition of pressure?

(A) $\dfrac{\text{area}}{\text{force}}$

(B) $\lim\limits_{\text{force}\to 0} \dfrac{\text{force}}{\text{area}}$

(C) $\lim\limits_{\text{area}\to 0} \dfrac{\text{force}}{\text{area}}$

(D) $\lim\limits_{\text{force}\to 0} \dfrac{\text{area}}{\text{force}}$

(E) $\lim\limits_{\text{area}\to 0} \dfrac{\text{area}}{\text{force}}$

The mathematical definition of pressure is:

$$\lim_{\text{area}\to 0} \frac{\text{force}}{\text{area}}$$

Answer is (C)

FLUIDS–5

For a fluid, viscosity is defined as the constant of proportionality between shear stress and what other variable?

(A) the time derivative of pressure
(B) the time derivative of density
(C) the spatial derivative of velocity
(D) the spatial derivative of density
(E) none of the above

By definition:

$$\tau = \mu \frac{dv}{dy}$$

Thus, viscosity (μ) is the constant of proportionality between the shear stress, τ, and the gradient (spatial derivative) of the velocity.

Answer is (C)

FLUIDS–6

Surface tension has which of the following properties?

    I.   It has units of force per unit length.
    II.   It exists whenever there is a density discontinuity.
    III.   It is strongly affected by pressure.

(A) I only   (B) II only   (C) III only
        (D) I and II   (E) I, II, and III

III is incorrect because pressure only slightly affects surface tension.
I and II are correct.

Answer is (D)

PROFESSIONAL PUBLICATIONS, INC. • Belmont, CA

FLUIDS–7

A leak from a faucet comes out in separate drops. Which of the following is the main cause of this phenomenon?

(A) gravity
(B) air resistance
(C) viscosity of the fluid
(D) shape of the faucet
(E) surface tension

Surface tension is caused by the molecular cohesive forces in a fluid. It is the main cause of the formation of the drops of water.

Answer is (E)

FLUIDS–8

The surface tension force, σ, of water in air is approximately 0.00518 lbf/ft. If the atmospheric pressure is 14.7 psia, what is the pressure inside a droplet 0.01″ in diameter?

(A) 14.53 psia   (B) 14.70 psia   (C) 14.78 psia
        (D) 14.87 psia   (E) 16.77 psia

For a spherical droplet:

$$\Delta p = \frac{2\sigma}{r}$$
$$= p_{in} - p_{out}$$
$$p_{in} = p_{out} + \frac{2\sigma}{r}$$
$$= 14.7 \frac{\text{lbf}}{\text{in}^2} + \frac{(4)\left(0.00518 \frac{\text{lbf}}{\text{ft}}\right)}{0.01 \text{ in}} \left(\frac{1 \text{ ft}}{12 \text{ in}}\right)$$
$$= (14.7 + 0.173) \,\text{psia}$$
$$= 14.87 \,\text{psia}$$

Answer is (D)

FLUIDS-9

Which of the following describes shear stress in a moving fluid?

(A) It is proportional to the absolute viscosity.
(B) It is proportional to the velocity gradient at the point of interest.
(C) It is non-existent.
(D) It is proportional to the fluid density.
(E) both A and B

$$\tau = \mu \frac{dv}{dy}$$

Thus, shear stress is proportional to the velocity gradient at a point, as well as the absolute viscosity.

Answer is (E)

FLUIDS-10

If the shear stress in a fluid varies linearly with the velocity gradient, which of the following describes the fluid?

(A) It is inviscid.
(B) It is a perfect gas.
(C) It is a Newtonian gas.
(D) It is at a constant temperature.
(E) It has a constant viscosity.

In order for shear stress to vary linearly with the velocity gradient, the fluid must be Newtonian.

Answer is (C)

FLUIDS-11

How are lines of constant pressure in a fluid related to the force field?

(A) They are parallel to the force field.
(B) They are perpendicular to the force field.
(C) They are at a 45° angle to the force field.
(D) They are perpendicular only to the force of gravity.
(E) They are unrelated to force fields.

Lines of constant pressure are always perpendicular to the direction of the force field.

Answer is (B)

FLUIDS-12

Which of the following is most accurate about a streamline?

(A) It is a path of a fluid particle.
(B) It is a line normal to the velocity vector everywhere.
(C) It is fixed in space in steady flow.
(D) It is defined for non-uniform flow only.
(E) It is perpendicular to the path line.

Streamlines are tangent to the velocity vectors at every point in the field. Thus, for a steady flow $\frac{d\mathbf{v}}{dt} = 0$, a streamline is fixed in space.

Answer is (C)

FLUIDS-13

Which of the following describes a streamline?

I.   It is a mathematical concept.
II.   It cannot be crossed by the flow.
III. It is a line of constant entropy.

(A) I only   (B) I and II   (C) II only
        (D) I and III   (E) I, II, and III

A streamline is a mathematical concept which defines lines that are tangential to the velocity vector. Therefore, no flow can cross a streamline. Entropy is not related to streamlines.

Answer is (B)

FLUIDS-14

The following figure shows several streamlines near the corner of two infinite plates. Which of the following could be the correct expression for the stream function, Ψ, of this potential flow?

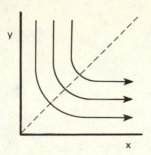

(A) $\Psi = x - y$
(B) $\Psi = 2xy$
(C) $\Psi = x$
(D) $\Psi = y$
(E) $\Psi = x + y$

Streamlines are graphs of constant values for the stream function. The graph shows hyperbolas that are of the form $axy = b$, where a and b are constants. Thus, of the choices shown, the stream function could only be $\Psi = 2xy$.

Answer is (B)

FLUIDS-15

What is the pressure at point A in the tank if $h = 2$ feet? ($g = 32.2$ ft/sec^2, and $\rho = 1.94$ slug/ft^3)

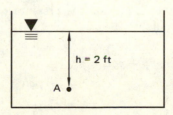

h = 2 ft

A

(A) 75 lbf/ft^2 (B) 85 lbf/ft^2 (C) 100 lbf/ft^2
         (D) 125 lbf/ft^2 (E) 150 lbf/ft^2

$$p = \rho gh$$

$$= \left(1.94 \, \frac{\text{slug}}{\text{ft}^3}\right)\left(32.2 \, \frac{\text{ft}}{\text{sec}^2}\right)\left(2\,\text{ft}\right)$$

$$= 125 \, \text{lbf/ft}^2$$

Answer is (D)

FLUIDS–16

A glass filled with a fluid is inverted. The bottom is open. What is the pressure at the closed end at point A?

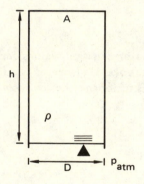

(A) p_{atm} (B) $p_{atm} + \rho gh$ (C) $p_{atm} - \rho gh$

(D) ρgh (E) none of the above

The pressure at point A, p, plus the pressure exerted by the water equals the pressure outside the glass.

$$p + \rho gh = p_{atm}$$
$$p = p_{atm} - \rho gh$$

Answer is (C)

FLUIDS–17

Find the pressure in the tank from the manometer readings shown in the illustration.

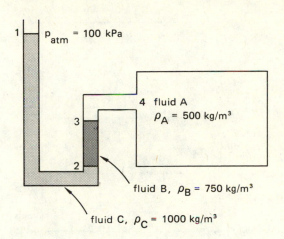

(A) 100 kPa   (B) 107 kPa   (C) 110 kPa

       (D) 118 kPa   (E) 122 kPa

$$p_2 - p_1 = \rho_C g(z_1 - z_2)$$
$$p_3 - p_2 = \rho_B g(z_2 - z_3)$$
$$p_4 - p_3 = \rho_A g(z_3 - z_4)$$
$$p_4 - p_1 = (p_4 - p_3) + (p_3 - p_2) + (p_2 - p_1)$$
$$p_4 = p_1 + g[\rho_C(z_1 - z_2) + \rho_B(z_2 - z_3) + \rho_A(z_3 - z_4)]$$
$$= 100,000 + 9.81\,[1000(1) + 750(-0.3) + 500(-0.1)]$$
$$= 100,000 + 9.81(725)$$
$$= 107,100 \text{ Pa}$$
$$= 107.1 \text{ kPa}$$

Answer is (B)

FLUIDS–18

In which fluid will a pressure of 700 kPa first be achieved?

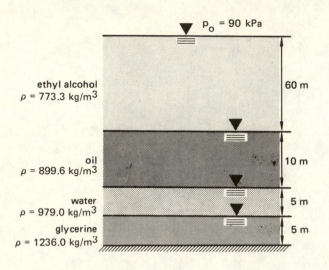

(A) ethyl alcohol   (B) oil   (C) water

                   (D) glycerin   (E) below glycerin

Let p_i be the maximum pressure that can be measured in fluid level i. If $p_i \geq 700$ kPa, then a pressure of 700 kPa can be measured at that level.

$$p_0 = 90 \text{ kPa}$$

$$p_1 = p_0 + \rho_1 g z_1$$

$$= 90 + \frac{(773.3)(9.81)(60)}{1000}$$

$$= 545.16 \text{ kPa}$$

$$p_1 < 700 \text{ kPa}$$

$$p_2 = p_1 + \rho_2 g z_2$$

$$= 545.16 + \frac{(899.6)(9.81)(10)}{1000}$$

$$= 633.41 \text{ kPa}$$

$$p_2 < 700 \text{ kPa}$$

$$p_3 = p_2 + \rho_3 g z_3$$

$$= 633.41 + \frac{(979.0)(9.81)(5)}{1000}$$

$$= 681.43 \text{ kPa}$$

$$p_3 < 700 \text{ kPa}$$

$$p_4 = p_3 + \rho_4 g z_4$$

$$= 681.43 + \frac{(1236.0)(9.81)(5)}{1000}$$

$$= 742.06 \text{ kPa}$$

$$p_4 > 700 \text{ kPa}$$

Thus, a pressure of 700 kPa occurs in the glycerin level.

Answer is (D)

FLUIDS–19

I_{xx} is the moment of inertia of the plane area about its centroidal x axis. How can I_{xx} be expressed?

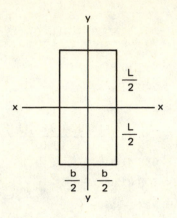

(A) $\dfrac{bL^3}{96}$

(B) $\dfrac{bL^3}{16}$

(C) $\dfrac{bL^3}{12}$

(D) $\dfrac{bL^3}{8}$

(E) $\dfrac{bL^3}{4}$

$$I_{xx} = \int y^2 dA$$

$$= \int_{-\frac{L}{2}}^{\frac{L}{2}} y^2 b\, dy$$

$$= b \cdot \left. \frac{y^3}{3} \right|_{-\frac{L}{2}}^{\frac{L}{2}}$$

$$= b \left[\frac{\left(\frac{L}{2}\right)^3}{3} - \frac{\left(\frac{-L}{2}\right)^3}{3} \right]$$

$$= \frac{b}{3} \left(\frac{2L^3}{8} \right)$$

$$= \frac{bL^3}{12}$$

Answer is (C)

FLUIDS–20

A circular window with a radius of 0.25 meter has its center three meters below the water's surface. Find the force acting on the window. The window is vertical.

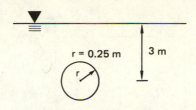

(A) 2.9 kN   (B) 5.8 kN   (C) 17.7 kN
          (D) 29.4 kN   (E) 92 kN

$$F = \bar{p}A$$

$$\bar{p} = \left(\frac{y_1 + y_2}{2}\right)(\rho g \sin \alpha)$$

$$y_1 = 2.75 \text{ m}$$

$$y_2 = 3.25 \text{ m}$$

$\alpha =$ the angle between the surface of the water and
        the surface of the window

$$= \frac{\pi}{2}$$

$$A = \pi(0.25)^2$$

$$F = \left(\frac{y_1 + y_2}{2}\right)(\rho g \sin \alpha)\left[\pi(0.25)^2\right]$$

$$= (3)(1000)(9.81)(1)(\pi)(0.25)^2$$

$$= 5770 \text{ N}$$

$$= 5.8 \text{ kN}$$

Answer is (B)

FLUIDS–21

Find the reaction moment per unit width, M_{dam}, acting at the base of the dam.

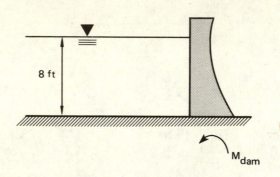

(A) 5325 lbf   (B) 5795 lbf   (C) 6656 lbf

         (D) 10,650 lbf   (E) 26,624 lbf

First, draw the free-body diagram.

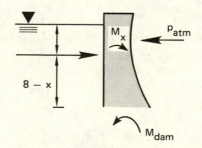

p_x = the pressure at point x

$$= p_{atm} + \frac{\rho g x}{g_c}$$

dF_x = the differential force at point x

$$= (p_x - p_{atm})\, dx$$

$$= \frac{\rho g x}{g_c} dx$$

dM_x = the differential moment on the dam

$$= F_x(x)(8 - x)$$

$$= \frac{\rho g}{g_c}(8 - x)dx$$

$M_{dam} = M_x$ when the system is in equilibrium

$$= \int dM_x$$

$$= \int_0^8 \frac{\rho g}{g_c}(x)(8 - x)dx$$

$$= \frac{\rho g}{g_c}\left[\frac{8x^2}{2} - \frac{x^3}{3}\right]_0^8$$

$$= \frac{\rho g}{g_c}\left(8^3\right)\left(\frac{1}{2} - \frac{1}{3}\right)$$

$$= \left(\frac{1}{6}\right)\left(\frac{\rho g}{g_c}\right)\left(8^3\right)$$

$$= \left(\frac{1}{6}\right)\left(\frac{(62.4)(32.2)}{32.2}\right)(8^3)$$

$$= 5325 \text{ lbf}$$

Answer is (A)

FLUIDS–22

Find the force per unit width, F_{dam}, that holds the dam upright.

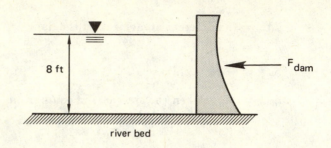

(A) 62.0 lbf/ft   (B) 499 lbf/ft   (C) 1997 lbf/ft

(D) 2114 lbf/ft   (E) 3994 lbf/ft

First, draw a free-body diagram.

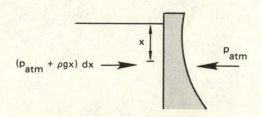

p_x = the pressure at point x

$$= p_{atm} + \frac{\rho g x}{g_c} - p_{atm}$$

$$= \frac{\rho g x}{g_c}$$

dF_x = the differential force at point x

$$= \frac{\rho g x}{g_c} dx$$

F_{dam} = total force exerted by the water

$$= \int_0^8 \frac{\rho g x}{g_c} dx$$

$$= \frac{\rho g x^2}{2 g_c} \Big|_0^8$$

$$= \frac{(62.4)(32.2)(8^2)}{(2)(32.2)}$$

$$= 1997 \text{ lbf/ft}$$

Answer is (C)

FLUIDS–23

Water is held in a tank by the sluice gate shown in the figure. What force per unit width of the dam must the latch supply to keep the gate closed?

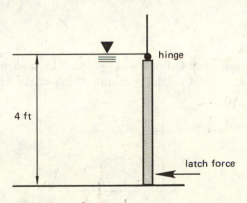

(A) 74 lbf/ft   (B) 125 lbf/ft   (C) 333 lbf/ft
      (D) 500 lbf/ft   (E) 1332 lbf/ft

Draw a free-body diagram.

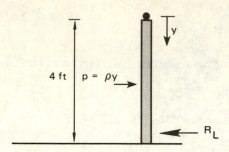

Use the coordinate system in the diagram. For the gate to stay in place, the sum of the moments around the hinge must be zero.

$$\sum M_{hinge} = 0$$

$$= 4(R_L) - \int_0^4 \frac{\rho g}{g_c} y^2 \, dy$$

$$4R_L = \frac{\rho g y^3}{3g_c}\bigg|_0^4$$

$$= \frac{(62.4)(32.2)(4)^3}{(3)(32.2)}$$

$$R_L = \frac{(62.4)(32.2)(4)^3}{(12)(32.2)}$$

$$= 333 \text{ lbf/ft}$$

$4 RL = \frac{62.4 \times 4 \times 4}{2} \times \frac{2}{3} \times 4$

$\therefore \quad RL = \frac{62.4 \times 16 \times 4}{3}$

$= \frac{998.4 \times 3}{3}$

$= 382.8$

$= 333 \; lbf/ft$

Answer is (C)

FLUIDS–24

A tank with one hinged wall is filled with water. The tank wall is held at a 30° angle by a belt. What is the tension per unit width of the tank in the belt?

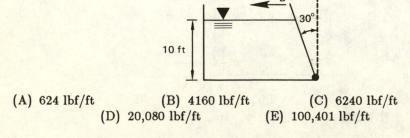

(A) 624 lbf/ft   (B) 4160 lbf/ft   (C) 6240 lbf/ft
           (D) 20,080 lbf/ft   (E) 100,401 lbf/ft

F_B = the force on the belt per unit width of the wall

w = width of the wall

for a tilted wall, $F_B = \dfrac{\overline{p}A}{w}$

$$\overline{p} = \rho\left(\dfrac{g}{g_c}\right)\overline{h}$$

$$\overline{h} = \dfrac{h_0 + h_1}{2}$$

$h_0 = 0$

h_1 = maximum height of the water measured along the surface of the wall

$$h_1 = \dfrac{10}{\cos 30°} = 11.55 \text{ ft}$$

$$
F_B = \dfrac{\left(\dfrac{\rho g}{g_c}\right)\left(\dfrac{0 + h_1}{2}\right)\left(h_1\, w\right)}{w}
$$

$$
= \left(\dfrac{\rho g}{g_c}\right)\left[\dfrac{(h_1)^2}{2}\right]
$$

$$
= \left(\dfrac{(62.4)(32.2)}{32.2}\right)\left[\dfrac{(11.55)^2}{2}\right]
$$

$$
= 4160 \text{ lbf/ft}
$$

Answer is (B)

(handwritten annotations in margin)

$\dfrac{wht}{L}$

$F = \overline{p}\,A$

$= \dfrac{11.55}{2} \times 62.4 \times 11.55 \times 1$

$h_1 = $

$\cos 30° = \dfrac{10}{hyp}$

$hyp = \dfrac{10}{\cos 30°}$

$F = \dfrac{awh}{2} \times$

$= \dfrac{62.4 \times (11.55)^2}{2}$

$= 4160$

FLUIDS-25

A tank of water has a rectangular panel at its lower left side, as shown in the figure. The location of the center of pressure on the panel is at the point P. Call the distance along the panel from the bottom of the tank to the center of pressure PA. Determine the length of PA.

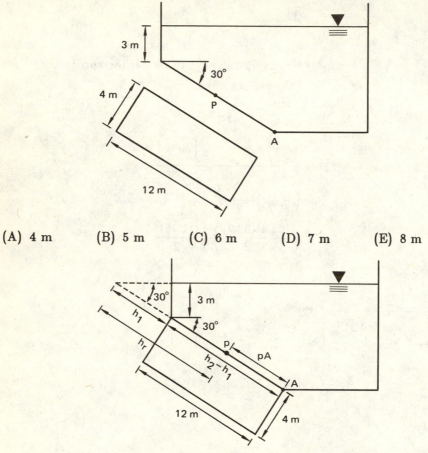

(A) 4 m   (B) 5 m   (C) 6 m   (D) 7 m   (E) 8 m

The distance along the surface of an object from the surface of the fluid to the center of pressure, h_r, is given by:

$$h_r = \frac{2}{3}\left(h_1 + h_2 - \frac{h_1 h_2}{h_1 + h_2}\right)$$

h_1 = distance along the surface of the object
        from the surface of the fluid to the object's upper edge

h_2 = distance along the surface of the object
        from the surface of the fluid to the object's lower edge

From the figure,

$$PA = h_2 - h_r$$

The plane is inclined at $30°$ below horizontal, has its upper edge at three meters (vertically) below the surface of the fluid, and is 12 meters long. Thus, the following can be determined:

$$h_1 = \frac{3}{\sin 30°}$$
$$= 6 \text{ m}$$
$$h_2 - h_1 = 12 \text{ m}$$
$$h_2 = h_1 + 12 = 6 + 12$$
$$= 18 \text{ m}$$
$$h_r = \frac{2}{3}\left(6 + 18 - \frac{(6)(18)}{6 + 18}\right)$$
$$= 13 \text{ m}$$
$$PA = 18 - 13$$
$$= 5 \text{ m}$$

Answer is (B)

FLUIDS-26

Determine the total force exerted on the curved surface described by the equation $y = x^2$. The width of the curved plate is two feet, and the specific weight of water is 62.4 lbf/ft^3.

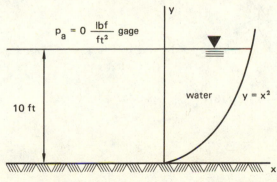

(A) 126 lbf

(B) 200 lbf

(C) 650 lbf

(D) 1064 lbf

(E) 2703 lbf

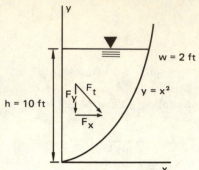

let w = width of the tank

$\mathbf{F}_t = \mathbf{F}_x + \mathbf{F}_y$

$\mathbf{F}_t = \sqrt{\mathbf{F}_x^2 + \mathbf{F}_y^2}$

$\mathbf{F}_y$ = weight of water in the portion of the tank above the curved surface

$$= \left(\frac{\rho g}{g_c}\right) V = \gamma V$$

$$= \gamma w \int_0^{\sqrt{10}} (10 - y)\,dx$$

$$= \gamma w \int_0^{\sqrt{10}} (10 - x^2)\,dx$$

$$= \gamma w \left(10x - \frac{x^3}{3}\right)_0^{\sqrt{10}}$$

$$= 62.4(2)\left(10\sqrt{10} - \frac{10\sqrt{10}}{3}\right)$$

$$= 2631 \text{ lbf}$$

$\mathbf{F}_x = \bar{p} A_x$

$$\bar{p} = \gamma\left(\frac{0 + h}{2}\right)$$

A_x = area of the tank perpendicular to the x axis

$$= wh$$

$$\mathbf{F}_y = 62.4\left(\frac{10}{2}\right)\big[(2)(10)\big]$$

$$= 624 \text{ lbf}$$

$$\mathbf{F}_t = \sqrt{2631^2 + 624^2}$$

$$= 2703 \text{ lbf}$$

Answer is (E)

FLUIDS–27

The stream potential, Φ, of a flow is given by $\Phi = 2xy - y$. Determine the stream function, Ψ, for this potential.

(A) $\Psi = x^2 - y^2 + C$
(B) $\Psi = x - x^2 + y^2 + C$
(C) $\Psi = x + x^2 - y^2 + C$
(D) $\Psi = x^2 + y^2 + C$
(E) $\Psi = x^2 + x - y^2 + C$

$$\text{Let} \quad u = \frac{dx}{dt} \quad v = \frac{dy}{dt}$$

$$\text{By definition,} \quad u = \frac{d\Phi}{dx}$$

$$= \frac{d\Psi}{dy}$$

$$v = \frac{d\Phi}{dy}$$

$$= -\frac{d\Psi}{dx}$$

$$u = 2y$$

$$v = 2x - 1$$

$$\frac{d\Psi}{dy} = 2y$$

$$\Psi = \int 2y \, dy + f(x)$$

$$= y^2 + f(x)$$

$$-\frac{df(x)}{dx} = v$$

$$= 2x - 1$$

$$f(x) = x - x^2 + C$$

$$\Psi = x - x^2 + y^2 + C$$

Answer is (B)

FLUIDS-28

Determine the average velocity through a circular section in which the velocity distribution is given as $v = v_{max}\left[1 - \left(\frac{r}{r_o}\right)^2\right]$. The distribution is symmetric with respect to the longitudinal axis, $r = 0$. r_o is the outer radius, and v_{max} is the velocity along the longitudinal axis.

(A) $\dfrac{v_{max}}{4}$ 　　　　 (B) $\dfrac{v_{max}}{3}$ 　　　　 (C) $\dfrac{v_{max}}{2}$

　　　　 (D) v_{max} 　　　　　　 (E) $2v_{max}$

$$
\begin{aligned}
v_{ave} &= \frac{1}{A}\int v\,dA \\
&= \frac{1}{\pi r_o^2}\int_0^{r_o} v_{max}\left[1 - \left(\frac{r}{r_o}\right)^2\right]2\pi r\,dr \\
&= \frac{2\pi v_{max}}{\pi r_o^2}\int_0^{r_o} r\left[1 - \left(\frac{r}{r_o}\right)^2\right]dr \\
&= \frac{2v_{max}}{r_o^2}\left(\frac{r^2}{2} - \frac{r^4}{4r_o^2}\right)\Bigg|_0^{r_o} \\
&= \frac{2v_{max}}{r_o^2}\left(\frac{r_o^2}{2} - \frac{r_o^2}{4}\right) \\
&= \frac{v_{max}}{2}
\end{aligned}
$$

Answer is (C)

FLUIDS-29

Under what conditions is mass conserved in fluid flow?

(A) The fluid is baratropic.
(B) The flow is isentropic.
(C) The flow is adiabatic.
(D) The fluid is incompressible.
(E) It is always conserved.

Mass is always conserved in fluid flow.

Answer is (E)

FLUIDS–30

What is the absolute velocity of a real fluid at a surface?

(A) the same as the bulk fluid velocity
(B) the velocity of the surface
(C) zero
(D) proportional to the smoothness of the surface
(E) proportional fo the kinematic viscosity of the fluid

For a real fluid (non-zero viscosity) there is no slip at the boundaries. In other words, the velocity of the surface is the same as the velocity of the fluid at the surface. Thus, (B) is true.

Choice (C) is true only if the velocity of the surface is zero.

Answer is (B)

FLUIDS–31

Which of the statements is true concerning the following continuity equation?

$$\frac{\partial \rho}{\partial t} + \frac{\partial (\rho u)}{\partial x} + \frac{\partial (\rho v)}{\partial y} + \frac{\partial (\rho w)}{\partial z} = 0$$

where $\rho = $ density

$u = $ velocity in the x direction

$v = $ velocity in the y direction

$w = $ velocity in the z direction

(A) It is valid only for incompressible flow.
(B) It is valid only for steady flow.
(C) It is derived from the principle of conservation of mass.
(D) It is derived from the principle of conservation of energy.
(E) It is valid only for pipe flow.

In essence, the continuity equation states that the mass flux entering a control volume is equal to the mass flux leaving the control volume plus the rate of accumulation of mass within the control volume. Thus, it is derived from the principle of conservation of mass. It is valid for all real and ideal fluids, and for all types of fluid flow.

Answer is (C)

FLUIDS–32

Which of the following satisfies the continuity equation? (u, v, and w are the components of velocity in the x, y, and z directions, respectively).

$$
\begin{aligned}
\text{I.} \qquad & u = x + 2y - t \\
& v = t - 2y + z \\
& w = t - 2x + z
\end{aligned}
$$

$$
\begin{aligned}
\text{II.} \qquad & u = y^2 - x^2 \\
& v = 2xy \\
& w = 2tz
\end{aligned}
$$

$$
\begin{aligned}
\text{III.} \qquad & u = x^2 - y^2 \\
& v = -2xy + ty \\
& w = -tz
\end{aligned}
$$

(A) I and II
(B) II and III
(C) I and III
(D) I, II, and III
(E) I or II or III

The continuity equation states that $\nabla \cdot \mathbf{V} = 0$. Check to see if this is true for each of the given flows.

$$
\begin{aligned}
\text{I:} \quad \frac{\partial u}{\partial x} + \frac{\partial v}{\partial y} + \frac{\partial w}{\partial z} &= 1 + (-2) + 1 \\
&= 0
\end{aligned}
$$

$$
\begin{aligned}
\text{II:} \quad \frac{\partial u}{\partial x} + \frac{\partial v}{\partial y} + \frac{\partial w}{\partial z} &= -2x + 2x + 2t \\
&= 2t \\
&\neq 0
\end{aligned}
$$

$$
\begin{aligned}
\text{III.} \quad \frac{\partial u}{\partial x} + \frac{\partial v}{\partial y} + \frac{\partial w}{\partial z} &= 2x + (-2x + t) - t \\
&= 0
\end{aligned}
$$

Thus, flows I and III both satisfy the continuity equation.

Answer is (C)

FLUIDS–33

A pipe has a diameter of 4″ at section AA, and a diameter of 2″ at section BB. For an ideal fluid flow, the velocity is given as 1 ft/sec at section AA. What is the flow velocity at section BB?

(A) 0.25 ft/sec   (B) 0.5 ft/sec   (C) 1.0 ft/sec

(D) 2.0 ft/sec   (E) 4 ft/sec

Use the continuity equation.

$$\text{mass flux through AA} = \text{mass flux through BB}$$
$$\rho A_1 v_1 = \rho A_2 v_2$$
$$A_1 = \text{area of section AA}$$
$$A_2 = \text{area of section BB}$$
$$v_1 = \text{velocity at section AA}$$
$$v_2 = \text{velocity at section BB}$$
$$= \frac{\rho A_1 v_1}{\rho A_2}$$
$$= \frac{A_1}{A_2} v_1$$
$$= \left(\frac{\pi(2)^2}{\pi(1)^2}\right)\left(1\right)$$
$$= 4 \text{ ft/sec}$$

Answer is (E)

FLUIDS-34

Consider the following two flows of water.

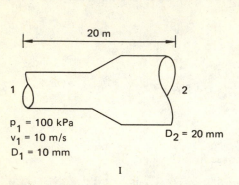

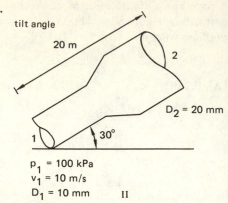

p_1 = 100 kPa
v_1 = 10 m/s
D_1 = 10 mm
D_2 = 20 mm

I

p_1 = 100 kPa
v_1 = 10 m/s
D_1 = 10 mm

D_2 = 20 mm

II

What is the relation between $v_2(I)$ and $v_2(II)$?

(A) $v_2(I) = v_2(II)$

(B) $v_2(I) = \dfrac{v_2(II)}{2}$

(C) $v_2(I) = 2v_2(II)$

(D) $v_2(I) = 4\,v_2(II)$

(E) $v_2(I) = \dfrac{v_2(II)}{4}$

From the continuity equation:

$$A_1 v_1 = A_2 v_2$$

$$v_2 = \left(\frac{A_1}{A_2}\right) v_1 \quad \text{independent of tilt angle}$$

$$v_2(I) = v_2(II)$$

Answer is (A)

FLUIDS-35

A mixing tank mixes two inlet streams containing salt. The salt concentration in stream 1 is 5% by weight, and in stream 2 it is 15% by weight. Stream 1 flows at 25 kg/s, and stream 2 flows at 10 kg/s. There is only one exit stream. Find the salt concentration in the exit stream.

(A) 5%   (B) 8%   (C) 11%   (D) 13%   (E) 15%

$$\sum_{\text{inlet}} \dot{m}_{\text{salt}} = \sum_{\text{outlet}} \dot{m}_{\text{salt}}$$

$$(0.05)(25) + (0.15)(10) = x(35)$$

$$x = \text{salt concentration in the exit stream}$$

$$= 0.079$$

$$= 8\%$$

Answer is (B)

FLUIDS–36

Water flowing with a velocity, v_1, in a pipe is turned to flow in the x direction, as shown in the figure. What is the relation between the y component of the force of the water jet on the inclined plate and the inclination angle?

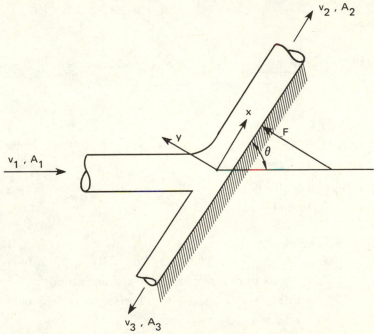

(A) $F_y = \rho A_1 v_1^2 \cos \theta$

(B) $F_y = \rho A_1 v_1 \sin \theta$

(C) $F_y = \rho A_1 v_1 \cos \theta$

(D) $F_y = \rho A_1 v_1^2 \sin \theta$

(E) $F_y = \rho A_1^2 v_1^2 \tan \theta$

Use the momentum equation in the y direction. The y component of the velocity v_1 is $v_y = v_1 \sin \theta$.

$$F_y = \dot{m}v_y$$
$$= (\rho A_1 v_1)(v_1 \sin \theta)$$
$$= \rho A_1 v_1^2 \sin \theta$$

Answer is (D)

FLUIDS-37

The vane shown in the figure deflects a jet of velocity v_{jet}, density ρ, and cross-sectional area A_{jet} by an angle of $40°$. Calculate F_h, the horizontal force on the vane.

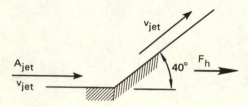

(A) $F_h = \rho A_{jet} v_{jet}^2$

(B) $F_h = \rho A_{jet} v_{jet}^2 \cos 40°$

(C) $F_h = \rho A_{jet} v_{jet}^2 (1 - \cos 40°)$

(D) $F_h = \rho A_{jet} v_{jet}^2 (1 - \sin 40°)$

(E) $F_h = \rho A_{jet} v_{jet}^2 (\cos 40° - 1)$

Use the momentum equation.

$$F_h = -(\text{rate of change of horizontal momentum})$$
$$= \dot{m}(\text{horizontal velocity in} - \text{horizontal velocity out})$$
$$\dot{m} = \rho A_{jet} v_{jet}$$
$$v_{jet} = \text{incoming horizontal velocity}$$
$$v_{jet} \cos 40° = \text{outgoing horizontal velocity}$$
$$F_h = \rho A_{jet} v_{jet} (1 - \cos 40°)$$

Answer is (C)

FLUIDS-38

A jet of velocity v_{jet}, cross-sectional area A_{jet}, and density ρ_{jet} impinges on a reversing vane. The vane is moving with velocity v_{vane}. What is the force, F_{vane}, exerted on the vane by the water?

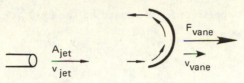

(A) $F_{vane} = 2\rho A_{jet} v_{jet}$
(B) $F_{vane} = \rho A_{jet} v_{jet}$
(C) $F_{vane} = 2\rho A_{jet} v_{vane}$
(D) $F_{vane} = 2\rho A_{jet} \left(v_{jet} - v_{vane}\right)^2$
(E) $F_{vane} = -2\rho A_{jet} \left(v_{jet} - v_{vane}\right)$

Use the momentum equation.

$$F_{vane} = \text{rate of change of momentum}$$
$$= \dot{m} \left(\Delta v\right)$$
$$\dot{m} = \rho A_{jet} \left(v_{jet} - v_{vane}\right)$$
$$F_{vane} = \rho A_{jet} \left(v_{jet} - v_{vane}\right) \times 2 \left(v_{jet} - v_{vane}\right)$$
$$= 2\rho A_{jet} \left(v_{jet} - v_{vane}\right)^2$$

Answer is (D)

FLUIDS-39

Oil (specific gravity $= 0.8$) flows at a constant rate of $1 \text{ m}^3/\text{s}$ through the circular nozzle shown. Calculate the force exerted to hold the nozzle in place.

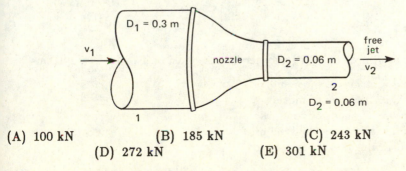

(A) 100 kN   (B) 185 kN   (C) 243 kN
        (D) 272 kN   (E) 301 kN

PROFESSIONAL PUBLICATIONS, INC. • Belmont, CA

$$\sum F_x = \dot{m}\Delta v$$

$$F_h - F_1 = \dot{m}(v_2 - v_1)$$

$$F_h = F_1 + \dot{m}(v_2 - v_1)$$

= horizontal force holding the nozzle in place

F_1 = pressure force exerted by the fluid

$$= pA_1$$

$$= 3000\left[\pi\left(\frac{0.3}{2}\right)^2\right]$$

$$= 212.07 \text{ N}$$

$$Q = vA \quad \text{from the continuity equation}$$

$$v = \frac{Q}{A}$$

$$v_1 = \frac{1}{\pi\left(\frac{0.3}{2}\right)^2}$$

$$= 14.15 \text{ m/s}$$

$$v_2 = \frac{1}{\pi\left(\frac{0.06}{2}\right)^2}$$

$$= 353.36 \text{ m/s}$$

$$\dot{m} = \rho Q$$

$$= (0.8)(1000)(1)$$

$$= 800 \text{ kg/s}$$

$$F_h = 212.07 + (800)(353.36 - 14.15)$$

$$= 271,600 \text{ N}$$

$$= 272 \text{ kN}$$

Answer is (D)

FLUIDS-40

What is the origin of the energy conservation equation used in flow systems?

(A) Newton's first law of motion
(B) Newton's second law of motion
(C) the first law of thermodynamics
(D) the second law of thermodynamics
(E) the third law of thermodynamics

The energy equation for fluid flow is based on the first law of thermodynamics, which states that the heat input into the system added to the work done on the system is equal to the change in energy of the system.

Answer is (C)

FLUIDS-41

Which of the following is the basis for Bernoulli's law for fluid flow?

(A) the principle of conservation of mass
(B) the principle of conservation of energy
(C) the continuity equation
(D) Fourier's law
(E) none of the above

Bernoulli's law is derived from the principle of conservation of energy.

Answer is (B)

FLUIDS-42

Under which of the following conditions is Bernoulli's equation valid?

(A) all points evaluated must be on the same streamline
(B) the fluid must be incompressible
(C) the fluid must be inviscid
(D) all of the above
(E) none of the above

Bernoulli's equation is valid only for incompressible, inviscid fluids. In order for Bernoulli's equation to be valid for two particular points, they must lie on the same streamline. Thus, (A), (B), and (C) are all valid conditions for Bernoulli's equation.

Answer is (D)

FLUIDS–43

Under certain flow conditions, the expression for the first law of thermodynamics for a control volume reduces to Bernoulli's equation:

$$gz_1 + \frac{v_1^2}{2} + \frac{p_1}{\rho} = gz_2 + \frac{v_2^2}{2} + \frac{p_2}{\rho}$$

Which combination of the following conditions is necessary and sufficient to reduce the first law for a control volume to Bernoulli's equation?

    I.   steady flow
    II.   incompressible fluid
    III.   no frictional losses of energy
    IV.   no heat transfer or change in internal energy

(A) I only   (B) I and II   (C) I and IV

      (D) I, III, and IV   (E) I, II, III, and IV

Bernoulli's equation is essentially a statement of conservation of energy for steady flow of an inviscid, incompressible fluid. Bernoulli's equation does not account for any frictional losses or changes in internal energy of the fluid. Thus, for Bernoulli's equation to be valid, I, II, III, and IV must all describe the flow.

> **Answer is (E)**

FLUIDS–44

Determine the velocity of the liquid at the exit, given that $h_1 = 5$ feet and $h_2 = 1$ foot.

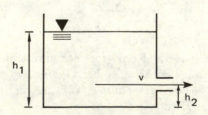

(A) 13 ft/sec   (B) 14 ft/sec   (C) 15 ft/sec
    (D) 16 ft/sec   (E) 17 ft/sec

Use Bernoulli's equation.

$$\rho g h_1 = \rho g h_2 + \frac{\rho v_2^2}{2} \quad \text{because } v_1 \text{ is zero}$$
$$v_2 = \sqrt{2g(h_1 - h_2)}$$
$$= \sqrt{2(32.2)(5-1)}$$
$$= 16 \text{ ft/sec}$$

Answer is (D)

FLUIDS-45

A pressure tank contains a fluid with weight density 81.5 lbf/ft^3. The pressure in the air space is 100 psia. Fluid exits to the atmosphere from the bottom of the tank. What is the exit velocity, v?

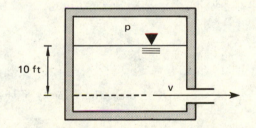

(A) 25.4 ft/sec   (B) 98.5 ft/sec   (C) 101.7 ft/sec

(D) 106.6 ft/sec   (E) 109.6 ft/sec

Apply Bernoulli's equation between the free surface and the exit.

$$\frac{p}{\rho} + gz_1 + \frac{v_1^2}{2} = \frac{p_{atm}}{\rho} + gz_2 + \frac{v^2}{2}$$

$v_1 = 0$ at the free surface

$z_2 = 0$ at the exit

$$\frac{p}{\rho} + gz_1 = \frac{p_{atm}}{\rho} + \frac{v_2^2}{2}$$

$$v_2 = \sqrt{2g\left(\frac{p - p_{atm}}{\rho g} + z_1\right)}$$

$$\rho g = 81.5 \frac{lbf}{ft^3}$$

$$= \sqrt{2\left(32.2\,\frac{ft}{sec^2}\right)\left[\frac{(100\,psi - 14.7\,psi)(144)\frac{in^2}{ft^2}}{81.5\,\frac{lbf}{ft^3}} + 10\,ft\right]}$$

$$= 101.7 \text{ ft/sec}$$

Answer is (C)

FLUIDS–46

Consider the holding tank shown. Calculate the velocity of the water exiting to the atmosphere.

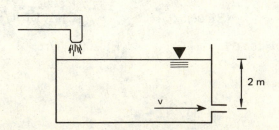

(A) 2.5 m/s   (B) 3.7 m/s   (C) 4.9 m/s

(D) 6.3 m/s   (E) 7.9 m/s

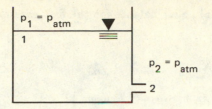

Apply Bernoulli's equation between the free surface (point 1) and the exit (point 2).

$$gz_1 + \frac{v_1^2}{2} + \frac{p_1}{\rho} = gz_2 + \frac{v_2^2}{2} + \frac{p_2}{\rho}$$

$$p_1 = p_2 \quad \text{(both are at atmospheric pressure)}$$

$$v_1 = 0 \quad \text{(the free surface is stationary)}$$

$$gz_1 = gz_2 + \frac{v_2^2}{2}$$

$$v_2 = \sqrt{2g(z_1 - z_2)}$$

$$= \sqrt{2(9.81)(2)}$$

$$= 6.3 \text{ m/s}$$

Answer is (D)

FLUIDS–47

Water is pumped at 1 m³/s to an elevation of 5 meters through a flexible hose using a 100% efficient pump rated at 100 kilowatts. Using the same length of hose, what size motor is needed to pump 1 m³/s of water to a tank with no elevation gain? In both cases both ends of the hose are at atmospheric pressure. Neglect kinetic energy effects.

(A) 18 kW   (B) 22 kW   (C) 37 kW

          (D) 43 kW   (E) 51 kW

From a mechanical power balance for the first case:

$$\dot{m}g\Delta z + \sum P_{friction} = P_{motor}$$

$$\sum P_{friction} = P_{motor} - \dot{m}g\Delta z$$

$$= P_{motor} - \rho Q g \Delta z$$

$$= 100,000 - (1000)(1)(9.81)(5)$$

$$= 100,000 - 49,000$$

$$= 51 \text{ kW}$$

In the second case, $\Delta z = 0$. Thus, a mechanical power balance yields the following:

$$\sum P_{friction} = P_{motor}$$

$$= 51\,\text{kW} \quad \text{because the same hose is used}$$

$$P_{motor} = 51 \text{ kW}$$

Answer is (E)

FLUIDS–48

The potential flow velocity distribution around a cylinder is:

$$U = 2U_\infty \sin\theta$$

The flow is in air, and the free stream conditions are atmospheric pressure and a free stream velocity of 100 ft/sec. What is the pressure at point A in the figure?

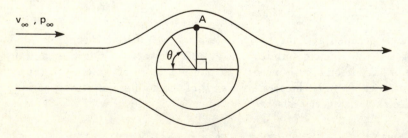

(A) 35.1 psi   (B) 14.94 psi   (C) 14.7 psi

(D) 14.45 psi   (E) − 20.4 psi

Apply Bernoulli's equation between the free stream and point A.

$$p_\infty + \frac{1}{2}\rho_\infty v_\infty^2 = p_A + \frac{1}{2}\rho U_A^2$$

$$p_{atm} + \frac{1}{2}\rho_{air}U_\infty^2 = p_A + \frac{1}{2}\rho_{air}U_A^2$$

$$U_A = 2U_\infty \sin 90^\circ$$

$$= 2U_\infty$$

$$p_A = p_{atm} + \rho_{air}\left(U_\infty^2 - 4U_\infty^2\right)$$

$$= p_{atm} - \frac{3}{2}\rho_{air}U_\infty^2$$

$$= 14.7 - \left(\frac{3}{2}\right)\left(0.00234\right)\left(100\right)^2$$

$$= 14.7\,\text{psi} - \left(35.1\frac{\text{lbf}}{\text{ft}^2}\right)\left(\frac{1\,\text{ft}^2}{144\,\text{in}^2}\right)$$

$$= 14.45\ \text{psi}$$

Answer is (D)

FLUIDS–49

Two tubes are mounted to the roof of a car. One tube points to the front of the car while the other points to the rear. The tubes are connected to a manometer filled with a fluid of specific gravity 0.62. When the height difference is 3″, what is the car's speed?

(A) 10.9 mph   (B) 43.8 mph   (C) 62 mph
             (D) 214.8 mph   (E) 239.8 mph

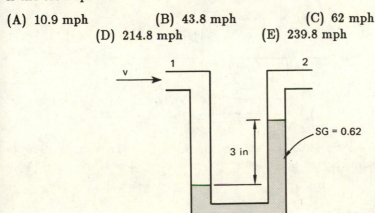

Apply Bernoulli's equation between the front tube (point 1) and the tube facing the rear (point 2).

$$p_1 + \frac{1}{2}\rho_{air}v^2 = p_2 + \rho_f gh$$

where $p_1 = p_2$ both are at atmospheric pressure

$\quad\quad\quad v =$ the speed of the car

$$= \sqrt{\frac{2\rho_f gh}{\rho_{air}}}$$

$$= \sqrt{\frac{2\left(0.62\right)\left(62.4\right)\left(\dfrac{3}{12}\right)}{0.00234}}$$

$$= 90.9 \text{ ft/sec}$$

$$= 62 \text{ mph}$$

Answer is (C)

FLUIDS–50

Water is flowing through a pipe with a manometer as shown:

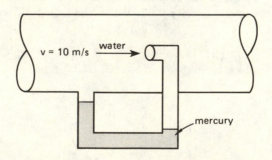

The density of mercury is 13,567 kg/m^3 and the velocity of the water is 10 m/s. Determine the reading on the manometer in centimeters of mercury.

(A) 40.6 cm   (B) 46.9 cm   (C) 57.1 cm

(D) 69.2 cm   (E) 80.0 cm

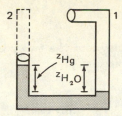

From Bernoulli's equation:

$$gz_1 + \frac{v_1^2}{2} + \frac{p_1}{\rho} = gz_2 + \frac{v_2^2}{2} + \frac{p_2}{\rho}$$

$$\Delta z = 0$$

$$p_1 - p_2 = \left(\frac{v_2^2 - v_1^2}{2}\right)\rho$$

$$v_1 = 0$$

$$v_2 = 10 \text{ m/s}$$

$$p_1 - p_2 = \left(\frac{100 - 0}{2}\right)(1000)$$

$$= 50,000 \text{ Pa}$$

$$= \rho_{Hg}\, g\, z_{Hg} - \rho_{H_2O}\, g\, z_{H_2O}$$

$$z_{Hg} = z_{H_2O}$$

$$50,000 = (\rho_{Hg} - \rho_{H_2O})g\, z_{Hg}$$

$$z_{Hg} = \frac{50,000}{9.81(13,567 - 1000)}$$

$$= 0.406 \text{ m}$$

$$= 40.6 \text{ cm}$$

Answer is (A)

FLUIDS-51

Given the following venturi meter and the two pressures shown, calculate the mass flow rate of water in the circular pipe.

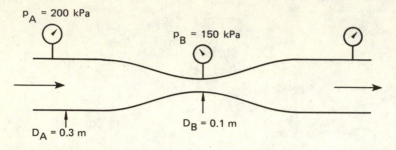

$P_A = 200$ kPa

$P_B = 150$ kPa

$D_B = 0.1$ m

$D_A = 0.3$ m

(A) 52 kg/s   (B) 61 kg/s   (C) 65 kg/s
          (D) 79 kg/s   (E) 100 kg/s

Use Bernoulli's equation along the streamline in the center of the pipe.

$$gz_A + \frac{v_A^2}{2} + \frac{p_A}{\rho} = gz_B + \frac{v_B^2}{2} + \frac{p_B}{\rho}$$

$$\Delta z = 0$$

$$\Delta p = -50\,\text{kPa}$$

$$\frac{v_B^2 - v_A^2}{2} = \frac{50,000}{1000}$$

$$v_B^2 - v_A^2 = 100\,\text{m}^2/\text{s}^2$$

from the continuity equation, $A_A v_A = A_B v_B$

$$v_A \pi (0.15)^2 = v_B \pi (0.05)^2$$

$$v_A = \left[\frac{(0.05)^2}{(0.15)^2} \right] v_B$$

$$= 0.111 v_B$$

$$v_B^2 - (0.111 v_B)^2 = 100$$

$$v_B^2 [1 - (0.111)^2] = 100$$

$$v_B = 10.06 \text{ m/s}$$

$$Q = \rho v_B A_B$$

$$= (1000)(10.06)(\pi)(0.05)^2$$

$$= 79 \text{ kg/s}$$

Answer is (D)

FLUIDS-52

What is the actual volumetric flow rate for the discharge of the tank shown? The coefficient of contraction for the orifice is 0.61, and the coefficient of velocity is 0.98.

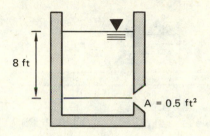

(A) 6.78 ft^3/sec   (B) 6.92 ft^3/sec   (C) 9.79 ft^3/sec

(D) 11.34 ft^3/sec   (E) 22.69 ft^3/sec

$$Q_{actual} = C_c A_{out} v_{out}$$
$$C_c = \text{coefficient of contraction}$$
$$A_{out} = \text{area of the outlet}$$
$$v_{out} = C_v \sqrt{2gh}$$
$$C_v = \text{coefficient of velocity}$$
$$h = \text{vertical distance from the exit to the fluid's surface}$$
$$Q_{actual} = C_c C_v A_{out} \sqrt{2gh}$$
$$= (0.61)(0.98)(0.5)\sqrt{2(32.2)(8)}$$
$$= 6.78 \text{ ft}^3/\text{sec}$$

Answer is (A)

FLUIDS–53

The upper plate is fixed, while the lower plate moves in the positive x direction at 0.5 m/s. The plate separation is 0.001 m, the fluid viscosity is 0.7 cp, and the velocity profile is linear. Calculate the shear stress, τ_{xy}, in the moving fluid.

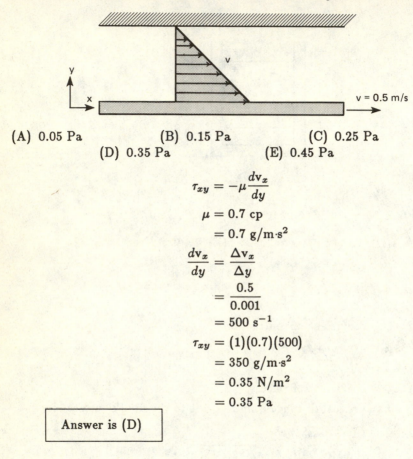

(A) 0.05 Pa   (B) 0.15 Pa   (C) 0.25 Pa
          (D) 0.35 Pa   (E) 0.45 Pa

$$\tau_{xy} = -\mu \frac{dv_x}{dy}$$

$$\mu = 0.7 \text{ cp}$$

$$= 0.7 \text{ g/m·s}^2$$

$$\frac{dv_x}{dy} = \frac{\Delta v_x}{\Delta y}$$

$$= \frac{0.5}{0.001}$$

$$= 500 \text{ s}^{-1}$$

$$\tau_{xy} = (1)(0.7)(500)$$

$$= 350 \text{ g/m·s}^2$$

$$= 0.35 \text{ N/m}^2$$

$$= 0.35 \text{ Pa}$$

Answer is (D)

FLUIDS–54

What are the units of Reynolds number for pipe flow?

(A) m/s   (B) ft^2/sec   (C) lbm/ft-sec^2
          (D) lbm/ft-sec   (E) none of the above

Reynolds number is dimensionless.

Answer is (E)

FLUIDS–55

Which of the following ratios is correct in providing a physical meaning for the Reynolds number, Re?

(A) $\mathrm{Re} = \dfrac{\text{buoyant forces}}{\text{inertial forces}}$

(B) $\mathrm{Re} = \dfrac{\text{viscous forces}}{\text{inertial forces}}$

(C) $\mathrm{Re} = \dfrac{\text{drag forces}}{\text{viscous forces}}$

(D) $\mathrm{Re} = \dfrac{\text{inertial forces}}{\text{viscous forces}}$

(E) Reynolds number has no physical meaning.

$$\mathrm{Re} = \frac{\rho v D}{\mu}$$

$$\propto \frac{\text{mass velocity}}{\mu}$$

$$\propto \frac{\text{inertial force}}{\text{viscous force}}$$

Answer is (D)

FLUIDS–56

Which of the following statements is incorrect?

(A) The Reynolds number is the ratio of the viscous force to the inertial force.

(B) Steady flows do not change with time at any point.

(C) The Navier-Stokes equation is the equation of motion for a viscous Newtonian fluid.

(D) Bernoulli's equation only holds on the same streamline.

(E) For a fluid at rest, the pressure is equal in all lateral directions.

Reynolds number is the ratio of the inertial forces to the viscous forces.

Answer is (A)

FLUIDS–57

Calculate the Reynolds number, Re, for water at 20 °C flowing in an open channel. The water is flowing at a volumetric rate of 200 gal/sec. The channel has a height of 4 feet and a width of 8 feet. At this temperature, water has a kinematic viscosity of 1.104×10^{-5} ft^2/sec.

(A) 600,000   (B) 800,000   (C) 1.0×10^6
          (D) 1.2×10^6 (E) 1.4×10^6

$$\text{Re} = \frac{\rho v D_e}{\mu}$$

$$= \frac{v D_e}{\nu}$$

$$D_e = 4 \times \left(\frac{\text{cross-sectional area}}{\text{wetted perimeter}} \right)$$

$$= 4 \left(\frac{(4)(8)}{4 + 8 + 4} \right)$$

$$= 4 \left(\frac{32}{16} \right)$$

$$= 8 \text{ ft}$$

$$Q = vA$$

$$v = \frac{Q}{A}$$

$$= \left(200 \, \frac{\text{gal}}{\text{sec}} \right) \left(\frac{1 \, \text{ft}^3}{7.481 \, \text{gal}} \right) \left(\frac{1}{32 \, \text{ft}^2} \right)$$

$$v = 0.835 \text{ ft/sec}$$

$$\text{Re} = \frac{(0.835)(8)}{1.104 \times 10^{-5}}$$

$$= 600,000$$

Answer is (A)

FLUIDS–58

A fluid with a kinematic viscosity of 2.5×10^{-5} ft^2/sec is flowing at 0.1 ft/sec from an orifice 3″ in diameter. How can the fluid be described?

(A) The fluid is completely turbulent.
(B) The fluid is in the transition zone.
(C) The fluid is laminar.
(D) The fluid's turbulence cannot be calculated from the information given.
(E) Turbulence cannot be calculated; it must be measured.

$$Re = \frac{vD}{\nu}$$

$$= \frac{(0.1)\left(\frac{3}{12}\right)}{2.5 \times 10^{-5}}$$

$$= 1000$$

A Reynolds number of 1000 means that the flow is well within the laminar region.

Answer is (C)

FLUIDS-59

The Reynolds number of a sphere falling in air is 1×10^6. If the sphere's radius is 1 ft, what is its velocity? ($\rho_{air} = 0.00234\,\text{slug/ft}^3$, $\mu_{air} = 3.8 \times 10^{-7}\,\text{lbf-sec/ft}^2$)

(A) 2.5 ft/sec   (B) 5.1 ft/sec   (C) 40.6 ft/sec
              (D) 81.2 ft/sec   (E) 162.4 ft/sec

$$Re = \frac{vD}{\nu} = \frac{\rho v D}{\mu}$$

$$v = \frac{Re\mu}{\rho D}$$

$$= \frac{(1 \times 10^6)(3.8 \times 10^{-7})}{(0.00234)(2)}$$

$$= 81.2 \text{ ft/sec}$$

Answer is (D)

FLUIDS-60

Which of the following is not true regarding the Blasius boundary layer solution?

(A) It is valid only for potential flow.
(B) It is valid for laminar flow
(C) It is an approximate solution.
(D) It permits one to calculate the skin friction on a flat plate.
(E) all of the above

The Blasius solution is an approximate solution. It is a solution to the boundary layer equations that use some simplifying assumptions. It is valid for laminar, viscous flow and permits the evaluation of shear stress and skin friction.

The Blasius solution or any other boundary layer concept has no meaning for potential flow.

Answer is (A)

FLUIDS-61

From the Blasius solution for laminar boundary layer flow, the average coefficient of skin friction is $C_f = \dfrac{1.328}{\sqrt{Re}}$. If air is flowing past a 33 feet long plate at a velocity of 100 ft/sec, what is the force on the plate?

(A) 0.00114 lbf   (B) 0.00344 lbf   (C) 0.086 lbf
          (D) 0.114 lbf   (E) 0.227 lbf

$$Re_L = \frac{\rho_{air} v_{air} L_{plate}}{\mu_{air}}$$

$$= \frac{(0.00234)(100)(33)}{3.8 \times 10^{-7}}$$

$$= 2.032 \times 10^7$$

$$C_f = \frac{1.328}{\sqrt{Re_L}}$$

$$= \frac{1.328}{\sqrt{2.032 \times 10^7}}$$

$$= 2.95 \times 10^{-4}$$

$$C_f = \frac{F}{\frac{1}{2}\rho_{air} v_{air}^2 L_{plate}}$$

$$f = C_f \left(\frac{1}{2}\rho_{air} v_{air}^2 L_{plate}\right)$$

$$= \left(2.95 \times 10^{-4}\right)\left[\frac{1}{2}\left(0.00234\right)\left(100\right)^2\left(33\right)\right]$$

$$= 0.114 \text{ lbf}$$

Answer is (D)

FLUIDS-62

Where does the Moody diagram for friction factors for pipe flow come from?

(A) calculations based on potential flow
(B) theoretical solutions of the Navier-Stokes equations
(C) experimental results for inviscid fluids
(D) experimental results for viscous fluids
(E) applying the principle of conservation of mass

Moody diagrams are experimental data plots. They are valid for viscous fluids.

Answer is (D)

FLUIDS-63

What is the friction factor for fully developed flow in a circular pipe where Reynolds number is 1000?

(A) 0.008　　　　　(B) 0.064　　　　　(C) 0.08
　　　　　(D) 0.10　　　　　(E) 0.64

For Re < 2000, the friction factor, f, is given by the following:

$$f = \frac{64}{Re}$$
$$= \frac{64}{1000}$$
$$= 0.064$$

Answer is (B)

FLUIDS-64

For pipe flow in the laminar flow region, how is the friction factor related to the Reynolds number?

(A) $f \propto \dfrac{64}{Re}$
(B) $f \propto \dfrac{1}{Re}$
(C) $f \propto Re$
(D) $f \propto Re^2$
(E) They are not directly related.

In the laminar region, $f = \dfrac{64}{\mathrm{Re}}$.

Answer is (A)

FLUIDS–65

Which of the following flow meters measure the average fluid velocity in a pipe rather than a point or local velocity?

I. venturi meter
II. pitot tube
III. impact tube
IV. orifice meter
V. hot wire anemometer

(A) I only   (B) II only   (C) I and IV
        (D) II and V   (E) I, II, and V

Of the five choices given, only venturi and orifice meters measure average velocity.

Answer is (C)

FLUIDS–66

What is the hydraulic radius of the semicircular channel shown?

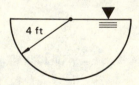

4 ft

(A) 2 ft   (B) 2.57 ft   (C) 4 ft
        (D) 6.14 ft   (E) 8 ft

$$r_h = \text{hydraulic radius}$$
$$= \frac{\text{cross-section area}}{\text{wetted perimeter}}$$
$$= \frac{\frac{1}{2}(\pi r^2)}{\pi r}$$
$$= \frac{r}{2}$$
$$= 2 \text{ ft}$$

Answer is (A)

FLUIDS–67

What is the hydraulic radius of the channel shown?

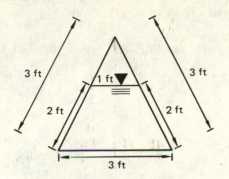

(A) 0.33 ft   (B) 0.43 ft   (C) 0.49 ft

(D) 1.0 ft   (E) 3.0 ft

$$r_h = \frac{\text{cross-sectional area}}{\text{wetted perimeter}}$$

$$= \frac{\left(\frac{3}{2}\right)\left(3\sin 60°\right) - \left(\frac{1}{2}\right)\left(3\sin 60°\right)}{7}$$

$$= 0.49 \text{ ft}$$

Answer is (C)

FLUIDS–68

For fully developed laminar flow of fluids through pipes, the average velocity is what fraction of the maximum velocity in the pipe?

(A) $\frac{1}{8}$ (B) $\frac{1}{4}$ (C) $\frac{1}{2}$

(D) $\frac{3}{4}$ (E) 1

For laminar flow in pipes:

$$\bar{v} = \frac{v_{max}}{2}$$

Answer is (C)

PROFESSIONAL PUBLICATIONS, INC. • Belmont, CA

FLUIDS-69

The flow rate of water through a cast iron pipe is 5000 gallons per minute. The diameter of the pipe is 1 foot, and the coefficient of friction is $f = 0.0173$. What is the pressure drop over a 100 foot length of pipe?

(A) 21.078 lbf/ft^2 (B) 23.78 lbf/ft^2 (C) 337.26 lbf/in^2

(D) 337.26 lbf/ft^2 (E) 488.65 lbf/ft^2

$$Q = 5000 \text{ gpm}$$

$$= 5000 \text{ gpm} \left(\frac{1 \text{ min}}{60 \text{ sec}} \right) \left(0.13368 \frac{\text{ft}^3}{\text{gal}} \right)$$

$$= 11.14 \text{ ft}^3/\text{sec}$$

$$v = \frac{Q}{A}$$

$$= \frac{11.14}{\pi d^2}$$

$$= \frac{11.14}{\pi \left(\frac{1}{2} \right)^2}$$

$$v = 14.18 \text{ ft/sec}$$

$$\Delta h = \text{head loss}$$

$$= f \left(\frac{l}{d} \right) \left(\frac{v_2}{2g} \right)$$

$$= \frac{\Delta p}{\rho g}$$

$$\Delta p = \frac{\rho g f l v^2}{2gd}$$

$$= \frac{\rho f v^2}{2d}$$

$$= \frac{(62.4)(0.0173)(100)(14.18)^2}{(2)(1)}$$

$$= 337.26 \text{ lbf/ft}^2$$

Answer is (D)

LUIDS–70

A cast iron pipe of equilateral triangular cross section with side length of 20.75″
as water flowing through it. The flow rate is 6000 gallons per minute, and the
riction factor for the pipe is $f = 0.017$. What is the pressure drop in a 100 foot
ection?

(A) 24.3 lbf/ft^2 (B) 48.7 lbf/ft^2 (C) 178.5 lbf/ft^2

(D) 309.7 lbf/ft^2 (E) 536.4 lbf/ft^2

$$D_e = 4 \times \frac{\text{cross-sectional area}}{\text{wetted perimeter}}$$

$$= 4 \frac{\left(\frac{1}{2}\right)\left(\frac{20.75}{12}\right)\left(\frac{\sqrt{3}}{2}\right)\left(\frac{20.75}{12}\right)}{3\left(\frac{20.75}{12}\right)}$$

$$= 1 \text{ ft}$$

$$Q = 6000 \text{ gpm}$$

$$= 6000 \text{ gpm} \left(\frac{1 \min}{60 \sec}\right)\left(0.13368 \frac{\text{ft}^3}{\text{gal}}\right)$$

$$= 13.37 \text{ ft}^3/\text{sec}$$

$$v = \frac{Q}{A}$$

$$= \frac{13.37}{\left(\frac{1}{2}\right)\left(\frac{20.75}{12}\right)\left(\frac{\sqrt{3}}{2}\right)\left(\frac{20.75}{12}\right)}$$

$$= 10.32 \text{ ft/sec}$$

$$\Delta h = \frac{\Delta p}{\rho g}$$

$$= \frac{f l v^2}{2 D_e g}$$

$$\Delta p = \frac{\rho f v^2}{2 D_e}$$

$$\Delta p = \frac{(62.4)(0.017)(100)(10.32)^2}{(2)(1)(32.2)}$$

$$= 178.5 \text{ lbf/ft}^2$$

Answer is (C)

FLUIDS-71

A circular cylinder 4 feet long and 3 feet in diameter is in an air stream. The flow velocity is 5 ft/sec perpendicular to the axis of the cylinder. Given that the coefficient of drag, C_D, on the cylinder is 1.3, and the density of air is 0.00234 slug/ft^3, what is the force on the cylinder?

(A) 0.09 lbf   (B) 0.11 lbf   (C) 0.46 lbf
          (D) 0.91 lbf   (E) 1.43 lbf

$$D = \frac{1}{2}\rho v^2 A C_D$$

$$= \frac{1}{2}\left(0.00234\right)\left(5\right)^2\left(3 \times 4\right)\left(1.3\right)$$

$$= 0.456 \text{ lbf}$$

Answer is (C)

FLUIDS-72

Air flows past a 2″ diameter sphere at 100 ft/sec. What is the drag force experienced by the sphere given that it has a coefficient of drag of 0.5 and that the density of the air is 0.00234 slug/ft^3?

(A) 0.041 lbf   (B) 0.064 lbf   (C) 0.128 lbf
          (D) 0.244 lbf   (E) 0.256 lbf

$$D = C_D \left(\frac{1}{2}\rho_{air} v_{air}^2 A\right)$$

$$= C_D \left[\frac{1}{2}\rho_{air} v_{air}^2 \pi \left(\frac{d_{sphere}}{2}\right)^2\right]$$

$$= \frac{C_D \rho_{air} v_{air}^2 \pi d_{sphere}^2}{8}$$

$$= \frac{\left(0.5\right)\left(0.00234\right)\left(100\right)^2 \pi \left(\frac{2}{12}\right)^2}{8}$$

$$= 0.128 \text{ lbf}$$

Answer is (C)

FLUIDS-73

A cylinder 10 feet long and 2 feet in diameter is suspended in air flowing at 8 ft/sec. The density of air is 0.00234 slug/ft^3, and the coefficient of drag of the sphere is 1.3. What is the drag on the cylinder?

(A) 0.311 lbf   (B) 0.39 lbf   (C) 1.95 lbf

(D) 3.89 lbf   (E) 4.989 lbf

$$D = C_D \left(\frac{1}{2}\rho v^2 A \right)$$
$$= \frac{(1.3)(0.00234)(8)^2(10)(2)}{2}$$
$$= 1.95 \text{ lbf}$$

Answer is (C)

FLUIDS-74

What is the terminal velocity of a 2" diameter aluminum sphere falling in air? Assume that the sphere has a coefficient of drag of 0.5, the density of aluminum is 5.12 slug/ft^3, and the density of air is 0.00234 slug/ft^3.

(A) 100 ft/sec   (B) 177 ft/sec   (C) 350 ft/sec

(D) 1000 ft/sec   (E) indeterminate

Let v_t be the terminal velocity. At terminal velocity:

$$\text{drag force} = \text{weight}$$

$$D = mg$$

$$mg = \frac{4}{3}\pi \left(\frac{d}{2}\right)^3 \rho_{alum}\, g$$

$$D = C_D \left(\frac{1}{2}\rho_{air} v_t^2 A\right)$$

$$\left(\frac{1}{2}\right) C_D \rho_{air} v_t^2 \pi \left(\frac{d}{2}\right)^2 = \frac{4}{3}\pi \left(\frac{d}{2}\right)^3 \rho_{alum}\, g$$

$$v_t = \sqrt{\frac{2(4\pi)\left(\dfrac{d}{2}\right)^3 \rho_{alum}\, g}{3 C_D \rho_{air} \pi \left(\dfrac{d}{2}\right)^2}}$$

$$= \sqrt{\frac{4 d \rho_{alum}\, g}{3 C_D \rho_{air}}}$$

$$= \sqrt{\frac{4\left(\dfrac{2}{12}\right)(5.12)(32.2)}{3(0.5)(0.00234)}}$$

$$= 177 \text{ ft/sec}$$

Answer is (B)

FLUIDS-75

In the real flow of air around a cylinder, the circulation is calculated to be 42.74 ft/sec. If the free stream velocity is 100 ft/sec, what is the lift generated per foot of the cylinder?

(A) 10 lbf/ft   (B) 42.74 lbf/ft   (C) 120 lbf/ft

(D) 322 lbf/ft   (E) indeterminate

The Kutta-Joukowsky theorem states that:

$$\frac{\text{lift}}{\text{foot}} = \rho_\infty \text{v}_\infty \Gamma$$

$$= (0.00234)(100)(42.74)$$

$$= 10 \text{ lbf/ft}$$

Answer is (A)

FLUIDS-76

The 5″ diameter cylinder shown below rotates at 3600 revolutions per minute. Air is flowing past the cylinder at 100 ft/sec. How much is the lift on the cylinder per unit length? The density of air is 0.00234 slug/ft³.

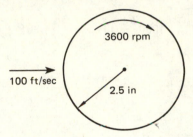

3600 rpm

100 ft/sec

2.5 in

(A) 0.15 lbf/ft   (B) 1.25 lbf/ft   (C) 11.75 lbf/ft
          (D) 20.75 lbf/ft   (E) 23.65 lbf/ft

From the Kutta-Joukowsky theorem:

$$\frac{\text{lift}}{\text{foot}} = \rho \text{v}_\infty \Gamma$$

$$\Gamma = \oint \mathbf{V} \cdot d\ell$$

$$= \int_0^{2\pi} (r\omega)(r\, d\theta)$$

$$= 2\pi r^2 \omega$$

$$= 2\pi \left(\frac{5}{24}\right)^2 \left[3600\left(\frac{2\pi}{60}\right)\right]$$

$$= 102.8 \text{ ft}^2/\text{sec}$$

$$\frac{\text{lift}}{\text{foot}} = (0.00234)(100)(102.8)$$

$$= 23.65 \text{ lbf/ft}$$

Answer is (E)

FLUIDS-77

A pump produces a head of 30 feet. The volumetric flow rate is 10 gallons per minute. The fluid pumped is oil with a specific gravity of 0.83. How much energy does the pump consume in one hour?

(A) 8.7 kJ   (B) 17.2 kJ   (C) 168.9 kJ

(D) 203.6 kJ   (E) 1.03 MJ

$$\text{power} = \Delta p\, Q$$

$$\Delta p = \text{change in pressure}$$

$$= \rho g h$$

$$= \left(0.83\right)\left(1000\frac{\text{kg}}{\text{m}^3}\right)\left(9.81\frac{\text{m}}{\text{s}^2}\right)\left(30\,\text{ft}\right)\left(\frac{0.3048\,\text{m}}{\text{ft}}\right)$$

$$= 74,377\ \text{Pa}$$

$$Q = 10\ \text{gpm}\left(0.003785\,\frac{\text{m}^3}{\text{gal}}\right)\left(\frac{1\,\text{min}}{60\,\text{sec}}\right)$$

$$= 6.31 \times 10^{-4}\,\text{m}^3/\text{s}$$

$$\text{energy} = \text{power} \times \text{time}$$

$$= \Delta p\, Q t$$

$$= (74,377)(6.31 \times 10^{-4})(3600)$$

$$= 168,900\ \text{J}$$

$$= 168.9\ \text{kJ}$$

Answer is (C)

FLUIDS-78

A pump has an efficiency of 65%. It is driven by a 0.75 horsepower motor. The pump produces a pressure rise of 120 Pa in water. What is the required flow rate?

(A) 3.03 m³/s   (B) 4.04 m³/s   (C) 4.55 m³/s

(D) 4.66 m³/s   (E) 6.32 m³/s

P_r = power supplied by the pump to the water

$\quad = \eta P_i$

where $\quad \eta$ = efficiency

$\qquad P_i$ = ideal power

$\qquad P_r = \Delta p \, Q$

where $\quad \Delta p$ = pressure rise

$\qquad Q$ = volumetric flow rate

$\Delta p \, Q = \eta P_i$

$Q = \dfrac{\eta P_i}{\Delta p}$

$$= \frac{\left(0.65\right)\left(0.75\right)\left(745.7\,\dfrac{\text{W}}{\text{hp}}\right)}{120}$$

$= 3.03 \text{ m}^3/\text{s}$

Answer is (A)

FLUIDS–79

The pressure drop across a turbine is 30 psi. The flow rate is 60 gallons per minute. Calculate the power output of the turbine.

(A) 0.41 hp (B) 1.05 hp (C) 2.54 hp
  (D) 6.30 hp (E) 8.35 hp

$$\Delta p = 30 \, \text{psi} \left(\frac{144 \, \text{in}^2}{\text{ft}^2}\right)$$

$$= 4320 \, \text{lbf/ft}^2$$

$$Q = 60 \, \text{gpm} \left(0.13368 \, \frac{\text{ft}^3}{\text{gal}}\right)\left(\frac{1 \, \text{min}}{60 \, \text{sec}}\right)$$

$$= 0.13368 \, \text{ft}^3/\text{sec}$$

$$\text{power} = \Delta p \, Q = (4320)(0.13368)$$

$$= 577.5 \, \frac{\text{ft-lbf}}{\text{sec}} \left(\frac{1 \, \text{hp}}{550 \, \dfrac{\text{ft-lbf}}{\text{sec}}}\right)$$

$$= 1.05 \, \text{hp}$$

Answer is (B)

5 THERMODYNAMICS

THERMODYNAMICS-1

Which of the following properties are intensive properties?

> I.   temperature
> II. pressure
> III. composition
> IV. mass

(A) I only   (B) IV only   (C) I and II

            (D) I and IV   (E) I, II, and III

An intensive property does not depend on the amount of material present. This is true for temperature, pressure, and composition.

Answer is (E)

THERMODYNAMICS-2

How many independent properties are required to completely fix the equilibrium state of a pure gaseous compound?

(A) 0   (B) 1   (C) 2   (D) 3   (E) 4

The number of properties needed to fix the state of a gaseous compound is given by Gibb's phase rule:

$$f = n - p + 2$$

where f = number of independently variable properties

n = number of components

p = number of phases

For a pure gas,

$$n = p = 1$$
$$f = 1 - 1 + 2$$
$$= 2$$

Answer is (C)

THERMODYNAMICS-3

Which of the following thermodynamic relations is incorrect?

(A) $TdS = dU + pdV$
(B) $TdS = dH - Vdp$
(C) $U = Q - W$
(D) $G = H - TS$
(E) $H = U - pV$

Enthalpy is given by the relation $H = U + pV$. Therefore, the relation in (E) is incorrect.

Answer is (E)

THERMODYNAMICS-4

If air is at a pressure, p, of 3200 lbf/ft², and at a temperature, T, of 800 °R, what is the specific volume, v? ($R = 53.3$ ft-lbf/lbm-°R, and air can be modeled as an ideal gas)

(A) 9.8 ft³/lbm   (B) 11.2 ft³/lbm   (C) 13.3 ft³/lbm

(D) 14.2 ft³/lbm   (E) 15.8 ft³/lbm

$$pv = RT$$

$$v = \frac{RT}{p}$$

$$= \frac{(53.3)(800)}{3200}$$

$$= 13.33 \text{ ft}^3/\text{lbm}$$

Answer is (C)

THERMODYNAMICS–5

Which of the following relations defines enthalpy?

(A) $h = u + \dfrac{p}{T}$

(B) $h = u + pV$

(C) $h = u + \dfrac{p}{V}$

(D) $h = pV + T$

(E) $h = \dfrac{p}{V} + \dfrac{u}{T}$

Enthalpy is given by $h = u + pV$.

Answer is (B)

THERMODYNAMICS–6

In a certain constant mass system, the conditions change from point 1 to point 2. How does the change in enthalpy for path A differ from the enthalpy change for path B in going from point 1 to point 2?

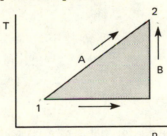

(A) $\Delta H_A > \Delta H_B$

(B) $\Delta H_A = \Delta H_B$

(C) $\Delta H_A < \Delta H_B$

(D) $\Delta H_A \div \left(\dfrac{P_1 + P_2}{2} \right) = \dfrac{\Delta H_B}{P_1}$

(E) Change cannot be determined from the information given.

Enthalpy is a state function. Therefore, its value depends only on the initial and final states, and not on the path taken between the two states. Thus, $\Delta H_A = \Delta H_B$.

Answer is (B)

THERMODYNAMICS-7

Steam at 1000 lbf/ft^2 pressure and 300 °R has a specific volume of 6.5 ft^3/lbm and a specific enthalpy of 9800 lbf-ft/lbm. Find the internal energy per pound mass of steam.

(A) 2500 lbf-ft/lbm   (B) 3300 lbf-ft/lbm   (C) 5400 lbf-ft/lbm

(D) 6900 lbf-ft/lbm   (E) 8040 lbf-ft/lbm

$$h = u + pV$$
$$u = h - pV$$
$$= 9800 - (1000)(6.5)$$
$$= 3300 \text{ lbf-ft/lbm}$$

Answer is (B)

THERMODYNAMICS-8

Which of the following is true for water at a reference temperature where enthalpy is zero?

(A) Internal energy is negative.
(B) Entropy is non-zero.
(C) Specific volume is zero.
(D) Vapor pressure is zero.
(E) pV energy (flow energy) is zero.

Typically, the saturation temperature (32 °F for water) is chosen as the enthalpic reference temperature. At that temperature, the water has a distinct (vapor) pressure and volume. Therefore, choices (C), (D), and (E) are false. Although there is no thermodynamic relationship between entropy and enthalpy, the values of enthalpy and entropy are commonly referenced to the same temperature. Thus, by convention, entropy is zero when enthalpy is zero. Therefore, (B) is also false.

The definition of enthalpy is the sum of internal energy and flow energy:

$$h = u + pV$$

If internal energy is zero and the flow energy, pV, is non-zero, then the internal energy must be negative.

Answer is (A)

THERMODYNAMICS–9

Which of the following is the triple point for the phase diagram given?

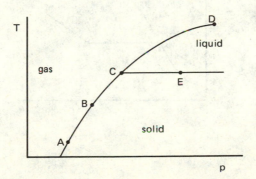

(A) A (B) B (C) C (D) D (E) E

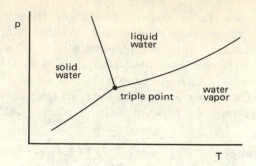

The triple point is the point at which the liquid, solid, and vapor states are all in equilibrium. Therefore, point C is the triple point.

Answer is (C)

THERMODYNAMICS–10

What does the dashed curve in the figure represent?

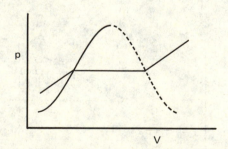

(A) an isobar
(B) an isotherm
(C) the saturated liquid line
(D) the saturated vapor line
(E) energy

By definition, the dashed line shown is the saturated vapor line.

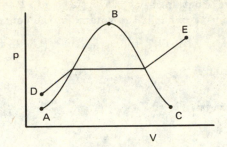

AB is the saturated liquid line.
B is the critical point.
BC is the saturated vapor line.
DE represents a possible path
    of the system when heated.

Answer is (D)

THERMODYNAMICS–11

In an ideal gas mixture of constituents i and j, what is the mole fraction x_i equal to?

(A) $\dfrac{T_i}{T_i + T_j}$ (B) $\dfrac{Z_i}{Z_i + Z_j}$ (C) $\dfrac{p_i}{p_i + p_j}$

    (D) $\dfrac{p_i V_i}{RT}$ (E) $\dfrac{n_i}{n_j}$

For an ideal gas, the mole fraction is equal to the partial pressure fraction.

$$x_i = \frac{p_i}{p_i + p_j}$$

Answer is (C)

THERMODYNAMICS–12

3.0 lbm of air are contained at 25 psia and 100 °F.   Given that $R_{air} = 53.35$ ft-lbf/lbm-°F, what is the volume of the container?

(A) 10.7 ft³   (B) 14.7 ft³   (C) 15 ft³

       (D) 20.6 ft³   (E) 24.9 ft³

Use the ideal gas law.

$$pV = mRT$$
$$T = (100 + 460)\,°R$$
$$V = \frac{mRT}{p}$$
$$= \frac{(3)(53.35)(560)}{3600}$$
$$= 24.9 \text{ ft}^3$$

> **Answer is (E)**

THERMODYNAMICS–13

The compressibility factor, z, is used for predicting the behavior of non-ideal gases. How is the compressibility factor defined relative to an ideal gas? (Subscript "c" refers to critical value.)

(A) $z = \dfrac{p}{p_c}$ (B) $z = \dfrac{pV}{RT}$ (C) $z = \dfrac{T}{T_c}$

    (D) $z = \left(\dfrac{T}{T_c}\right)\left(\dfrac{p_c}{p}\right)$ (E) $z = \dfrac{RT}{pV}$

For real gases, the compressibility factor, z, is a dimensionless constant given by $pV = zRT$. Therefore, $z = \dfrac{pV}{RT}$.

> **Answer is (B)**

THERMODYNAMICS-14

On what plane is the Mollier diagram plotted?

(A) p-V (B) p-T ✓(C) h-s

     (D) s-u (E) u-h

The axes for a Mollier diagram are enthalpy and entropy (h-s).

> Answer is (C)

THERMODYNAMICS-15

How is the quality, x, of a liquid-vapor mixture defined?

(A) the fraction of the total volume that is saturated vapor
(B) the fraction of the total volume that is saturated liquid
(C) the fraction of the total mass that is saturated vapor
(D) the fraction of the total mass that is saturated liquid
(E) none of the above

The quality of the liquid-vapor mixture is defined as the fraction of the total mass that is saturated vapor.

> Answer is (C)

THERMODYNAMICS-16

What is the expression for the heat of vaporization?

$$h_g = \text{enthalpy of the saturated vapor}$$
$$h_f = \text{enthalpy of the saturated liquid}$$

(A) h_g (B) h_f (C) $h_g - h_f$

     (D) $h_g^2 - h_f^2$ (E) h_g^2

The heat of vaporization, h_{fg}, is the difference between the enthalpy of the saturated vapor and the enthalpy of the saturated liquid. Thus, $h_{fg} = h_g - h_f$.

> Answer is (C)

THERMODYNAMICS–17

From the steam tables, determine the average constant pressure specific heat (c_p) of steam at 10 kPa and 45.8 °C.

(A) 1.79 kJ/kg·°C   (B) 10.28 kJ/kg·°C   (C) 30.57 kJ/kg·°C

(D) 100.1 kJ/kg·°C   (E) 121.8 kJ/kg·°C

$$\Delta h = c_p \Delta T$$

$$c_p = \frac{\Delta h}{\Delta T}$$

From the steam tables:

$$\begin{aligned} \text{at } 47.7\,°C \qquad & h = 2588.1 \text{ kJ/kg} \\ \text{at } 43.8\,°C \qquad & h = 2581.1 \text{ kJ/kg} \end{aligned}$$

$$c_p = \frac{2588.1 - 2581.1}{47.7 - 43.8}$$

$$= 1.79 \text{ kJ/kg·°C}$$

Answer is (A)

THERMODYNAMICS–18

A 10 m³ vessel initially contains 5 m³ of liquid water and 5 m³ of saturated water vapor at 100 kPa. Calculate the internal energy of the system using the steam tables.

(A) 5×10^5 kJ   (B) 8×10^5 kJ   (C) 1×10^6 kJ

(D) 2×10^6 kJ   (E) 3×10^6 kJ

from the steam tables, $\quad v_f = 0.001043 \text{ m}^3/\text{kg}$

$$v_g = 1.6940 \text{ m}^3/\text{kg}$$

$$u_f = 417.3 \text{ kJ/kg}$$

$$u_g = 2506 \text{ kJ/kg}$$

$$m_{\text{vap}} = \frac{V_{\text{vap}}}{v_g}$$

$$= \frac{5}{1.694}$$

$$= 2.95 \text{ kg}$$

$$m_{\text{liq}} = \frac{V_{\text{liq}}}{v_f}$$

$$= \frac{5}{0.001043}$$

$$= 4794 \text{ kg}$$

$$u = u_f m_{\text{liq}} + u_g m_{\text{vap}}$$

$$= (417.3)(4794) + (2506.1)(2.95)$$

$$= 2.01 \times 10^6 \text{ kJ}$$

Answer is (D)

THERMODYNAMICS–19

A vessel with a volume of 1 cubic meter contains liquid water and water vapor in equilibrium at 600 kPa. The liquid water has a mass of 1 kg. Using the steam tables, calculate the mass of the water vapor.

(A) 0.99 kg   (B) 1.57 kg   (C) 1.89 kg

       (D) 2.54 kg   (E) 3.16 kg

From the steam tables at 600 kPa:

$$v_f = 0.001101 \text{ m}^3/\text{kg}$$

$$v_g = 0.3157 \text{ m}^3/\text{kg}$$

$$V_{\text{tot}} = m_f v_f + m_g v_g$$

$$m_g = \frac{V_{\text{tot}} - m_f v_f}{v_g}$$

$$= \frac{1 - (1)(0.001101)}{0.3157}$$

$$= 3.16 \text{ kg}$$

Answer is (E)

THERMODYNAMICS–20

Calculate the entropy of steam at 60 psia with a quality of 0.6.

(A) 0.4274 BTU/lbm-°R
(B) 0.7303 BTU/lbm-°R
(C) 1.1577 BTU/lbm-°R
(D) 1.2172 BTU/lbm-°R
(E) 1.6446 BTU/lbm-°R

From the steam tables at 60 psia:

$$s_f = 0.4274 \text{ BTU/lbm-°R}$$
$$s_{fg} = 1.2172 \text{ BTU/lbm-°R}$$
$$s = s_f + x s_{fg}$$
$$\text{where} \quad x = \text{quality}$$
$$s = 0.4274 + (0.6)(1.2172)$$
$$= 1.1577 \text{ BTU/lbm-°R}$$

Answer is (C)

THERMODYNAMICS–21

If 1 lbm of steam at 14.7 psia, 63% quality is heated isentropically, at what pressure will it reach the saturated vapor state?

(A) 56 psia   (B) 313 psia   (C) 1852 psia
          (D) 2585 psia   (E) indeterminate

Use the steam tables.

$$p_1 = 14.7 \, \text{psia}$$
$$s_{f1} = 0.3122 \text{ BTU/lbm-°R}$$
$$s_{fg1} = 1.4447 \text{ BTU/lbm-°R}$$
$$s_1 = s_{f1} + 0.63 s_{fg1}$$
$$= 0.3122 + (0.63)(1.4447)$$
$$= 1.2224 \text{ BTU/lbm-°R}$$

Now, find p_2 such that $s_{g2} = 1.2224$ BTU/lbm-°R. Interpolating from the steam tables, $p_2 \approx 2585$ psia.

Answer is (D)

THERMODYNAMICS–22

The first law of thermodynamics is based on which of the following principles?

(A) conservation of mass
(B) the enthalpy-entropy relationship
(C) action-reaction
(D) conservation of energy
(E) the entropy-temperature relationship

The first law of thermodynamics is based on the principle of conservation of energy.

Answer is (D)

THERMODYNAMICS–23

The general energy equation for an open system involves the following five terms:

I.   accumulation of energy
II.   net energy transfer by work (standard sign convention)
III. net energy transfer by heat (standard sign convention)
IV.   transfer of energy in by mass flow
V.   transfer of energy out by mass flow

Using the standard sign conventions, what is the proper arrangement of these terms for the general energy equation satisfying the first law of thermodynamics?

(A) I = −II + III + IV − V
(B) I = II + III + IV + V
(C) I = II + III + IV − V
(D) I = II − III − IV + V
(E) I = −II − III − IV − V

The first law of thermodynamics states that the total change in energy (I) is equal to the energy in (IV) minus the energy out (V) minus the work done on the system (II) plus the heat transferred to the system (III). Thus, I = − II + III + IV − V.

Answer is (A)

THERMODYNAMICS-24

In a reversible process, the state of a system changes from state 1 to state 2 as shown on the p-V diagram. What does the shaded area on the diagram represent?

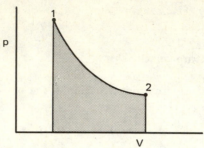

(A) free energy change
(B) heat transfer
(C) enthalpy change
(D) work done by the system
(E) entropy change

For a reversible process, the work done by the system is given by the following:

$$W = \int_1^2 p \, dV$$

Therefore, the shaded area represents the work done by the system.

Answer is (D)

THERMODYNAMICS-25

What is the value of the work done for a closed, reversible, isometric system?

(A) zero
(B) positive
(C) negative
(D) positive or negative
(E) cannot be determined

$$W = \int p \, dV$$

However, an isometric system is a system which has a constant volume ($dV = 0$). Therefore, the work done by the system is zero.

Answer is (A)

THERMODYNAMICS–26

The expansion of a gas through a plug at a high enough pressure results in a temperature rise, while at lower pressures a temperature drop occurs. The Joule-Thompson coefficient, μ_{JT}, is defined as the ratio of the change in temperature to the change in pressure. The temperature at which μ_{JT} changes from positive to negative is called the inversion temperature. When μ_{JT} is negative, which of the following statements is true?

(A) Gases may be liquified.
(B) No liquification is possible.
(C) Only trace liquification is possible.
(D) Liquification can be obtained only with a catalyst.
(E) none of the above

When $\mu_{JT} < 0$, then $\dfrac{\partial T}{\partial p} < 0$. Thus, a pressure rise is accompanied by a temperature drop. Therefore, a gas may be liquified.

Answer is (A)

THERMODYNAMICS–27

A 5 m^3 vessel initially contains 50 kg of liquid water and saturated water vapor at a total internal energy of 27,300 kJ. Calculate the heat requirement to vaporize all of the liquid.

(A) 100,000 kJ   (B) 200,00 kJ   (C) 300,000 kJ
          (D) 400,000 kJ   (E) 500,000 kJ

An expression for the first law of thermodynamics is:

$$Q = U_2 - U_1$$
$$U_1 = 27,300 \text{ kJ}$$

Find μ_2 in the steam tables at 100% vapor and $v_g = \dfrac{5}{50} = 0.10$ m^3/kg.

The final state is at $p = 2.00$ MPa and $\mu_g = 2600$ kJ/kg.

$$U_2 = (2600 \text{ kJ/kg}) \left(50 \text{ kg} \right)$$

$$= 130,000 \text{ kJ}$$
$$Q = 130,000 - 27,300$$
$$= 103,000$$
$$\approx 100,000 \text{ kJ}$$

Answer is (A)

THERMODYNAMICS-28

Find the change in internal energy of 5 lbm of oxygen gas when the temperature changes from 100 °F to 120 °F. $c_v = 0.157$ BTU/lbm-°R.

(A) 14.7 BTU   (B) 15.7 BTU   (C) 16.8 BTU
          (D) 147 BTU   (E) 157 BTU

$$\Delta U = mc_v \Delta T$$
$$\Delta T = 120 - 100$$
$$= 20 \,°F$$
$$= 20 \,°R$$
$$\Delta U = (5)(0.157)(20)$$
$$= 15.7 \text{ BTU}$$

Answer is (B)

THERMODYNAMICS-29

Water (specific heat $c_v = 4.2$ kJ/kg · K) is being heated by a 1500 W heater. What is the rate of change in temperature of 1 kg of the water?

(A) 0.043 K/s   (B) 0.179 K/s   (C) 0.357 K/s
          (D) 1.50 K/s   (E) 3.499 K/s

$$\dot{Q} = mc_v (\dot{\Delta T})$$
$$\dot{\Delta T} = \frac{Q}{mc_v}$$
$$= \frac{1500}{(1)(4200)}$$
$$= 0.357 \text{ K/s}$$

Answer is (C)

THERMODYNAMICS-30

One kilogram of water ($c_v = 4.2$ kJ/kg·K) is heated by 300 BTU of energy. What is the change in temperature, in K?

(A) 17.9 K   (B) 71.4 K   (C) 73.8 K
          (D) 75.4 K   (E) 125.2 K

$$mc_v\Delta T = Q$$

$$\Delta T = \frac{Q}{mc_v}$$

$$Q = 300 \text{ BTU} \left(1.055 \frac{\text{kJ}}{\text{BTU}}\right)$$

$$= 316.5 \text{ kJ}$$

$$\Delta T = \frac{316.5}{(1)(4.2)}$$

$$= 75.4 \text{ K}$$

Answer is (D)

THERMODYNAMICS-31

Determine the change in enthalpy per lbm of nitrogen gas as its temperature changes from 500 °F to 200 °F. ($c_p = 0.2483$ BTU/lbm-°R)

(A) −74.49 BTU/lbm   (B) −72.68 BTU lbm   (C) −68.47 BTU/lbm

(D) 63.78 BTU/lbm   (E) 84.48 BTU/lbm

$$\Delta h = c_p\Delta T$$

$$= (0.2483)(200 - 500)$$

$$= -74.49 \text{ BTU/lbm}$$

Answer is (A)

THERMODYNAMICS-32

Calculate the change in enthalpy as 1 kg of nitrogen is heated from 1000 K to 1500 K, assuming the nitrogen is an ideal gas at a constant pressure. The temperature-dependent specific heat of nitrogen is:

$$c_p = 39.06 - 512.79T^{-1.5} + 1072.7T^{-2} - 820.4T^{-3}$$

c_p is in kJ/kmole·K, and T is in K.

(A) 600 kJ   (B) 700 kJ   (C) 800 kJ

(D) 900 kJ   (E) 1000 kJ

$$c_p \equiv \left(\frac{\partial H}{\partial T} \right)_P$$

$$\partial H = c_p \partial T$$

$$H = \int c_p dT$$

$$= \int_{1000}^{1500} \left(39.06 - 512.79T^{-1.5} + 1072.7 - 820.4T^{-3} \right) dT$$

$$= \left[39.06T + 1025.6T^{-0.5} + 1072.7 + 410.2T^{-2} \right]_{1000}^{1500}$$

$$= 58,616 - 39,091$$

$$= 19,525 \, \frac{kJ}{kgmole} \left(1\,kg \right) \left(\frac{1\,kgmole}{28\,kg} \right)$$

$$= 697.3 \text{ kJ}$$

$$\Delta H \approx 700 \text{ kJ}$$

Answer is (B)

THERMODYNAMICS-33

What is the resulting pressure when one pound of air at 15 psia and 200 °F is heated at constant volume to 800 °F?

(A) 15 psia   (B) 28.6 psia   (C) 36.4 psia

(D) 52.1 psia   (E) 102.8 psia

$$\frac{T_1}{p_1} = \frac{T_2}{p_2}$$

$$p_2 = \frac{p_1 T_2}{T_1}$$

$$= \frac{(15)(1260)}{660}$$

$$= 28.6 \text{ psia}$$

Answer is (B)

THERMODYNAMICS-34

What horsepower is required to isothermally compress 800 ft³ of air per minute from 14.7 psia to 120 psia?

(A) 28 hp   (B) 108 hp   (C) 256 hp

          (D) 13,900 hp   (E) 17,000 hp

For an isothermal process:

$$W = -p_1 V_1 \ln \frac{p_1}{p_2}$$

$$= p_1 V_1 \ln \frac{p_2}{p_1}$$

$$= \left(14.7 \, \text{psia}\right) \left(144 \, \frac{\text{in}^2}{\text{ft}^2}\right) \left(800 \, \text{ft}^3\right) \ln \frac{120}{14.7}$$

$$= 3.56 \times 10^6 \, \text{ft-lbf}$$

$$\text{power} = \frac{dW}{dt}$$

$$= \left(\frac{3.56 \times 10^6 \, \text{ft-lbf}}{60 \, \text{sec}}\right) \left(\frac{1 \, \text{hp}}{550 \, \frac{\text{ft-lbf}}{\text{sec}}}\right)$$

$$= 108 \, \text{hp}$$

Answer is (B)

THERMODYNAMICS-35

Calculate the work done by a system in which 1 kgmole of water completely evaporates at 100 °C and 1 atmosphere pressure.

(A) 1000 kJ   (B) 2130 kJ   (C) 2490 kJ

          (D) 3050 kJ   (E) 4200 kJ

$$W = \int_1^2 pdV$$
$$= p(V_2 - V_1)$$
$$p = 1\,atm = 101.3\,kPa$$

from the steam tables, $v_f = 0.001044\,m^3/kg$

$$v_g = 1.673\,m^3/kg$$

the molecular weight of water is $M_{H_2O} = 18.016\,kg/kgmole$

$$V_1 = v_f M_{H_2O} m$$
$$= (0.001044)(18.016)(1)$$
$$= 0.01881\,m^3$$
$$V_2 = v_g M_{H_2O} m$$
$$= (1.673)(18.016)(1)$$
$$= 30.141\,m^3$$
$$W = (101,300)(30.141 - 0.01881)$$
$$= 3.05 \times 10^6\,J$$
$$= 3050\,kJ$$

Answer is (D)

The following situation applies to Thermodynamics questions 36 to 38.

Five moles of water vapor at 100 °C and 1 atmosphere pressure are compressed isobarically to form liquid at 100 °C. The process is reversible, and the ideal gas laws apply.

THERMODYNAMICS-36

Compute the initial volume of the vapor.

(A) 123 l (B) 133 l (C) 143 l

         (D) 153 l (E) 163 l

Use the ideal gas law.

$$pV = nR^*T$$
$$V = \frac{nR^*T}{p}$$
$$= \frac{(5)(0.0821)(373)}{1}$$
$$= 153\,l$$

Answer is (D)

THERMODYNAMICS-37

Compute the work, in joules, done on the system.

(A) −15.5 MJ   (B) −10.5 MJ   (C) −15.5 kJ

        (D) 10.5 MJ   (E) 15.5 MJ

$$W = -p(V_2 - V_1)$$

$$\text{from the steam tables,} \quad v_f = 0.001044 \text{ m}^3/\text{kg}$$

$$M_{H_2O} = 18.016 \text{ kg/kgmole}$$

$$V_2 = m M_{H_2O} v_f$$

$$= (5)(18.016)(0.001044)$$

$$= 0.094 \text{ m}^3$$

$$W = -(101,325)(0.094 - 152.9)$$

$$= 1.548 \times 10^7 \text{ J}$$

$$= 15.5 \text{ MJ}$$

Answer is (E)

THERMODYNAMICS-38

Determine the heat for condensation, Q, for the amount of water given. The heat of vaporization is 2257 kJ/kg.

(A) −203.3 MJ   (B) −40.66 MJ   (C) 203.3 kJ

        (D) 40.66 MJ   (E) 203.3 MJ

$$h_{fg} = 2257 \text{ kJ/kg}$$

$$\text{heat of condensation} = -h_{fg}$$

$$Q = \Delta H$$

$$= m(-h_{fg})$$

$$= (5)(18.016)(-2257)$$

$$= -203.3 \text{ MJ}$$

Answer is (A)

THERMODYNAMICS-39

What is the equation for the work done by a constant temperature system?

(A) $W = mRT \ln(V_2 - V_1)$

(B) $W = mR(T_2 - T_1) \ln \dfrac{V_2}{V_1}$

(C) $W = mRT \ln \dfrac{V_2}{V_1}$

(D) $W = RT \ln \dfrac{V_2}{V_1}$

(E) $W = \ln \dfrac{V_2}{V_1}$

$$W = \int_1^2 p\,dV$$

$$p = \frac{mRT}{V}$$

$$W = \int_1^2 \frac{mRT}{V}\,dV$$

$$= mRT \ln V \Big|_1^2$$

$$W = mRT \ln \frac{V_2}{V_1}$$

Answer is (C)

THERMODYNAMICS-40

Twenty grams of oxygen gas (O_2) are reversibly compressed at a constant temperature of 30 °C to 5% of their original volume. What work is done on the system?

(A) 824 cal   (B) 924 cal   (C) 944 cal

(D) 1124 cal   (E) 1144 cal

$$W = -\int_{V_2}^{V_1} p\,dV$$

$$= mRT \ln \frac{V_1}{V_2}$$

$$R = \left(1.98 \frac{\text{cal}}{\text{gmole} \cdot \text{K}}\right)\left(\frac{32\,\text{g}}{\text{gmole}}\right)$$

$$= 0.0619 \frac{\text{cal}}{\text{g} \cdot \text{K}}$$

$$W = (20)(0.0619)(303) \ln\left(\frac{1}{0.05}\right)$$

$$= 1124 \text{ cal}$$

Answer is (D)

THERMODYNAMICS–41

Helium ($R = 0.4968$ BTU/lbm-°R) is compressed isothermally from 14.7 psia and 68 °F. The compression ratio is 4. Calculate the work done by the gas.

(A) −1454 BTU/lbm   (B) −364 BTU/lbm   (C) −187 BTU/lbm
(D) 46.7 BTU/lbm   (E) 363 BTU/lbm

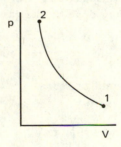

$$W = \int_1^2 p\,dV$$

$$= RT \ln \frac{V_2}{V_1}$$

$$W = (0.4968)(460 + 68) \ln\left(\tfrac{1}{4}\right)$$

$$= -364 \text{ BTU/lbm}$$

Answer is (B)

THERMODYNAMICS–42

Gas is enclosed in a cylinder with a weighted piston as the top boundary. The gas is heated and expands from a volume of 0.04 m^3 to 0.10 m^3 at a constant pressure of 200 kPa. Calculate the work done by the system.

(A) 8 kJ   (B) 10 kJ   (C) 12 kJ

       (D) 14 kJ   (E) 16 kJ

At constant pressure:

$$
\begin{aligned}
W &= \int_1^2 p\,dV \\
&= p(V_2 - V_1) \\
&= 200,000(0.10 - 0.04) \\
&= 12,000 \text{ J} \\
&= 12 \text{ kJ}
\end{aligned}
$$

constant
F
↓

gas

Answer is (C)

THERMODYNAMICS–43

A piston-cylinder system contains a gas which expands under a constant pressure of 1200 lbf/ft^2. If the piston is displaced 12″ during the process, and the piston diameter is 24″, what is the work done by the gas on the piston?

(A) 1768 ft-lbf   (B) 1890 ft-lbf   (C) 2387 ft-lbf

       (D) 3768 ft-lbf   (E) 4000 ft-lbf

The work is done at constant pressure.

$$
\begin{aligned}
W &= \int_1^2 p\,dV \\
&= p\Delta V \\
\Delta V &= A(\Delta L) \\
&= \pi(1)^2(1) \\
&= 3.14 \text{ ft}^3 \\
W &= (1200)(3.14) \\
&= 3768 \text{ ft-lbf}
\end{aligned}
$$

Answer is (D)

THERMODYNAMICS–44

Gas is enclosed in a cylinder with a weighted piston as the top boundary. The gas is heated and expands from a volume of 0.04 m^3 to 0.10 m^3. The pressure varies such that $pV = \text{constant}$, and the initial pressure is 200 kPa. Calculate the work done by the system.

(A) 6.80 kJ   (B) 7.33 kJ   (C) 9.59 kJ
         (D) 12.0 kJ   (E) 17.33 kJ

The work done by the system on the piston is given as follows:

$$W = \int_1^2 p dV$$
$$pV = k$$
$$\quad = p_1 V_1$$
$$p = \frac{p_1 V_1}{V}$$
$$W = p_1 V_1 \int_1^2 \frac{dV}{V}$$
$$\quad = p_1 V_1 \ln \frac{V_2}{V_1}$$
$$\quad = (200,000)(0.04) \ln \left(\frac{0.01}{0.04} \right)$$
$$\quad = 7330 \text{ J}$$
$$\quad = 7.33 \text{ kJ}$$

Answer is (B)

THERMODYNAMICS–45

Steam flows into a turbine at a rate of 10 kg/s, and 10 kilowatts of heat are lost from the turbine. Ignoring elevation and kinetic energy effects, calculate the power output from the turbine.

	inlet conditions	exit conditions
pressure	2.0 MPa	0.1 MPa
temperature	350 °C	—
quality	—	100%

(A) 4000 kW   (B) 4375 kW   (C) 4625 kW
         (D) 4973 kW   (E) 5200 kW

Use the first law of thermodynamics.

$$P = \dot{W}$$
$$= \dot{m}(h_i - h_e) + Q$$

from the steam tables, $h_i = 3137.0 \ \text{kJ/kg}$

$$h_e = 2675.5 \ \text{kJ/kg}$$
$$\dot{W} = (10)(3137 - 2675) + 10$$
$$= 4625 \ \text{kW}$$

Answer is (C)

THERMODYNAMICS–46

How does an adiabatic process compare to an isentropic process?

(A) adiabatic: heat transfer = 0; isentropic: heat transfer $\neq$ 0
(B) adiabatic: heat transfer $\neq$ 0; isentropic: heat transfer = 0
(C) adiabatic: reversible; isentropic: not reversible
(D) both: heat transfer = 0; isentropic: reversible
(E) none of the above

An adiabatic process is one in which there is no heat flow. It is not necessarily reversible. An isentropic process has no heat flow and is reversible.

Answer is (D)

THERMODYNAMICS–47

What is true about the polytropic exponent, n, for a perfect gas undergoing an isobaric process?

(A) $n > 0$ (B) $n < 0$ (C) $n \to \infty$

(D) $n = 0$ (E) none of the above

For an isobaric process,

$$p_1 = p_2 \qquad \text{(I)}$$

For a polytropic process,

$$p_1 V_1^n = p_2 V_2^n \qquad \text{(II)}$$

Equation (I) can be derived from (II) only if $n = 0$.

Answer is (D)

THERMODYNAMICS-48

In an isentropic compression, $p_1 = 100$ psia, $p_2 = 200$ psia, $V_1 = 10$ in^3, and $\gamma = 1.4$. Find V_2.

(A) 3.509 in^3 (B) 4.500 in^3 (C) 5.000 in^3
     (D) 6.095 in^3 (E) 7.077 in^3

For an isentropic proces the following is true:

$$\frac{p_1}{p_2} = \left(\frac{V_2}{V_1}\right)^{\gamma}$$

$$V_2 = \left(\frac{p_1}{p_2}\right)^{1/\gamma} \left(V_1\right)$$

$$= \left(\frac{100}{200}\right)^{1/1.4} \left(10\right)$$

$$= 6.095 \text{ in}^3$$

Answer is (D)

THERMODYNAMICS-49

In an adiabatic, isentropic process, $p_1 = 200$ psi, $p_2 = 300$ psi, and $T_1 = 700\,^\circ$R. Find T_2, using $\gamma = 1.4$.

(A) 576 $^\circ$R   (B) 590 $^\circ$R   (C) 680 $^\circ$R
     (D) 786 $^\circ$R   (E) 800 $^\circ$R

For an isentropic process:

$$T_2 = T_1 \left(\frac{p_2}{p_1}\right)^{(\gamma-1)/\gamma}$$

$$= \left(700\right)\left(\frac{300}{200}\right)^{(1.4-1)/1.4}$$

$$= 786\,°R$$

Answer is (D)

THERMODYNAMICS-50

Air undergoes an isentropic compression from 14.7 psia to 180.6 psia. If the initial temperature is 68 °F and the final temperature is 621.5 °F, calculate the work done by the gas.

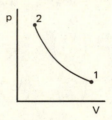

(A) −138.2 BTU/lbm   (B) −94.8 BTU/lbm   (C) 0 BTU/lbm

(D) 94.8 BTU/lbm   (E) 138.2 BTU/lbm

$$\text{for air,} \quad c_v = 0.1714 \text{ BTU/lbm-°R}$$
$$\gamma = 1.4$$
$$W = c_v(T_1 - T_2)$$
$$= (0.1714)(528 - 1081.2)$$
$$= -94.8 \text{ BTU/lbm}$$

Answer is (B)

THERMODYNAMICS–51

Nitrogen is expanded isentropically. Its temperature changes from 620 °F to 60 °F. Find the pressure ratio (p_1/p_2).

(A) 0.08   (B) 12.9   (C) 26.2

        (D) 3547   (E) indeterminate

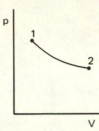

For an isentropic process:

$$\frac{p_1}{p_2} = \left(\frac{T_1}{T_2} \right)^{\gamma/\gamma-1}$$

$$\gamma = 1.4$$

$$\frac{p_1}{p_2} = \left[\frac{(620 + 460)\,°R}{(60 + 460)\,°R} \right]^{1.4/(1.4-1)}$$

$$= 12.9$$

Answer is (B)

THERMODYNAMICS–52

Nitrogen is expanded isentropically. Its temperature changes from 620 °F to 60 °F. The volumetric ratio is $V_2/V_1 = 6.22$, and the value of R for nitrogen is 0.787 BTU/lbm-°R. What is the work done by the gas?

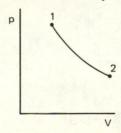

(A) −1112.7 BTU/lbm
(B) −99.22 BTU/lbm
(C)   0 BTU/lbm
(D)   99.22 BTU/lbm
(E)   1112.7 BTU/lbm

$$W = \frac{R(T_1 - T_2)}{\gamma - 1}$$
$$= \frac{(0.0787)(620 - 60)}{0.4}$$
$$= 99.22 \text{ BTU/lbm}$$

Answer is (D)

THERMODYNAMICS–53

For the cycle shown, what is the work done on the system?

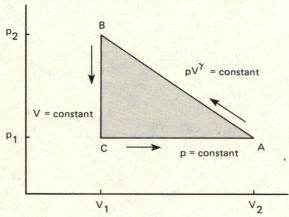

(A) 0
(B) area enclosed by the cycle in T-V space
(C) $R(T_B - T_C)$
(D) $S_B - S_A$
(E) area enclosed by the cycle in p-V space

$$W = \oint p dV$$
$$= W_A^B - W_C^A$$
$$= \frac{R(T_B - T_A)}{\gamma - 1} - \frac{R(T_A - T_C)}{\gamma - 1}$$
$$= \left(\frac{R}{\gamma - 1}\right)\left(T_B + T_C - 2T_A\right)$$
$$W \neq R(T_B - T_C)$$

Thus, (A), (B), (C), and (D) are incorrect. However, $\oint p dV$ is the area enclosed in p-V space. Therefore, (E) is correct.

Answer is (E)

THERMODYNAMICS–54

An isobaric steam generating process starts with saturated liquid at 20 psia. The change in entropy is equal to the initial entropy. What is the change in enthalpy during the process? (Hint: Not all of the liquid is vaporized.)

(A) −230.4 BTU/lbm   (B) −196.2 BTU/lbm   (C) 0 BTU/lbm

(D) 196.2 BTU/lbm   (E) 230.4 BTU/lbm

$$\Delta h = x h_{fg}$$
$$x = \text{quality}$$
$$h_{fg} = 960.2 \text{ BTU/lbm}$$
$$s_{\text{initial}} = s_f \quad \text{at 20 psia}$$
$$= 0.3359 \text{ BTU/lbm-}^\circ\text{R}$$
$$s_{\text{final}} = s_{\text{initial}} + \Delta s$$
$$= 2 s_{\text{initial}}$$
$$= 0.6718 \text{ BTU/lbm-}^\circ\text{R}$$
$$= s_f + x s_{fg}$$
$$x = \frac{s_{\text{final}} - s_f}{s_{fg}}$$
$$= \frac{0.6718 - 0.3359}{1.3963}$$
$$= 24.05\%$$
$$\Delta h = (0.24)(960.2)$$
$$= 230.4 \text{ BTU/lbm}$$

Answer is (E)

THERMODYNAMICS–55

A cylinder and piston arrangement contains saturated water vapor at 110 °C. The vapor is compressed in a reversible adiabatic process until the pressure is 1.6 MPa. Determine the work done by the system per kilogram of water.

(A) −637 kJ/kg   (B) −509 kJ/kg   (C) −432 kJ/kg

(D) −330 kJ/kg   (E) −213 kJ/kg

Use the first law of thermodynamics.

$$Q = U_1 - U_2 + W$$
$$Q = 0$$
$$W = U_1 - U_2$$

State 1 is specified. Because the process is reversible and adiabatic, $s_1 = s_2$. Thus, it is possible to find state 2. To do so, use the steam tables.

$$\text{at } 110\,°C \quad s_1 = 7.2387 \text{ kJ/kg·K}$$
$$u_1 = 2518.1 \text{ kJ/kg}$$
$$\text{at } 1.6 \text{ MPa} \quad s_2 = 7.2374 \text{ kJ/kg·K}$$
$$u_2 = 2950.1 \text{ kJ/kg}$$
$$T_2 = 400\,°C$$
$$w = 2518.1 - 2950.1$$
$$= -432 \text{ kJ/kg}$$

Answer is (C)

THERMODYNAMICS–56

During an adiabatic, internally reversible process, what is true about the change in entropy?

(A) It is always zero.
(B) It is always less than zero.
(C) It is always greater than zero.
(D) It is temperature-dependent.
(E) It is pressure-dependent.

The second law of thermodynamics means that an adiabatic, reversible process always has a zero change in entropy.

Answer is (A)

THERMODYNAMICS–57

For an irreversible process, what is true about the total change in entropy of the system and surroundings?

(A) $dS = \dfrac{dQ}{T}$ (B) $dS = 0$ (C) $dS > 0$

      (D) $dS < 0$ (E) none of the above

For an irreversible process, it is always true that

$$dS = dS_{\text{system}} + dS_{\text{surroundings}} > 0$$

> **Answer is (C)**

THERMODYNAMICS–58

For which type of process does the equation $dQ = TdS$ hold?

(A) irreversible   (B) isothermal   (C) reversible

      (D) isobaric   (E) isometric

$$TdS = dH - Vdp$$
$$dH = dU + pdV + Vdp$$
$$TdS = dU + pdV$$
$$dW = pdV$$
$$TdS = dU + dW$$

For a reversible process, there is no work lost ($\eta = 100\%$).

$$\text{Thus,} \quad dQ = dU + dW$$
$$= TdS \quad \text{for a reversible process}$$

> **Answer is (C)**

THERMODYNAMICS–59

Which of the following is true for any process?

(A) $\Delta S_{\text{surroundings}} + \Delta S_{\text{system}} \geq 0$
(B) $\Delta S_{\text{surroundings}} + \Delta S_{\text{system}} \leq 0$
(C) $\Delta S_{\text{surroundings}} + \Delta S_{\text{system}} < 0$
(D) $\Delta S_{\text{surroundings}} + \Delta S_{\text{system}} > 0$
(E) $\Delta S_{\text{surroundings}} + \Delta S_{\text{system}} = 0$

The total entropy either increases or, for a reversible process, remains the same. Therefore, the total change in entropy is always greater than or equal to zero.

Answer is (A)

THERMODYNAMICS-60

In a reversible process, the state of a system changes from state 1 to state 2, as shown on the T-S diagram. What does the shaded area on the diagram represent?

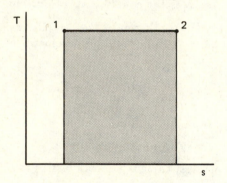

(A) free energy change
(B) heat transfer
(C) enthalpy change
(D) work
(E) volume change

For a reversible process, $Q = \int T dS$. Thus, the shaded area represents the heat transfer.

Answer is (B)

THERMODYNAMICS–61

Helium is compressed isothermally from 14.7 psia and 68 °F. The compression ratio is 4. Calculate the change in entropy of the gas given that $R_{He} = -0.4961$ BTU/lbm-°R.

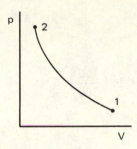

(A) −2.76 BTU/lbm-°R
(B) −0.689 BTU/lbm-°R
(C) 0 BTU/lbm-°R
(D) 0.689 BTU/lbm-°R
(E) 2.76 BTU/lbm-°R

for an isothermal process, $\quad \Delta s = R_{He} \ln \dfrac{V_2}{V_1}$

$$= 0.4961 \ln \left(\frac{1}{4} \right)$$

$$= -0.689 \text{ BTU/lbm-°R}$$

Note: $\Delta s < 0$. In this compression, the entropy of the helium is decreasing. However, the total entropy, $\Delta s_{He} + \Delta s_{surroundings}$, cannot decrease.

Answer is (B)

THERMODYNAMICS–62

For an ideal gas, what is the specific molar entropy change during an isothermal process in which the pressure changes from 200 kPa to 150 kPa?

(A) 2.00 J/mole·K   (B) 2.39 J/mole·K   (C) 2.79 J/mole·K
(D) 3.12 J/mole·K   (E) 3.49 J/mole·K

For an ideal gas:

$$\Delta s = c_p \ln \frac{T_2}{T_1} - R^* \ln \frac{p_2}{p_1}$$

$T_1 = T_2$ for an isothermal process, therefore,

$$\ln \frac{T_2}{T_1} = 0$$

$$\Delta s = -R^* \ln \frac{p_2}{p_1}$$

$$= -(8.314) \ln \left(\frac{150}{200} \right)$$

$$= 2.39 \text{ J/mole·K}$$

Answer is (B)

THERMODYNAMICS-63

In the p-V diagram shown, heat addition occurs between points 1 and 2. Given that $c_v = 0.336$ BTU/lbm-°R, what is the entropy produced during this step?

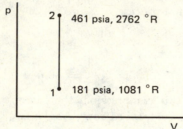

(A) −0.167 BTU/lbm-°R
(B) 0 BTU/lbm-°R
(C) 0.234 BTU/lbm-°R
(D) 0.315 BTU/lbm-°R
(E) indeterminate

$$\Delta s = c_v \ln \frac{T_2}{T_1} + R \ln \frac{V_2}{V_1}$$

$$V_2 = V_1$$

$$\Delta s = c_v \ln \frac{T_2}{T_1}$$

$$= (0.336) \ln \left(\frac{2762}{1081} \right)$$

$$= 0.315 \text{ BTU/lbm-°R}$$

Answer is (D)

THERMODYNAMICS–64

200 g of water are heated from 5 °C to 100 °C and vaporized at a constant pressure. The heat of vaporization of water at 100 °C is 539.2 cal/g. The heat capacity at constant pressure, c_p, is 1.0 cal/g·K. Determine the total change in entropy.

(A) 248.2 cal/K   (B) 298.2 cal/K   (C) 348.0 cal/K

     (D) 398.2 cal/K   (E) 448.2 cal/K

$$\Delta S = \Delta S_{\text{heat}} + \Delta S_{\text{vaporization}}$$

$$\Delta s_{\text{heat}} = s_2 - s_1$$

$$= \int_{T_1}^{T_2} \frac{c_p}{T} dT - R \ln \frac{p_2}{p_1}$$

$$p_1 = p_2$$

$$\Delta s_{\text{heat}} = \int_{T_1}^{T_2} \frac{c_p}{T} dT$$

$$= c_p \ln \frac{T_2}{T_1}$$

$$= (1) \ln \left(\frac{373}{278} \right)$$

$$= 0.2940 \text{ cal/g·K}$$

$$\Delta S_{\text{heat}} = (\Delta s_{\text{heat}})(m)$$

$$= (0.2940)(200)$$

$$= 58.8 \text{ cal/K}$$

$$\Delta s_{\text{vaporization}} = \frac{h_{fg}}{T_{\text{vap}}}$$

$$= \frac{539.2}{373}$$

$$= 1.446 \text{ cal/g·K}$$

$$\Delta S_{\text{vaporization}} = (\Delta s_{\text{vaporization}})(m)$$

$$= (1.446)(200)$$

$$= 289.2 \text{ cal/K}$$

$$\Delta S = 58.5 + 289.2$$

$$= 348 \text{ cal/K}$$

Answer is (C)

6

POWER CYCLES

POWER CYCLES-1

What kind of process occurs between points 3 and 4?

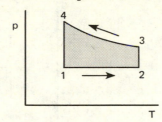

(A) adiabatic
(B) isentropic
(C) isobaric
(D) isothermal
(E) There is insufficient information to determine the process type.

> The process between points 3 and 4 is indeterminate. All that can be
> said about the process is that it is neither isobaric nor isothermal.

> Answer is (E)

POWER CYCLES-2

Which of the following thermodynamic cycles is the most efficient?

(A) Brayton
(B) Rankine
(C) Carnot
(D) combined Brayton-Rankine
(E) Rankine with reheat

No cycle is more efficient than the Carnot cycle, because it is completely reversible.

Answer is (C)

POWER CYCLES-3

For the reversible heat engine shown, which area on the corresponding T-S diagram represents the work done by the system?

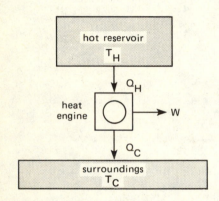

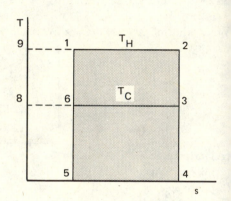

(A) work $= 0$ (B) $1-2-4-5$ (C) $6-3-4-5$

(D) $1-2-3-6$ (E) $2-3-8-9$

Use the first law of thermodynamics.

$$W = Q_H - Q_C$$
$$= T_H \Delta S - T_C \Delta S$$
$$= (1-2-4-5) - (3-4-5-6)$$
$$= 1-2-3-6$$

Answer is (D)

POWER CYCLES–4

The ideal, reversible Carnot cycle involves four basic processes. What type of processes are they?

(A) all isothermal
(B) all adiabatic
(C) all isentropic
(D) two adiabatic and two isentropic
(E) two isothermal and two isentropic

By definition, a Carnot cycle consists of two isothermal processes and two isentropic processes.

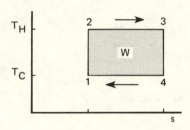

Answer is (E)

POWER CYCLES–5

An ideal, reversible Carnot cycle is represented on a T-S diagram below.

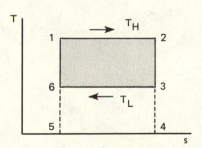

The efficiency of the cycle is represented by which of the following ratios of areas?

(A) $\dfrac{1-2-3-6}{1-2-4-5}$

(B) $\dfrac{1-2-4-5}{1-2-3-6}$

(C) $\dfrac{3-4-5-6}{1-2-4-5}$

(D) $\dfrac{1-2-4-5}{3-4-5-6}$

(E) $\dfrac{1-2-3-6}{3-4-5-6}$

The efficiency, η, is defined as follows:

$$\eta = \frac{W_{net}}{Q_H}$$
$$W_{net} = Q_{net}$$
$$= 1 - 2 - 3 - 6$$
$$Q_H = 1 - 2 - 4 - 5$$
$$\eta = \frac{1 - 2 - 3 - 6}{1 - 2 - 4 - 5}$$

Answer is (A)

POWER CYCLES-6

Which of the following T-S diagrams may be that of a Carnot cycle?

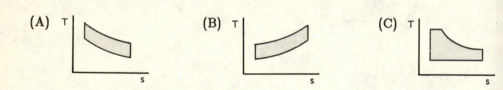

A Carnot cycle has two isothermal processes (horizontal lines on a T-S diagram) and two isentropic processes (vertical lines on a T-S diagram). The only diagram that has both of these properties is the diagram in choice (D).

Answer is (D)

POWER CYCLES–7

Which of the following are representations of a Carnot cycle?

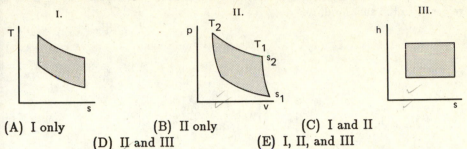

I.   II.   III.

(A) I only   (B) II only   (C) I and II
        (D) II and III   (E) I, II, and III

A Carnot cycle has two isentropic processes and two isothermal processes. Diagram I has two isentropic processes, but no isothermal processes. Diagram II has two isentropic processes and two isothermal
processes. When h is constant, T is constant. Diagram III, therefore,
also has two isentropic processes and two isothermal processes. Thus,
II and III both represent a Carnot cycle.

    Answer is (D)

POWER CYCLES–8

Consider the $T\text{-}S$ diagram of a Carnot cycle in the figure. What is the amount
of total work done in one cycle?

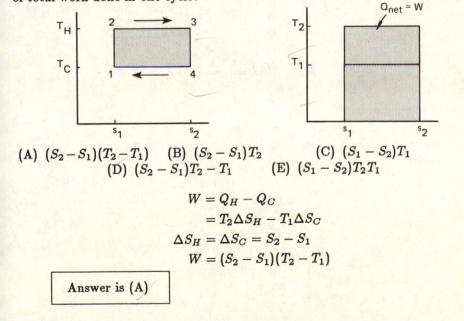

(A) $(S_2 - S_1)(T_2 - T_1)$ (B) $(S_2 - S_1)T_2$ (C) $(S_1 - S_2)T_1$
        (D) $(S_2 - S_1)T_2 - T_1$ (E) $(S_1 - S_2)T_2 T_1$

$$W = Q_H - Q_C$$
$$= T_2 \Delta S_H - T_1 \Delta S_C$$
$$\Delta S_H = \Delta S_C = S_2 - S_1$$
$$W = (S_2 - S_1)(T_2 - T_1)$$

    Answer is (A)

POWER CYCLES-9

Consider the T-S diagram of a Carnot cycle shown. What amount of heat is rejected to the surroundings?

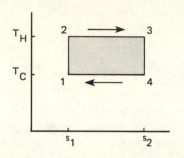

(A) $T_H(S_3 - S_2)$ (B) $T_H(S_2 - S_1)$ (C) $T_C(S_3 - S_2)$

(D) $T_C(S_2 - S_1)$ (E) $T_C(S_4 - S_3)$

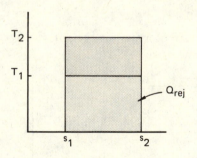

The rejected heat is equal to the area under the lower branch of the cycle.

$$dQ = TdS$$
$$Q = T_C(S_2 - S_1)$$

Answer is (D)

POWER CYCLES–10

What is the temperature difference of the cycle if the entropy difference is ΔS, and the work done is W?

(A) $W - \Delta S$ (B) $\dfrac{W}{\Delta S}$ (C) $\dfrac{\Delta S}{W}$

(D) $W(\Delta S)$ (E) $\Delta S - W$

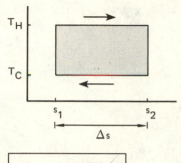

$W = $ shaded area
$$= (\Delta T)(\Delta S)$$
$$\Delta T = \frac{W}{\Delta S}$$

Answer is (B)

POWER CYCLES–11

In the Carnot cycle shown, the net amount of heat put into the system is equal to the total amount of work done by the system. However, it cannot be stated that the heat put into the system between states 1 and 2 is equal to the work done between states 1 and 2. What is the reason for this?

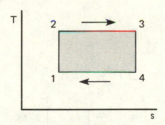

(A) The process is adiabatic.

(B) The process is not adiabatic.

(C) The second law states that the amount of energy put into the system is equal to the amount taken out of the system.

(D) The first law states that $dQ = dU + dW$. Since $dU \neq 0$, $dQ \neq 0$.

(E) $\int T dS = W$

PROFESSIONAL PUBLICATIONS, INC. • Belmont, CA

The first law states that $dQ = dU + dW$. Between states 1 and 2, $\Delta U \neq 0$. Therefore, $dQ_{12} \neq dW_{12}$.

Answer is (D)

POWER CYCLES-12

For the Carnot cycle shown, helium $(\gamma = \dfrac{5}{3})$ is the gas used. Given that $\dfrac{V_B}{V_A} = 2$ and $\dfrac{T_A}{T_D} = 1.9$, calculate $\dfrac{p_C}{p_A}$.

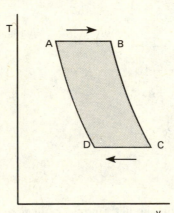

(A) 0.0633   (B) 0.0725   (C) 0.180   (D) 0.262   (E) 0.385

$$\frac{T_A}{T_D} = \frac{T_B}{T_C} = 1.9$$

$$\frac{p_B}{p_A} = \left(\frac{V_A}{V_B}\right)^{5/3}$$

$$\frac{p_C}{p_B} = \left(\frac{T_C}{T_B}\right)^{\frac{\gamma}{\gamma-1}}$$

$$\frac{p_C}{p_A} = \left(\frac{p_C}{p_B}\right)\left(\frac{p_B}{p_A}\right)$$

$$= \left(\frac{T_C}{T_B}\right)^{5/2}\left(\frac{V_A}{V_B}\right)^{5/3}$$

$$= \left(\frac{1}{1.9}\right)^{5/2}\left(\frac{1}{2}\right)^{5/3}$$

$$= 0.0633$$

Answer is (A)

POWER CYCLES-13

A Carnot engine operates between 800 °R and 1000 °R. What is its thermal efficiency?

(A) 20% (B) 30% (C) 40% (D) 50% (E) 60%

$$\eta_t = 1 - \frac{T_C}{T_H}$$
$$= 1 - \frac{800}{1000}$$
$$= 0.2$$
$$= 20\%$$

Answer is (A)

POWER CYCLES-14

For a heat engine operating between two temperatures $(T_2 > T_1)$, what is the maximum efficiency attainable?

(A) $1 - \dfrac{T_2}{T_1}$ (B) $1 - \dfrac{T_1}{T_2}$ (C) $\dfrac{T_1}{T_2}$

(D) $1 - \left(\dfrac{T_1}{T_2}\right)^\gamma$ (E) $0.5\left(\dfrac{T_1}{T_2}\right)$

The maximum efficiency attainable is the Carnot efficiency.

$$\eta = \frac{W}{Q_{in}}$$
$$= \frac{Q_{in} - Q_{out}}{Q_{in}}$$
$$= 1 - \frac{T_1}{T_2}$$

Answer is (B)

POWER CYCLES–15

Which of the following is not an advantage of a superheated, closed Rankine cycle over an open Rankine cycle?

(A) lower equipment costs
(B) increased efficiency
(C) increased turbine work output
(D) increased turbine life
(E) increased boiler life

Choice (E) is not an advantage because a superheated Rankine cycle has higher boiler heat temperatures that decrease boiler life.

Answer is (E)

POWER CYCLES–16

Which of the following statements regarding Rankine cycles is not true?

(A) Use of a condensible vapor in the cycle increases the efficiency of the cycle.

(B) The temperatures at which energy is transferred to and from the working liquid are less separated than in a Carnot cycle.

(C) Superheating increases the efficiency of a Rankine cycle.

(D) In practical terms, the susceptibility of the engine materials to corrosion is not a key limitation on the operating efficiency.

(E) Improvement in efficiency can generally be achieved by increasing the temperature at which heat is received.

Corrosion is a principal limitation on the use of higher temperatures for this type of engine. Thus, the susceptibility of engine materials to corrosion does limit operating efficiency.

Answer is (D)

POWER CYCLES–17

What type of power cycle does the following diagram illustrate?

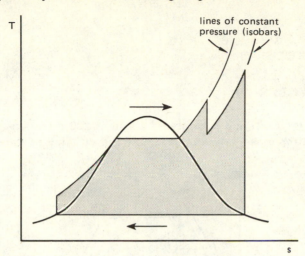

(A) a Carnot cycle
(B) an idealized Rankine cycle with reheat
(C) an idealized Diesel cycle
(D) an idealized Stirling cycle
(E) an idealized Brayton cycle

The diagram shows an idealized Rankine cycle with reheat.

Answer is (B)

POWER CYCLES–18

For the steam Rankine cycle shown, determine the enthalpy, h, at state 3.

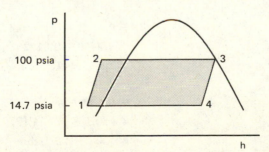

(A) 180.1 BTU/lbm   (B) 298.6 BTU/lbm   (C) 889.2 BTU/lbm
      (D) 1150.5 BTU/lbm   (E) 1187.8 BTU/lbm

State 3 is saturated steam at 100 psia. From the steam tables:

$$h_3 = h_g$$
$$= 1187.8 \text{ BTU/lbm}$$

Answer is (E)

POWER CYCLES-19

In a Rankine cycle, state 3 is saturated steam at 200 psia. Assuming that the turbine is isentropic and $p_4 = 14.7$ psia, find the enthalpy at state 4.

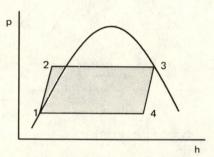

(A) 180.1 BTU/lbm   (B) 970.4 BTU/lbm   (C) 1008.8 BTU/lbm
(D) 1150.5 BTU/lbm   (E) 1199.3 BTU/lbm

$$p_3 = 200 \text{ psia}$$

from the steam tables, $s_3 = s_g$
$$= 1.5466 \text{ BTU/lbm-°R}$$

$$s_4 = s_3$$
$$= 1.5466 \text{ BTU/lbm-°R}$$

$$x = \text{quality}$$

$$s_4 = s_f + x s_{fg}$$
$$= 0.3122 + (x)(1.4447)$$

$$x = \frac{1.5466 - 0.3122}{1.4447}$$
$$= 0.854$$

$$h_4 = h_f + x h_{fg}$$
$$= 180.1 + (0.854)(970.4)$$
$$= 1008.8 \text{ BTU/lbm}$$

Answer is (C)

POWER CYCLES–20

n a steam Rankine cycle, state 1 is saturated liquid at 14.7 psia. State 2 is high
pressure liquid at 100 psia. How much work is required to pump 1 lbm of water
from state 1 to state 2?

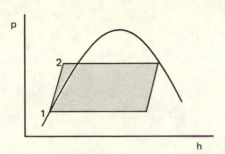

A) −118 BTU   (B) 0.129 BTU   (C) 0.264 BTU

(D) 18.5 BTU   (E) 118 BTU

$$p_1 = 14.7 \, \text{psia}$$
$$v_f = 0.01672 \, \text{ft}^3/\text{lbm}$$
$$W = m(h_2 - h_1)$$
$$= mv\Delta p$$
$$= \left(1\,\text{lbm}\right)\left(0.01672 \, \frac{\text{ft}^3}{\text{lbm}}\right)\left(100\,\text{psia} - 14.7\,\text{psia}\right)\left(\frac{144\,\text{in}^2}{\text{ft}^2}\right)$$
$$= 205.38 \, \text{ft-lbf}$$
$$= 0.264 \, \text{BTU}$$

Note: State 2 is not a saturated state.

Answer is (C)

POWER CYCLES-21

In a steam Rankine cycle, saturated liquid at 14.7 psia is pumped to 200 psia. If the pump were isentropic, the enthalpy of state 2 would be 180.67 BTU/lbm. The isentropic efficiency of the pump is 60%. What is the enthalpy of state 2?

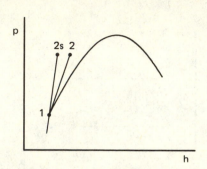

(A) 108.4 BTU/lbm   (B) 179.5 BTU/lbm   (C) 180.1 BTU/lbm

(D) 180.7 BTU/lbm   (E) 181.1 BTU/lbm

$$\eta \equiv \frac{W_s}{W}$$

$$= \frac{m\Delta h_s}{m\Delta h}$$

$$= \frac{\Delta h_s}{\Delta h}$$

$$= \frac{(h_2 - h_1)_s}{h_2 - h_1}$$

$$h_2 - h_1 = \frac{(h_2 - h_1)_s}{\eta}$$

$$h_2 = h_1 + \frac{(h_2 - h_1)_s}{\eta}$$

from the steam tables, $h_1 = 180.1$ BTU/lbm

$$h_2 = 180.1 + \frac{180.67 - 180.1}{0.6}$$

$$= 181.1 \text{ BTU/lbm}$$

Answer is (E)

POWER CYCLES–22

Steam in a Rankine cycle is expanded from a 200 psia saturated vapor state to 20 psia. The turbine has an efficiency of 0.8. What is the enthalpy of state 4?

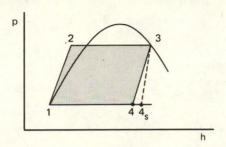

(A) 960.2 BTU/lbm   (B) 986.1 BTU/lbm   (C) 1028.7 BTU/lbm

(D) 1062.8 BTU/lbm   (E) 1156.4 BTU/lbm

Call the isentropic state 4_s.

$$s_3 = s_{f,4s} + x s_{fg,4s}$$
$$1.5466 = 0.3359 + x(1.3963)$$
$$x = \frac{1.5466 - 0.3359}{1.3063}$$
$$= 86.7\%$$
$$h_{4s} = h_{f,4s} + x h_{fg,4s}$$
$$= 146.2 + (0.867)(960.2)$$
$$h_{4s} = 1028.69 \text{ BTU/lbm}$$
$$\eta = \frac{h_s - h_4}{(h_3 - h_4)_s}$$
$$h_4 = h_3 - \eta(h_3 - h_4)_s$$
$$= 1199.3(0.8)(1199.3 - 1028.7)$$
$$= 1062.8 \text{ BTU/lbm}$$

Answer is (D)

POWER CYCLES-23

Which of the following set of reversible processes describes an ideal Otto cycle?

    I.   adiabatic compression, isometric heat addition,
           adiabatic expansion, isometric heat rejection

    II.   isothermal compression, isobaric heat addition,
           isothermal expansion, isobaric heat rejection

(A) I only
(B) II only
(C) I and II in succession
(D) II and I in succession
(E) none of the above

    An Otto cycle is defined by the set of reversible processes in I.

               Answer is (A)

POWER CYCLES-24

In the power stroke $(3 \rightarrow 4)$ of the ideal Otto cycle shown, what is the entropy change?

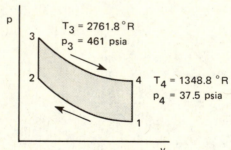

$T_3 = 2761.8\,°R$
$p_3 = 461$ psia

$T_4 = 1348.8\,°R$
$p_4 = 37.5$ psia

(A) -0.128 BTU/lbm-°R
(B) 0 BTU/lbm-°R
(C) 0.128 BTU/lbm-°R
(D) 0.179 BTU/lbm-°R
(E) indeterminate

    In the idealized Otto cycle, the expansion process is considered to be reversible and adiabatic (isentropic). For an isentropic process, $\Delta S = 0$.

               Answer is (B)

POWER CYCLES–25

The compression ratio of an ideal air Otto cycle is 6:1. p_1 is 14.7 psia, T_1 is 68 °F. Find the pressure and temperature at state 2.

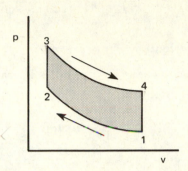

(A) 180.6 psig, 139 °F
(B) 180.6 psia, 139 °F
(C) 180.6 psia, 1081 °R
(D) 180.6 psig, 1081 °F
(E) 180.6 psia, 1081 °F

The process from state 1 to state 2 is an isentropic compression, with $\gamma = 1.4$ for air.

$$\frac{p_2}{p_1} = \left(\frac{V_1}{V_2}\right)^{\gamma}$$
$$= (6)^{1.4}$$
$$p_2 = p_1(6)^{1.4}$$
$$= (14.7)(6)^{1.4}$$
$$= 180.6 \text{ psia}$$
$$\frac{T_2}{T_1} = \left(\frac{V_1}{V_2}\right)^{\gamma-1}$$
$$= (6)^{0.4}$$
$$T_2 = T_1(6)^{0.4}$$
$$= (68 + 460)(6)^{0.4}$$
$$= 1081.2 \text{ °R}$$
$$T_2 = 1081.2 \text{ °R}, \quad p_2 = 180.6 \text{ psia}$$

Answer is (C)

POWER CYCLES–26

What is the efficiency of an Otto cycle with a compression ratio of 6:1? The gas used is air.

(A) 0.167   (B) 0.191   (C) 0.488   (D) 0.512   (E) 0.809

$$\text{by definition,} \quad \eta = 1 - \frac{1}{r^{\gamma-1}}$$

$$\text{where} \quad r = \text{compression ratio}$$

$$\text{for air} \quad \gamma = 1.4$$

$$\eta = 1 - \frac{1}{(6)^{0.4}}$$

$$= 0.512$$

Answer is (D)

POWER CYCLES–27

The cycle shown in the diagram can be described as follows:

the process from 1 to 2: adiabatic, isentropic compression
the process from 2 to 3: isobaric heat addition
the process from 3 to 4: adiabatic, isentropic compression
the process from 4 to 1: isometric heat rejection

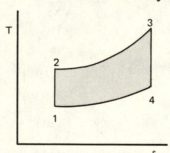

Which of the following is the name of this cycle?

(A) Otto   (B) Carnot   (C) Diesel
          (D) Rankine   (E) Brayton

The cycle described above is, by definition, a Diesel cycle.

Answer is (C)

POWER CYCLES–28

A device produces 37.5 joules per cycle. There is one power stroke per cycle. Calculate the power output if the device is run at 45 rpm.

(A) 4.69 W   (B) 14.063 W   (C) 28.125 W
     (D) 275.625 W   (E) 1.6875 MW

$$\text{power} = \frac{\text{energy}}{\text{time}}$$

$$= \left(37.5 \, \frac{\text{J}}{\text{cycle}}\right) \left(45 \, \frac{\text{cycle}}{\text{min}}\right) \left(\frac{1 \, \text{min}}{60 \, \text{s}}\right)$$

$$= 28.125 \, \text{W}$$

Answer is (C)

POWER CYCLES–29

A steam generator produces saturated steam at 100 psia from saturated liquid at 14.7 psia. If the heat source is a bath at 340 °F that provides 800 BTU/lbm, which of the following is true?

(A) The device cannot work because the pressure gradient is greater than zero.
(B) The device can work, but is inefficient.
(C) The device violates the first law of thermodynamics.
(D) The device violates the second law of thermodynamics.
(E) For the device to work, the steam produced must be superheated.

$$\Delta Q = \text{heat output}$$

$$= h_2 - h_1$$

$$= 1187.8 - 180.1$$

$$= 1007.7 \, \text{BTU/lbm}$$

However, it is given that the heat input is 800 BTU/lbm. This would mean that the heat output is greater than the heat input. Therefore, the first law of thermodynamics is violated.

$$\Delta s = s_2 - s_1 - \frac{Q}{T}$$

$$= 1.6036 - 0.3122 - \frac{800}{(340 + 460)}$$

$$= 1.6036 - 0.3122 - 1$$

$$= 0.2914 \, \text{BTU/lbm-°R}$$

Since $\Delta s > 0$, the second law of thermodynamics is not violated.

Answer is (C)

POWER CYCLES-30

A device that is meant to extract power from waste process steam starts with steam of 75% quality at 100 psia. The exit conditions of the steam are 70% quality at 14.7 psia. Which of the following statements are true?

I.   This device violates the first law of thermodynamics.
II. This device violates the second law of thermodynamics.
III. The device generates positive net power.
IV. The device generates no net power.

(A) I   (B) II   (C) I and II   (D) III   (E) IV

$$W = -(h_{final} - h_{initial})$$
$$h = h_g + xh_{fg}$$
$$h_{final} = 180.1 + (0.7)(970.4)$$
$$= 859.4 \text{ BTU/lbm}$$
$$h_{initial} = 298.6 + (0.75)(889.2)$$
$$= 965.5 \text{ BTU/lbm}$$
$$W = 965.5 - 859.4$$
$$= 106.1 \text{ BTU/lbm}$$

Thus, the device generates positive net power and the first law of thermodynamics is not violated.

$$\Delta s = s_{final} - s_{initial}$$
$$s = s_f + xs_{fg}$$
$$s_{final} = 0.3122 + (0.7)(1.4447)$$
$$= 1.3235 \text{ BTU/lbm-°R}$$
$$s_{initial} = 0.4745 + (0.75)(1.1291)$$
$$= 1.3213 \text{ BTU/lbm-°R}$$
$$\Delta s = 1.3235 - 1.3213$$
$$= 0.0022 \text{ BTU/lbm-°R}$$

Since $\Delta s > 0$, the second law of thermodynamics is not violated. Therefore, statements I, II, and IV are all false. Only statement III is true.

Answer is (D)

POWER CYCLES–31

An engineer devises a scheme for extracting some power from waste process steam. The steam enters the device at 100 psia and quality 75%, and exits at 14.7 psia and 65% quality. Which of the following statements are true?

I. The device produces 155 BTU/lbm of work.
II. The device violates the second law of thermodynamics.
III. The device violates the first law of thermodynamics.

(A) I   (B) II   (C) III

(D) I and II   (E) II and III

$$W = h_{\text{initial}} - h_{\text{final}}$$
$$h = h_g + x h_{fg}$$
$$h_{\text{initial}} = 298.6 + (0.75)(889.2)$$
$$= 965.5 \text{ BTU/lbm}$$
$$h_{\text{final}} = 180.1 + (0.65)(970.4)$$
$$= 810.9 \text{ BTU/lbm}$$
$$W = 965.5 - 810.9$$
$$= 154.6 \text{ BTU/lbm}$$

Thus, the device produces 155 BTU/lbm of work without violating the first law of thermodynamics.

$$\Delta s = s_{\text{final}} - s_{\text{initial}}$$
$$s = s_f + x s_{fg}$$
$$s_{\text{initial}} = 0.4745 + (0.75)(1.291)$$
$$= 1.3213 \text{ BTU/lbm-°R}$$
$$s_{\text{final}} = 0.3122 + (0.65)(1.4447)$$
$$= 1.2513 \text{ BTU/lbm-°R}$$
$$\Delta s = 1.2513 - 1.3213$$
$$= -0.07 \text{ BTU/lbm-°R}$$

Since $\Delta s < 0$, the device violates the second law of thermodynamics. Thus, I and II are true, but III is false.

Answer is (D)

POWER CYCLES–32

An engine burns a liter of fuel each 12 minutes. The fuel has a specific gravity of 0.8, and a heating value of 45 MJ/kg. The engine has an efficiency of 25%. What is the brake horsepower of the engine?

(A) 12.5 hp   (B) 15.63 hp   (C) 16.76 hp

          (D) 20.95 hp   (E) 50 hp

$$Q = \text{fuel flow rate}$$
$$= \left(\frac{1\,l}{12\,\text{min}}\right)\left(\frac{1\,\text{m}^3}{1000\,l}\right)\left(\frac{1\,\text{min}}{60\,\text{s}}\right)$$
$$= 1.389 \times 10^{-6}\,\text{m}^3/\text{s}$$

$$P_i = \text{power input to the engine}$$
$$= \dot{m}(\text{heating value})$$
$$= \rho Q(\text{heating value})$$
$$= (0.8)\left(1000\,\frac{\text{kg}}{\text{m}^3}\right)\left(1.389 \times 10^{-6}\,\frac{\text{m}^3}{\text{s}}\right)\left(45\,\frac{\text{MJ}}{\text{kg}}\right)$$
$$= 0.05\,\text{MJ/s}$$
$$= 50\,\text{kW}$$

$$P_o = \text{power output (actual power)}$$
$$= \eta P_i$$
$$= (0.25)(50)$$
$$= (12.5\,\text{kW})\left(\frac{1\,\text{hp}}{0.746\,\text{kW}}\right)$$
$$= 16.76\,\text{hp}$$

Answer is (C)

POWER CYCLES–33

A Carnot refrigerator operates between two reservoirs. One reservoir is at a higher temperature, T_H, and the other is at a cooler temperature, T_C. What is the coefficient of performance, COP, of the refrigerator?

(A) $\dfrac{T_H}{T_C}$ (B) $T_H - \dfrac{T_C}{T_H}$ (C) $1 - \dfrac{T_C}{T_H}$

          (D) $\dfrac{T_H}{T_C} - T_C$ (E) $\dfrac{T_C}{T_H - T_C}$

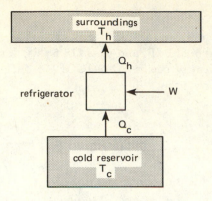

$$\text{COP} = \frac{Q_C}{W}$$

$$= \frac{Q_C}{Q_H - Q_C}$$

$$= \frac{T_C}{T_H - T_C}$$

Answer is (E)

POWER CYCLES-34

A refrigeration system produces 150 BTU/lbm of cooling. In order to have a rating of 1 ton of refrigeration, what must be the mass flow rate of the vapor? (1 ton of refrigeration = 12,000 BTU/hr, approximately the rate required to freeze 1 ton of ice in a day.)

(A) 2.2 lbm/hr   (B) 15 lbm/hr   (C) 80 lbm/hr

      (D) 360 lbm/hr   (E) 800 lbm/hr

$$\dot{m} = \frac{12,000 \text{ BTU/hr}}{150 \text{ BTU/hr}}$$

$$= 80 \text{ lbm/hr}$$

Answer is (C)

POWER CYCLES-35

A vapor compression refrigerator cycle is shown. The fluid used is Freon-12. State 1 is saturated vapor at 0 °F, and p_2 is 114.5 psia. If the isentropic efficiency is 0.8 and the coefficient of performance (COP) is 4, then what is the cooling produced by 2 lbm of vapor?

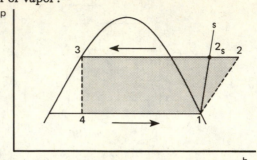

(A) 15.9 BTU   (B) 31.8 BTU   (C) 63.6 BTU
        (D) 81.4 BTU   (E) 107.3 BTU

State 1 is saturated vapor at 0 °F, state 2 is at 114.5 psia, and the entropy of state 2 is the same as that of state 1.

$$h_1 = 77.27 \text{ BTU/lbm}$$
$$s_1 = 0.16888 \text{ BTU/lbm-°R}$$
$$s_{2_s} = 0.16888 \text{ BTU/lbm-°R}$$
$$p_2 = 114.5 \text{ psia}$$
$$h_{2_s} = 88 \text{ BTU/lbm}$$
$$W_s = h_{2_s} - h_1$$
$$= 88 - 77.27$$
$$= 10.73 \text{ BTU/lbm}$$
$$W = \frac{W_s}{\eta}$$
$$= \frac{10.73}{0.8}$$
$$= 13.41 \text{ BTU/lbm}$$
$$\text{COP} = \frac{Q}{W}$$
$$Q = (4)(13.41)$$
$$= 53.65 \text{ BTU/lbm}$$
$$\text{for 2 lbm,} \quad Q_{tot} = (2)(53.65)$$
$$= 107.3 \text{ BTU}$$

Answer is (E)

7

CHEMISTRY

CHEMISTRY-1

The mole is a basic unit of measurement in chemistry. Which of the following is not equal to or the same as one mole of the substance indicated?

(A) 22.4 liters of nitrogen (N_2) gas at STP
(B) 6.02×10^{23} oxygen (O_2) molecules
(C) 12 g of carbon atoms
(D) 1 g of hydrogen (H) atoms
(E) 16 g of oxygen (O_2) molecules

Oxygen has a molar mass of 16 g/mole. Therefore, one mole of O_2 has a mass of 32 g.

Answer is (E)

CHEMISTRY-2

Which one of the following is standard temperature and pressure (STP)?

(A) 0 K and one atmosphere pressure
(B) 0 °F and zero pressure
(C) 32 °F and zero pressure
(D) 0 °C and one atmosphere pressure
(E) 100 °C and zero pressure

By definition, standard temperature and pressure (STP) is 0 °C and 1 atm pressure.

Answer is (D)

PROFESSIONAL PUBLICATIONS, INC. • Belmont, CA

CHEMISTRY-3

An ideal gas at 0.60 atmosphere and 87 °C occupies 0.450 liter. How many moles are in the sample? ($R^* = 0.0821 \, l \cdot \text{atm/mole} \cdot \text{K}$)

(A) 0.0002 mole   (B) 0.0091 mole   (C) 0.0198 mole

          (D) 0.0378 mole   (E) 0.0894 mole

Use the ideal gas law.

$$pV = nR^*T$$
$$n = \frac{pV}{R^*T}$$
$$= \frac{(0.60)(0.45)}{(0.0821)(360)}$$
$$= 0.0091 \text{ mole}$$

Answer is (C)

CHEMISTRY-4

At STP a gas occupies 0.213 liters. How many moles are there in this sample of gas?

(A) 0.0089 mole   (B) 0.0095 mole   (C) 0.0890 mole

          (D) 0.0950 mole   (E) 0.9000 mole

$$pV = nR^*T$$
$$n = \frac{pV}{R^*T}$$

at STP $p = 1 \text{ atm}$
$$T = 273 \text{ K}$$
$$n = \frac{(1)(0.213)}{(0.0821)(273)}$$
$$= 0.00950 \text{ mole}$$

Answer is (B)

CHEMISTRY–5

An ideal gas is contained in a vessel of unknown volume at a pressure of 1 atmosphere. The gas is released and allowed to expand into a previously evacuated bulb whose volume is 0.500 liter. Once equilibrium has been reached, the temperature remains the same while the pressure is recorded as 500 millimeters of mercury. What is the unknown volume, V, of the first bulb?

(A) 0.853 l (B) 0.961 l (C) 1.069 l
      (D) 1.077 l (E) 1.385 l

For an ideal gas at a constant temperature, the following holds true:

$$p_1 V_1 = p_2 V_2$$
$$p_1 = 1\,\text{atm} = 760\text{ mm Hg}$$
$$760\,V_1 = 500(0.5 + V_1)$$
$$= 250 + 500\,V_1$$
$$260\,V_1 = 250$$
$$V_1 = 0.961\ l$$

Answer is (B)

CHEMISTRY–6

What is the combined volume of 1.0 g of hydrogen gas (H_2) and 10.0 g of helium gas (He) when confined at 20 °C and 5 atmospheres?

(A) 4.4 l (B) 12.4 l (C) 14.4 l
      (D) 16.4 l (E) 24.2 l

Use the ideal gas law.

$$pV = nR^*T$$
$$V = \frac{n_{tot} R^* T}{p}$$
$$n_{tot} = n_{H_2} + n_{He}$$
$$= (1.0\,\text{g})\left(\frac{1\,\text{mole } H_2}{2\,\text{g } H_2}\right) + (10.0\,\text{g})\left(\frac{1\,\text{mole He}}{4\,\text{g He}}\right)$$
$$= 0.5 + 2.5$$
$$= 3.0\text{ mole}$$
$$V = \frac{(3)(0.0821)(293)}{5}$$
$$= 14.4\ l$$

Answer is (C)

CHEMISTRY-7

The valve between a 9 liter tank containing gas at 5 atmospheres and a 6 liter tank containing gas at 10 atmospheres is opened. What is the equilibrium pressure obtained in the two tanks at constant temperature? Assume ideal gas behavior.

(A) 5 atm   (B) 6 atm   (C) 7 atm

(D) 8 atm   (E) 9 atm

$$p_{tot} = p_1 + p_2$$
where p_n = partial pressure of gas n

For an ideal gas at a constant temperature, the following holds true:

$$p_i V_i = p_f V_f$$
$$p_f = p_i \left(\frac{V_i}{V_f} \right)$$
$$p_1 = 5 \left(\frac{9}{15} \right)$$
$$= 3 \text{ atm}$$
$$p_2 = 10 \left(\frac{6}{15} \right)$$
$$= 4 \text{ atm}$$
$$p_{tot} = 3 + 4$$
$$= 7 \text{ atm}$$

Answer is (C)

CHEMISTRY-8

A bicycle tire has a volume of 600 cm^3. It is inflated with CO_2 to a pressure of 80 psi at 20 °C. How many grams of CO_2 are contained in the tire?

(A) 3.83 g   (B) 4.83 g   (C) 5.98 g

(D) 6.43 g   (E) 7.80 g

$$pV = nR^*T$$

$$n = \frac{pV}{R^*T}$$

$$p = \left(80\,\text{psi}\right)\left(\frac{1\,\text{atm}}{14.7\,\text{psi}}\right)$$

$$= 5.44\,\text{atm}$$

$$V = 600\,\text{cm}^3\left(\frac{1\,l}{1000\,\text{cm}^3}\right) = 0.6\,l$$

$$T = 20 + 273 = 293\text{ K}$$

$$n = \frac{(5.44)(0.6)}{(0.0821)(293)}$$

$$= 0.136\text{ mole}$$

$$\text{MWT}_{CO_2} = \text{molecular weight of } CO_2$$

$$= 12 + (16)(2)$$

$$= 44\text{ g/mole}$$

$$m = \text{mass of } CO_2 \text{ in the tire}$$

$$= n(\text{MWT}_{CO_2})$$

$$= (0.136)(44)$$

$$= 5.98\text{ g}$$

Answer is (C)

CHEMISTRY–9

On a hot day, the temperature rises from 50 °F early in the morning to 99 °F in the afternoon. What is the ratio of the concentration (in moles/ft^3) of helium in a spherical balloon in the afternoon to the concentration of helium in the balloon in the morning?

(A) 0.51   (B) 0.69   (C) 0.91   (D) 1.10   (E) 1.98

$$pV = nR^*T$$

$$c = \text{concentration}$$

$$= \frac{n}{V}$$

$$= \frac{p}{R^*T}$$

c_1 = concentration of helium in the balloon in the morning

c_2 = concentration of helium in the balloon in the afternoon

$$\frac{c_2}{c_1} = \frac{T_1}{T_2}$$

$$= \frac{(50 + 460)\,^\circ R}{(99 + 460)\,^\circ R}$$

$$= 0.912$$

Answer is (C)

CHEMISTRY–10

When 0.5 g of a liquid is completely evaporated and collected in a 1 liter manometer, the pressure is 0.25 atmosphere and the temperature is 27 °C. Assuming ideal gas behavior, determine the molecular weight. The gas constant is $R^* = 0.0821\, l \cdot \text{atm/mole} \cdot \text{K}$.

(A) 2 g   (B) 2.2 g   (C) 12.3 g   (D) 49.2 g   (E) 64.0 g

$$pV = nR^*T$$

$$\text{MWT} = \text{molecular weight}$$

$$n = \frac{m}{\text{MWT}}$$

$$= \frac{mR^*T}{pV}$$

$$= \frac{(0.5)(0.0821)(300)}{(0.25)(1)}$$

$$= 49.2 \text{ g}$$

Answer is (D)

CHEMISTRY–11

Two hundred milliliters of oxygen gas (O_2) are collected over water at 23 °C and a pressure of 1 atmosphere. What volume would the oxygen occupy dry at 273 K and 1 atmosphere?

(A) 179.3 ml   (B) 184.4 ml   (C) 190.9 ml

        (D) 194.5 ml   (E) 200 ml

At 23 °C, the vapor pressure of water is 0.0277 atm. Find the pressure of the oxygen assuming ideal gas behavior.

$$p_{O_2} = 1 - 0.0277$$
$$= 0.9723 \text{ atm} = p_1$$
$$\frac{p_1 V_1}{T_1} = \frac{p_2 V_2}{T_2}$$
$$V_2 = \left(\frac{p_1}{p_2}\right)\left(\frac{T_2}{T_1}\right) V_1$$
$$= \left(\frac{0.9723}{1}\right)\left(\frac{273}{296}\right)(200)$$
$$= 179.3 \text{ ml}$$

Answer is (A)

CHEMISTRY–12

Eight grams of Ag_2O (solid) are heated to produce oxygen gas (O_2) as follows:

$$2Ag_2O \longrightarrow 4Ag + O_2$$

The oxygen gas is collected at 35 °C over water (water vapor pressure at 35 °C is 0.0555 atmosphere). Given that the barometric pressure is 1 atmosphere, what (wet) volume of O_2 is collected?

(A) 414.9 ml   (B) 424.5 ml   (C) 434.5 ml

        (D) 444.9 ml   (E) 454.9 ml

$$n_{Ag_2O} = \text{number of moles of } Ag_2O$$

$$= (8\,g) \left(\frac{1\,\text{mole}}{[2(108) + 16]\,g} \right)$$

$$= 0.034$$

$$n_{O_2} = \left(\frac{1}{2} \right) \left(0.034 \right)$$

$$= 0.017 \text{ mole}$$

$$T = 35 + 273$$

$$= 308 \text{ K}$$

$$p_{O_2} = p_{bar} - p_{H_2O}$$

$$= 1 - 0.0555$$

$$= 0.945 \text{ atm}$$

$$pV = nR^*T$$

$$V_{O_2} = \frac{n_{O_2} R^* T}{p_{O_2}}$$

$$= \frac{(0.017)(0.0821)(308)}{0.945}$$

$$= 0.4549 \; l$$

$$= 454.9 \text{ ml}$$

Answer is (E)

CHEMISTRY-13

A total of 0.1 grams of water is produced in a closed container at 40 °C. The container holds 500 cm³. The pressure is atmospheric pressure, and the vapor pressure of water at 40 °C is 55.3 torr. Is there any liquid in the container when it is in equilibrium? If so, how much is there?

(A) No, there is no liquid present.

(B) yes, $\frac{1}{2}$ of the water

(C) yes, $\frac{2}{3}$ of the water

(D) yes, $\frac{3}{4}$ of the water

(E) yes, $\frac{7}{8}$ of the water

Use the ideal gas law to determine how much of the H_2O is gas.

$$pV = nR^*T$$
$$= \frac{mR^*T}{MWT}$$

MWT = molecular weight of the water

$$m = \frac{pV(MWT)}{R^*T}$$

$$MWT = 2 + 16$$
$$= 18\,g$$

$$p = \frac{55.3}{760} \quad \text{1 TORR} = 760\ atm$$

$$= 0.073\,atm$$

$$V = \frac{500}{1000}\ cm^3 \quad cm^3/liters$$

$$= 0.5\,l$$

$$m = \frac{(18)(0.073)(0.5)}{(0.0821)(313)}$$

$$= 0.025\,g$$

The remainder of the H_2O is liquid.

$$\text{mass of liquid} = 0.1 - 0.025$$
$$= 0.075\,g$$

$$\text{fraction that is liquid} = \frac{m_{liq}}{m_{tot}}$$

$$= \frac{0.075}{0.1}$$

$$= \frac{3}{4}$$

Answer is (D)

CHEMISTRY–14

Which of the following statements is true for a real gas, but not for an ideal gas?

(A) The total volume of molecules in a gas is nearly the same as the volume of the gas as a whole.

(B) $pV = nR^*T$

(C) Collisions between gas molecules are perfectly elastic with no net decrease in kinetic energy.

(D) No attractive forces exist between the molecules of a gas.

(E) An increase in temperature causes an increase in the kinetic energy of the gas.

Real gases consist of molecules of finite volume that are widely separated in space.

Answer is (A)

CHEMISTRY–15

The following statements are made with regard to the boiling point of a liquid. Which statement is false?

(A) A nonvolatile substance having zero vapor pressure in solution (for example, sugars or salts) has no true boiling point.

(B) The boiling point is the temperature at which the vapor pressure of a liquid equals the applied pressure on the liquid.

(C) Combinations of liquids having different boiling points can be separated by slowly raising the temperature to draw off each fraction (by fractional distillation).

(D) The total vapor pressure of a solution is equal to the sum of the partial pressures of the components of the solution.

(E) At high elevations, water boils at a lower temperature because of a reduction in the surface tension of the water.

A liquid boils when its vapor pressure is equal to the pressure of the surroundings. The lower boiling temperature at high elevations is due to the reduced atmospheric pressure, not to a change in the surface tension of a liquid.

Answer is (E)

CHEMISTRY-16

The critical point for a mixture occurs for which of the following cases?

(A) The vapor and liquid have a single form.
(B) The liquid has no absorbed gas.
(C) The vapor phase is stable.
(D) The liquid is completely vaporized.
(E) none of the above

> The critical point for a mixture occurs when the vapor and the liquid
> have a form that is stable for a "critical temperature and critical
> pressure." It is both a liquid and a vapor with no boundaries and a
> uniform composition (a single form). A few substances have a triple
> point at which a solid, a liquid, and a gas are in equilibrium.

> **Answer is (A)**

CHEMISTRY-17

How is "molality" defined?

(A) the number of moles of solute in 1000 g of solvent
(B) the number of moles of solute in 1 l of solution
(C) the number of gram-formula weights of solute per liter
(D) the number of gram-equivalent weights of solute in one liter of solution
(E) the number of grams of solute per liter of solution

> Molality is defined as the number of moles of solute per 1000 g of
> solvent. Note: choice (B) is the definition of molarity, choice (C) is the
> definition of formality, and choice (D) is the definition of normality.

> **Answer is (A)**

CHEMISTRY-18

L is a nonvolatile, non-electrolytic liquid. A solid, S, is added to L to form a
solution that just boils at 1 atmosphere pressure. The vapor pressure of pure L
is 850 torr. What is the mole fraction of liquid L in the solution?

(A) 64.3%   (B) 79.4%   (C) 85.7%
            (D) 89.4%   (E) 91.5%

A liquid boils when its vapor pressure equals the pressure of its surroundings. Thus, the vapor pressure of the solution is 760 torr. From Raoult's law:

$$p_{solution} = p_{solvent}(\text{mole fraction of solvent})$$
$$760 = (850)(\text{mole fraction of L})$$
$$\text{mole fraction of L} = \frac{760}{850}$$
$$= 0.894$$
$$= 89.4\%$$

Answer is (D)

CHEMISTRY-19

Which of the following postulates does Bohr's model of the hydrogen atom involve?

(A) The electron in an atom has an infinite range of motion allowed to it.

(B) When an atom changes from a low energy state to a high energy state, it emits a quantum of radiation whose energy is equal to the difference in energy between the two states.

(C) In any of its energy states, the electron moves in a circular orbit about the nucleus.

(D) The states of allowed electron motion are those in which the angular momentum of the electron is an integral multiple of $\hbar/\pi$.

(E) all of the above

Bohr's model of the hydrogen atom involve the following postulates:

1. Each atom has only certain definite stationary states of motion allowed to it.
2. A quantum of energy is emitted when an atom changes from a higher energy state to a lower energy state.
3. The states of allowed electron motion are those in which the angular momentum of the electron is an integral multiple of $\hbar/2\pi$.

Thus, the only choice that is correct is (C).

Answer is (C)

CHEMISTRY–20

Which of the following diagrams best depicts the electron configuration of carbon?

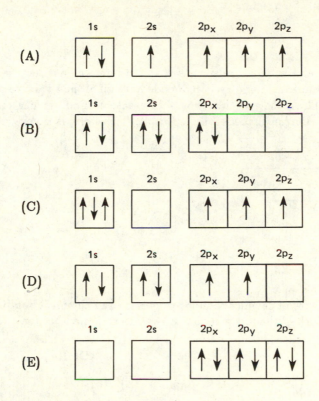

Carbon has a total of six electrons. Electrons position themselves in orbitals according to the following rules:

1. There is a maximum of two electrons per orbital.
2. Electrons in the same orbital have different spins $\left(\pm\frac{1}{2}\right)$.
3. Electrons fill up empty orbitals before moving into the same orbital as another electron.

Thus, (D) gives the correct electron configuration of carbon.

Answer is (D)

CHEMISTRY–21

Which of the following elements and compounds is unstable in its pure form?

(A) sodium (Na)
(B) helium (He)
(C) carbon dioxide (CO_2)
(D) hydrochloric acid (HCl)
(E) neon (Ne)

Helium and neon are inert gases and, therefore, not very reactive. Hydrochloric acid and carbon dioxide have all of their valence orbitals filled. Thus, they are also not very reactive. Sodium has only one valence electron that is easily ionizable. Therefore, it is very reactive.

Answer is (A)

CHEMISTRY–22

Two major types of chemical bonds are observed in chemical bonding: ionic and covalent. Which of the following has a bond that is the least ionic in character?

(A) NaCl   (B) CH_4 (C) H_2

(D) H_2O (E) Na_2CO_3

The electronegativity difference between two similar atoms is zero. Therefore, the H_2 bond is completely covalent. It has no ionic bond characteristics.

Answer is (C)

CHEMISTRY-23

Which of the following statements is false?

(A) It is not possible for bonds between a pair of atoms to be different (for example, difference in bond length or bond energy) in different compounds.

(B) The bond length for a pair of atoms is the point of lowest energy.

(C) The electrostatic repulsion between two nuclei increases as the atoms are brought together.

(D) The repulsion between two nuclei increases as their charge increases.

(E) The dissociation energy of a particular type of bond is largely independent of the molecule in which the bond occurs.

It is possible for bonds between a pair of atoms to be different in different compounds. For example, there is more than one type of carbon—carbon bond, although each type remains fairly consistent.

Answer is (A)

CHEMISTRY-24

Which of the following statements is false?

(A) For a diatomic molecule, the bond dissociation energy is the change in the enthalpy of the reaction when the diatomic molecule is separated into atoms.

(B) The average bond energy is the approximate energy required to break a bond in any compound in which it occurs.

(C) The energy released when a gaseous molecule is formed from its gaseous atoms can be estimated using average bond energies.

(D) ΔH is negative when energy is absorbed in the formation of a compound from its elements.

(E) Bond energies can be used to estimate the energies of chemical reactions.

ΔH is negative for the formation of a compound from elements when energy is released in the process.

Answer is (D)

CHEMISTRY–25

Which of the following is (are) the correct Lewis structure for sulfur dioxide?

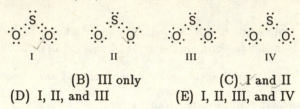

I   II   III   IV

(A) I only   (B) III only   (C) I and II
         (D) I, II, and III   (E) I, II, III, and IV

Sulfur and oxygen each have 6 valence electrons. Thus, there are a total of 18 valence electrons in SO_2. Therefore, there is one single S—O bond and one double S=O bond. The Lewis structure of sulfur dioxide is as follows:

$$.\ddot{O} = \ddot{S} - \ddot{O}:$$

Answer is (C)

CHEMISTRY–26

The molecule methane, CH_4, is often represented by the following structural formula:

$$H - \overset{\overset{\displaystyle H}{|}}{\underset{\underset{\displaystyle H}{|}}{C}} - H$$

What is the actual geometric shape of the molecule?

(A) linear
(B) square planar
(C) planar, but not 90° bond angles
(D) pyramidal
(E) tetrahedral

The structure of methane is as follows:

$$\overset{\displaystyle H}{\underset{\displaystyle H \quad \underset{\displaystyle H}{|} \quad H}{\overset{|}{C}}}$$

In the tetrahedral structure, bond angles are maximized and repulsions minimized, with bond angles of 109°.

Answer is (E)

CHEMISTRY-27

SO_3 has a structural formula represented as a resonance hybrid:

Which of the following is a true statement about the meaning of such a structure?

A) One-third of the SO_3 molecules exist as each of the three structures shown.

B) The true structure is a combination of the three with each $S \doteq O$ bond identical to another.

C) The molecule fluctuates between the three structures.

D) The arrows indicate equilibrium where an actual chemical reaction is taking place.

E) none of the above

The true structure is a combination with each bond identical, somewhere between a single and a double bond.

Answer is (B)

CHEMISTRY-28

Which of the following chemical equations is incorrect?

(A) $S + Fe \rightarrow FeS$
(B) $ZnSO_4 + Na_2S \rightarrow ZnS + Na_2SO_4$
(C) $H_2SO_4 + ZnS \rightarrow ZnSO_4 + H_2S$
(D) $Al_2S_3 + 6H_2O \rightarrow 2Al(OH)_3 + 3H_2S$
(E) $ZnS + O_2 \rightarrow SO_2 + ZnO$

The equation in choice (E) does not balance. It needs $\frac{3}{2}O_2$ on the left side. The equation $ZnS + \frac{3}{2}O_2 \rightarrow SO_2 + ZnO$ is correct.

Answer is (E)

CHEMISTRY–29

Na_2CO_3 reacts with HCl, but not by the stoichiometry implied in the following unbalanced chemical equation:

$$Na_2CO_3 + HCl \longrightarrow NaCl + H_2O + CO_2$$

What is the smallest possible whole number coefficient for Na_2CO_3 in the balanced equation?

(A) 1   (B) 2   (C) 4   (D) 5   (E) 10

The simplest balanced equation is:

$$Na_2CO_3 + 2HCl \longrightarrow 2NaCl + H_2O + CO_2$$

Thus, the smallest whole number coefficient for Na_2CO_3 is 1.

Answer is (A)

CHEMISTRY–30

Which of the following is the result of the reaction given below?

$$5SO_2 + 2KMnO_4 + 2H_2O \longrightarrow ?$$

(A) $2MnSO_4 + K_2SO_4 + 2H_2SO_4$
(B) $2MnSO_4 + K_2SO_2 + HSO_4 + H_2O$
(C) $2MnSO_4 + K_2SO_4 + H_2SO_4$
(D) $MnSO_4 + 2K_2SO_4 + 2H_2SO_4$
(E) $MnSO_4 + K_2SO_4 + 2H_2SO_4$

Only the products listed in choice (A) would balance the elements on the right and left sides of the equation.

Answer is (A)

CHEMISTRY-31

What is the balanced form of the equation listed below?

$$H_2O + P_4 + OH^- \longrightarrow PH_3 + H_2PO_2^-$$

(A) $4OH^- + 4P_4 + H_2O \longrightarrow 6H_2PO_2^- + 2PH_3$
(B) $P_4 + H_2O \longrightarrow H_2PO_2^- + 3PH_3$
(C) $8OH^- + 2P_4 + 2H_2O \longrightarrow H_2PO_2^- + PH_3$
(D) $6OH^- + 3P_4 + 6H_2O \longrightarrow 6H_2PO_2^- + 2PH_3$
(E) $3OH^- + P_4 + 3H_2O \longrightarrow 3H_2PO_2^- + PH_3$

The two half reactions are:

$$8OH^- + P_4 \longrightarrow 4H_2PO_2^- + 4e^-$$
$$12e^- + 12H_2O + P_4 \longrightarrow 4PH_3 + 12OH^-$$

Multiplying the top equation by 3 and adding the two equations together yields:

$$12OH^- + 4P_4 + 12H_2O \longrightarrow 12H_2PO_2^- + 4PH_3$$

In order to reduce the equation to the lowest whole number coefficients, divide by 4.

$$3OH^- + P_4 + 3H_2O \longrightarrow 3H_2PO_2^- + PH_3$$

Answer is (E)

CHEMISTRY-32

Which of the following chemical reactions relates to the softening procedure in water purification?

(A) $CO_2 + Ca(OH)_2 \longrightarrow CaCO_3 + H_2O$
(B) $Ca(HCO_3)_2 + Ca(OH)_2 \longrightarrow 2CaCO_3 + 2H_2O$
(C) $2H_2O + O_2 \longrightarrow 2H_2O_2$
(D) $NaOH + HCl \longrightarrow NaCl + H_2O$
(E) $Na_2O_2 + H_2O \longrightarrow 2NaOH + \frac{1}{2}O_2$

PROFESSIONAL PUBLICATIONS, INC. • Belmont, CA

Choice (B) gives the chemical reaction for adding lime to hard water in order to remove calcium salts. The resulting calcium carbonate precipitate can be removed by sedimentation.

Answer is (B)

CHEMISTRY-33

A substance is oxidized when which of the following occurs?

(A) It turns red.
(B) It becomes more negative.
(C) It loses electrons.
(D) It gives off heat.
(E) It absorbs energy.

By definition, a substance is oxidized when it loses electrons.

Answer is (C)

CHEMISTRY-34

In order to assign oxidation states in polyatomic molecules, which of the following rules is followed?

(A) The oxidation of all elements in any allotropic form is zero.

(B) The oxidation state of oxygen is always -2.

(C) The oxidation state of hydrogen is always $+1$.

(D) All other oxidation states are chosen such that the algebraic sum of the oxidation states for the ion or molecule is zero.

(E) All of the above rules are used.

Choice (B) is false because it does not take into account the peroxides in which the oxidation state of O is -1. Choice (C) is false because it does not account for hydrogen combined with metals, where its oxidation state is -1. Choice (D) is wrong because the sum of the oxidation states should equal the net charge on the ion or molecule. Thus, only choice (A) is correct.

Answer is (A)

CHEMISTRY-35

What is the oxidation state of nitrogen in NO_3^-?

(A) -1 (B) $+1$ (C) $+3$ (D) $+5$ (E) $+7$

The oxidation state of O is -2, and the net charge on the ion is -1. The oxidation state of nitrogen is given as follows:

$$3(\text{oxidation state of O}) + (\text{oxidation state of N}) = -1$$
$$3(-2) + (\text{oxidation state of N}) = -1$$
$$\text{oxidation state of N} = +5$$

> Answer is (D)

CHEMISTRY-36

What is the oxidation number of Cr in the dichromate ion $(Cr_2O_7)^{-2}$?

(A) -1 (B) 0   (C) 2.5   (D) 6   (E) 7

The oxidation number of O is -2. Therefore, the oxidation number of O_7 is -14. The charge on the ion is -2, so the charge on Cr_2 is 12. Thus, the oxidation number of Cr is 6.

> Answer is (D)

CHEMISTRY-37

Given the following information, determine the oxidation state of nitric acid, HNO_3.

oxidation state	formula	name
1	$HClO$	hypochlorous acid
3	$HClO_2$	chlorous acid
5	$HClO_3$	chloric acid
7	$HClO_4$	perchloric acid
3	HNO_2	nitrous acid

(A) 1   (B) 2   (C) 3   (D) 4   (E) 5

The outer shell of oxygen is 2 electrons short of being full (inert gases have a full shell). Thus, the oxidation number of oxygen is 2 for both ions. By adding another oxygen atom to nitrous acid, the oxidation level is increased by 2. This situation compares directly with that of $HClO_2$ and $HClO_3$. Thus, the oxidation state of nitric acid is 5.

Answer is (E)

CHEMISTRY–38

Which are the oxidizing and reducing agents in the following reaction?

$$2CCl_4 + K_2CrO_4 \longrightarrow 2Cl_2CO + CrO_2Cl_2 + 2KCl$$

(A) oxidizing agent: chromium; reducing agent: chlorine
(B) oxidizing agent: oxygen; reducing agent: chlorine
(C) oxidizing agent: chromium; reducing agent: oxygen
(D) oxidizing agent: chlorine; reducing agent: carbon
(E) There are no oxidizing or reducing agents in this reaction.

The oxidation state of chromium is 6 in each compound. Carbon remains with a +4 oxidation state throughout the reaction. The oxidation states of both chlorine and oxygen remain the same throughout this reaction. Thus, nothing is oxidized or reduced in the reaction.

Answer is (E)

CHEMISTRY–39

A volumetric analysis of a gaseous mixture is as follows:

CO_2	12%
O_2	4%
N_2	82%
CO	2%

What is the percentage of CO on a mass basis?

(A) 1.0%   (B) 1.2%   (C) 1.5%
(D) 1.9%   (E) 2.0%

name	vol. %	mole frac.		mol. wt.		mass
CO_2	12	0.12	×	44	=	5.28
O_2	4	0.04	×	32	=	1.28
N_2	82	0.82	×	28	=	22.96
CO	2	0.02	×	28	=	0.56

The total mass of the mixture is 30.08 kg. Thus, the mass percentage of CO is given as follows:

$$\text{mass \% of CO} = \frac{0.56}{30.08}$$
$$= 1.9\%$$

Answer is (D)

CHEMISTRY-40

What is the empirical formula for a compound that has the following composition by mass:

element	mass %
Si	30.2
O	8.59
F	61.2

(A) $SiOF_4$　　　　(B) Si_2OF_4　　　　(C) Si_2OF_6

　　　　(D) Si_3OF_6　　　　(E) $Si_3O_2F_6$

element	mass %	mass (based on 100 g)	moles	mole %
Si	30.2	30.2	1.075	22.2
O	8.59	8.59	0.537	11.1
F	61.2	61.2	3.221	66.6

Find the smallest whole number ratio of the mole percentage of each element to that of oxygen.

$$\frac{Si}{O} = 2$$

$$\frac{F}{O} = 6$$

$$\frac{O}{O} = 1$$

Therefore, the simplest formula is Si_2OF_6.

Answer is (C)

CHEMISTRY–41

The following equation describes the decomposition of potassium chlorate to produce oxygen gas.

$$2KClO_3 \longrightarrow 2KCl \text{ (solid)} + 3O_2 \text{ (gas)}$$

How many grams of $KClO_3$ must be used to produce 4.00 liters of O_2 (gas) measured at 7400 torr and 30 °C.

(A) 108 g   (B) 118 g   (C) 128 g

              (D) 138 g   (E) 148 g

$1 \, atm = \dfrac{1 \, Torr}{760 \, \dfrac{Torr}{atm}}$

$$p = \frac{7400}{760} = 9.74 \, \text{atm}$$

$$V = 4 \, l$$

$$T = 30 + 273 = 303 \, \text{K}$$

$$pV = nR^*T$$

$$n = \frac{pV}{R^*T}$$

$$= \frac{(9.74)(4)}{(0.0821)(303)}$$

$$= 1.566 \, \text{mole}$$

$$\frac{\text{no. of moles } KClO_3}{\text{no. of moles of } O_2} = \frac{2}{3}$$

$$\text{no. of moles } KClO_3 \text{ needed} = \left(\frac{2}{3}\right)\left(1.566\right)$$

$$= 1.04 \, \text{mole}$$

$$MWT_{KClO_3} = 39.1 + 35.5 + (16)(3)$$

$$= 123 \, \text{g/mole}$$

$$\text{number of grams } KClO_3 = \left(1.04 \, \text{mole}\right)\left(123 \, \frac{\text{g}}{\text{mole}}\right)$$

$$= 128 \, \text{g}$$

Answer is (C)

CHEMISTRY-42

Determine which of the statements is true, given the following facts.

1. A 40 liter sample of H_2 (gas) at 10 °C and 740 torr is added to a 75 liter sample of O_2 (gas) at 20 °C and 730 torr.

2. The mixture is ignited to produce water.

(A) There is an excess of O_2 greater than 0.2 mole.
(B) There is an excess of H_2 greater than 0.2 mole.
(C) There is H_2O only.
(D) There is an excess of H_2 less than 0.2 mole.
(E) There is an excess of O_2 less than 0.2 mole.

The stoichiometric equation is:

$$H_2 + \tfrac{1}{2}O_2 \longrightarrow H_2O$$

The number of moles of each gas initially present is:

$$n_{H_2} = \frac{pV}{R*T}$$

$$= \frac{\left(\frac{740}{760}\right)(40)}{(283)(0.0821)}$$

$$= 1.68 \text{ mole}$$

$$n_{O_2} = \frac{pV}{R*T}$$

$$= \frac{\left(\frac{730}{760}\right)(75)}{(293)(0.0821)}$$

$$= 2.99 \text{ mole}$$

For each mole of H_2O formed, 0.5 mole of O_2 and 1 mole of H_2 are required. The oxygen necessary to completely react with 1.68 moles of H_2 is given by:

$$n = \frac{1.68}{2}$$

$$= 0.84 \text{ mole}$$

Therefore, there is an excess of O_2. The amount of O_2 extra is: $2.99 - 0.84 = 2.15$ moles.

Answer is (A)

CHEMISTRY–43

If 2.25 g of pure calcium metal are converted to 3.13 g of pure CaO, what is the atomic weight of calcium? The atomic weight of oxygen is 16 g/mole.

(A) 28.3 g/mole   (B) 32.5 g/mole   (C) 36.7 g/mole
        (D) 40.9 g/mole   (E) 45.1 g/mole

The stoichiometric equation is:

$$Ca + O \longrightarrow CaO$$

Thus, one mole of oxygen and one mole of calcium are required to make one mole of CaO.

$$n_O = \frac{3.13 - 2.25}{16}$$
$$= \frac{0.880}{16}$$
$$n_O = 0.055 \text{ mole}$$
$$n_{Ca} = 0.055$$
$$= \frac{2.25}{\text{atomic weight of Ca}}$$
$$\text{atomic weight of Ca} = \frac{2.25}{0.055}$$
$$= 40.9 \text{ g/mole}$$

Answer is (D)

CHEMISTRY–44

Methane, CH_4, burns to form CO_2 and H_2O according to the equation

$$CH_4 + 2O_2 \longrightarrow CO_2 + 2H_2O$$

How many grams of CO_2 will theoretically be formed when a mixture of 50 g of CH_4 and 100 g of O_2 is ignited?

(A) 34.4 g   (B) 68.8 g   (C) 103.1 g
        (D) 137.5 g   (E) 171.9 g

$$n = \frac{m}{MWT}$$

where m = mass of compound

MWT = molecular weight of compound

$$n_{CH_4} = \frac{50}{12 + (4)(2)}$$

$$= 3.125 \text{ mole}$$

$$n_{O_2} = \frac{100}{(2)(16)}$$

$$= 3.125 \text{ mole}$$

Since 1 mole of CH_4 and 2 moles of O_2 are needed for each mole of CO_2 formed, O_2 is the limiting reactant.

$$\frac{\text{no. moles } CO_2 \text{ formed}}{\text{no. moles } O_2 \text{ ignited}} = \frac{1}{2}$$

$$n_{CO_2} = \left(3.125\right)\left(\frac{1}{2}\right)$$

$$= 1.563 \text{ mole}$$

$$m_{CO_2} = (n_{CO_2})(MWT_{CO_2})$$

$$= (1.563)(44)$$

$$= 68.75 \text{ g}$$

Answer is (B)

CHEMISTRY–45

Determine the mole percent of CO_2 in the products of combustion of C_8H_{18} when 200% theoretical air is used.

(A) 5.5%   (B) 6.5%   (C) 7.5%
          (D) 8.5%   (E) 9.5%

The formula for theoretical air is: $O_2 + 3.76N_2$. For 200% theoretical air, the stoichiometric equation is:

$$C_8H_{18} + 25(O_2 + 3.76N_2) \longrightarrow 8CO_2 + 9H_2O + 12.5O_2 + 94N_2$$

The mole percent of CO_2 is given by the ratio of the number of moles of CO_2 formed to the total number of moles formed.

$$\%CO_2 = \frac{8}{8 + 9 + 12.5 + 94}$$

$$= 6.5\%$$

Answer is (B)

CHEMISTRY–46

What volume of O_2 at 298 K and 1 atmosphere is required for complete combustion of 10 liters of C_2H_6 (gas) at 500 K and 1 atmosphere? The combustion equation is:

$$7O_2 + 2C_2H_6 \longrightarrow 6H_2O + 4CO_2$$

(A) 15.6 l (B) 19.1 l (C) 20.7 l

            (D) 22.4 l (E) 35.0 l

Assuming ideal gas behavior for C_2H_6 at 500 K:

$$n_{C_2H_6} = \frac{pV}{R*T}$$
$$= \frac{(1)(10)}{(0.0821)(500)}$$
$$= 0.24 \text{ mole}$$
$$\frac{n_{O_2}}{n_{C_2H_6}} = \frac{7}{2}$$
$$n_{O_2} = \left(\frac{7}{2}\right)(0.24)$$
$$= 0.85 \text{ mole}$$

The volume of one mole of ideal gas at STP (standard temperature and pressure) is 22.4 l. Therefore, assuming ideal gas behavior, the volume of O_2 required at 298 K is:

$$V_{O_2} = n_{O_2}\left(\frac{V_{298}}{V_{STP}}\right)V_{STP}$$
$$\frac{V_{298}}{V_{STP}} = \frac{T_{298}}{T_{STP}}$$
$$= \frac{298}{273}$$
$$= 1.09$$
$$V_{O_2} = (0.85)(1.09)(22.4)$$
$$= 20.7 l$$

Answer is (C)

CHEMISTRY–47

One gram of gas made up of carbon and hydrogen combusts to give 3.30 g of CO_2 and 1.125 g of H_2O. What is the empirical formula of the compound?

(A) CH   (B) CH_3 (C) C_2H_3
        (D) C_2H_5 (E) C_3H_5

The stoichiometric equation is:

$$C_xH_y + \left(x + \frac{y}{2}\right) O_2 \longrightarrow xCO_2 + \frac{y}{2}H_2O$$

$$MWT_{CO_2} = 44 \text{ g/mole}$$

$$MWT_{H_2O} = 18 \text{ g/mole}$$

$$x = \text{moles of C} = \text{moles of } CO_2$$

$$= \frac{3.3}{44}$$

$$= 0.0750 \text{ mole}$$

$$y = \text{moles of H} = 2 \times (\text{moles of } H_2O)$$

$$= 2\left(\frac{1.125}{18}\right)$$

$$= 0.125 \text{ mole}$$

$$\frac{x}{y} = \frac{\text{C atoms}}{\text{H atoms}}$$

$$= \frac{\text{moles of C}}{\text{moles of H}}$$

$$= \frac{0.075}{0.125}$$

$$= \frac{3}{5}$$

Thus, the empirical formula of the gas is C_3H_5.

Answer is (E)

CHEMISTRY–48

Combustion of 13.02 g of a compound (C_xH_y) gives 40.94 g CO_2 and 16.72 g of H_2O. Determine the empirical formula of the compound.

(A) CH   (B) CH_2 (C) CH_4
        (D) CH_2O (E) C_6H_6O

The stoichiometric equation for combustion is:

$$C_xH_y + \left(x + \frac{y}{2}\right)O_2 \longrightarrow xCO_2 + \frac{y}{2}H_2O$$

$$n = \frac{m}{MWT}$$

where m = mass of compound

MWT = molecular weight

$$n_{CO_2} = \frac{40.94}{44}$$

$$= 0.93 \text{ mole}$$

$$= n_C$$

therefore, $n_C = 0.93$ mole

$$n_{H_2O} = \frac{16.72}{18}$$

$$= 0.93 \text{ mole}$$

$$= \frac{n_H}{2}$$

$$n_H = 1.86 \text{ mole}$$

$$\frac{n_C}{n_H} = \frac{0.93}{1.86}$$

$$= \frac{1}{2}$$

Therefore, the empirical formula for the compound is CH_2.

Answer is (B)

CHEMISTRY–49

When 0.01 mole of a substance consisting of O, H, and C is burned, the following products are obtained:

1.   896 cm³ of CO_2 at standard temperature and pressure (STP)
2.   0.72 g of water

It is also found that the ratio of oxygen mass to the mass of H plus C in the substance is $\frac{4}{7}$. What is the chemical formula of the substance? One mole of CO_2 has a volume of 22,400 cm³ at STP.

(A) CHO_2 (B) $C_4H_6O_2$ (C) CH_2O_2

          (D) $C_4H_8O_2$ (E) $C_2H_4O_2$

The stoichiometric equation is:

$$C_x H_y O_z + \left(x + \frac{y}{4} - \frac{z}{2}\right) O_2 \longrightarrow x CO_2 + \frac{y}{2} H_2O$$

$$\frac{n_1}{V_2} = \frac{n_1}{V_1}$$

$$n_C = V_C \left(\frac{n_{STP}}{V_{STP}}\right)$$

$$= (896)\left(\frac{1}{22,400}\right)$$

$$= 0.04 \text{ mole}$$

$$n_{H_2O} = \frac{m_{H_2O}}{MWT_{H_2O}}$$

$$= \frac{0.72}{18}$$

$$= 0.04 \text{ mole}$$

$$= \frac{n_H}{2}$$

$$n_H = 0.08 \text{ mole}$$

Thus, there are 0.04 mole C and 0.08 mole H in 0.01 mole of the substance $C_x H_y O_z$. For 1 mole of $C_x H_y O_z$, there are $x = \frac{0.04}{0.01} = 4$ moles of C and $y = \frac{0.08}{0.01} = 8$ moles of H.

$$\frac{\text{weight of O}}{\text{weight of H + weight of C}} = \frac{16z}{(8)(1) + (4)(12)}$$

$$= \frac{4}{7}$$

$$\frac{16z}{56} = \frac{4}{7}$$

$$z = 2$$

Thus, the formula is $C_4 H_8 O_2$.

Answer is (D)

CHEMISTRY–50

Determine the melting point of sodium chloride, given that the heat of melting is 30 kJ/mole, and the associated entropy change is 28 J/mole·K.

(A) 373 K   (B) 879 K   (C) 933 K

(D) 1071 K   (E) 2000 K

For the phase change:

$$\Delta G = \Delta H - T_m \Delta S = 0$$

$$T_m = \frac{\Delta H}{\Delta S}$$

$$= \frac{30,000}{28}$$

$$= 1071 \text{ K}$$

Answer is (D)

CHEMISTRY–51

The temperature of 100 g of liquid water at $0\,^\circ$C is raised by 1 °C. How many calories are consumed?

(A) 0 cal   (B) 4.18 cal   (C) 80 cal

(D) 100 cal   (E) 1000 cal

By definition, 1 calorie is the energy needed to heat 1 g of liquid water by 1 °C. Therefore, the heat needed to heat 100 g of water by 1 °C is:

$$\left(100\,\text{g}\right)\left(1\,^\circ\text{C} - 0\,^\circ\text{C}\right)\left(1\,\frac{\text{cal}}{\text{g}\cdot^\circ\text{C}}\right) = 100 \text{ cal}$$

Answer is (D)

CHEMISTRY–52

If 50 cm^3 of ice at 0 °C are added to 100 g of water at 20 °C, how much ice is left unmelted? Assume that there is no spurious heat loss. The density of ice is 0.92 g/cm^3, and the heat of fusion of ice is 1.44 kcal/mole at 0 °C.

(A) 12.83 cm^3 (B) 18.83 cm^3 (C) 22.83 cm^3

(D) 38.83 cm^3 (E) 42.83 cm^3

The ice will melt until the temperature of the water reaches 0 °C. The number of calories necessary to lower the water temperature from 20 °C to 0 °C is:

$$q = mc_p\Delta T$$
$$= (100)(1)(20)$$
$$= 2000 \text{ cal}$$

The calories necessary to melt all the ice are:

$$q = \rho V (\text{MWT}) h_f$$
$$= \left(0.92\right)\left(50\right)\left(\frac{1}{18}\right)\left(1440\right)$$
$$= 3680 \text{ cal}$$

Thus, the amount of ice unmelted is:

$$V = \left(50\right)\left(\frac{3680 - 2000}{3680}\right)$$
$$= 22.83 \text{ cm}^3$$

Answer is (C)

CHEMISTRY-53

Determine the final temperature when 10 g of copper and 20 g of lead at $-100\,°C$ are added to 50 g of H_2O at 50 °C. Disregard spurious heat losses. The atomic weight of copper is 63.55 g/mol, and the specific heat of lead = 0.032 cal/g·°C = 0.134 J/g·°C.

(A) 33.21 °C   (B) 38.21 °C   (C) 39.21 °C

          (D) 45.21 °C   (E) 49.21 °C

The law of Dulong and Petit is:

$$\text{atomic weight} \left(\frac{\text{g}}{\text{mole}}\right) \times \text{specific heat} \left(\frac{\text{cal}}{\text{g} \cdot {}^\circ\text{C}}\right) = 6.4 \frac{\text{cal}}{\text{mole} \cdot {}^\circ\text{C}} = 26.8 \frac{\text{J}}{\text{mole} \cdot {}^\circ\text{C}}$$

Since there are no spurious heat losses, the heat loss by H_2O equals the heat gained by Cu and Pb.

$$q = mc_p \Delta T$$

$$\text{where} \quad m = \text{mass (in grams)}$$
$$c_p = \text{specific heat capacity}$$
$$\Delta T = \text{change in temperature}$$

$$(50)(1)(323 - T_f) = \left[10 \left(\frac{6.4}{63.55}\right) + 20(0.032)\right]\left(T_f - 173\right)$$

$$T_f = 318.21\,{}^\circ\text{C}$$
$$= 45.21\,{}^\circ\text{C}$$

Answer is (D)

CHEMISTRY–54

A bomb calorimeter is used to determine thermal properties. Calculate the molar enthalpy of the combustion of glucose when 2.22 g of glucose are ignited, and the water in the well-insulated calorimeter rises in temperature from 18.00 °C to 23.19 °C. Assume that the water absorbs all of the heat given off.

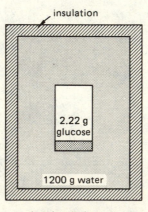

MW glucose
= 180

bomb calorimeter

(A) 102 kcal/mole   (B) 323 kcal/mole   (C) 506 kcal/mole
(D) 729 kcal/mole   (E) 1000 kcal/mole

$$n = \frac{m}{\text{MWT}}$$

$$n_{glucose} = \frac{2.22}{180}$$

$$= 0.0123 \text{ mole}$$

$$q = mc_p\Delta T$$

$$q_{H_2O} = (1200)(1)(23.19 - 18.00)$$

$$= 6228 \text{ cal}$$

$$q_{glucose} = 6228 \text{ cal}$$

$$\text{molar enthalpy} = \frac{6228}{0.0123}$$

$$= 506.34 \text{ kcal/mole}$$

Answer is (C)

CHEMISTRY–55

Calculate the standard heat of reaction, $\Delta \hat{H}^\circ$, per mole of C_6H_6 for the following reaction.

$$C_6H_6(g) \longrightarrow 3C_2H_2(g)$$

(A) −650.2 kJ/mole
(B) −597.4 kJ/mole
(C) 597.4 kJ/mole
(D) 650.2 kJ/mole
(E) 750.1 kJ/mole

Use the enthalpy of formation tables.

$$n_{C_6H_6} = 1 \text{ mole}$$
$$n_{C_2H_2} = 3 \text{ mole}$$
$$\hat{h}^\circ_{C_6H_6} = 82,923 \text{ J/gmole}$$
$$\hat{h}^\circ_{C_2H_2} = 226,757 \text{ J/gmole}$$
$$\hat{H}^\circ_R = \sum_i n_i \hat{h}^\circ_i \Big|_{reactants}$$
$$= 82,923$$
$$= 82.9 \text{ kJ/mole}$$
$$\hat{H}^\circ_P = \sum_i n_i \hat{h}^\circ_i \Big|_{products}$$
$$= (3)(226,757)$$
$$= 680,300$$
$$= 680.3 \text{ kJ/mole}$$
$$\Delta \hat{H}^\circ = \hat{H}^0_R - \hat{H}^0_P$$
$$= 82.9 - 680.3$$
$$= -597.4 \text{ kJ/mole}$$

Answer is (B)

CHEMISTRY-56

The heats of reaction for 3 equations are as follows:

1. $-2C_2H_2 - 5O_2 + 4CO_2 + 2H_2O$ $\qquad = -620,000 \text{ cal/mole}$
2. $\quad - O_2 + CO_2 \qquad -C \quad = - 96,960 \text{ cal/mole}$
3. $\quad - O_2 \qquad +2H_2O \quad -2H_2 = -136,800 \text{ cal/mole}$

What is the heat of formation of C_2H_2?

(A) 4.14 kcal/mole   (B) 45.7 kcal/mole   (C) 47.7 kcal/mole

(D) 95.7 kcal/mole   (E) 97.7 kcal/mole

CHEMISTRY 7-37

Adding $(-\text{eq. 1}) + 4(\text{eq. 2}) + (\text{eq. 3})$ gives the formation of 2 moles of C_2H_2.

$$2C_2H_2 + 5O_2 - 4CO_2 - 2H_2O \qquad\qquad = \quad 620,000$$
$$-4O_2 + 4CO_2 \qquad\qquad -4C \quad = -387,840$$
$$-\ O_2 \qquad\quad +2H_2O \qquad -2H_2 = -136,800$$

Therefore, $2C_2H_2 - 4C - 2H_2 = 95,360$

Because H_2 and C are at the standard reference state, $\hat{h}_C = 0$ and $\hat{h}_{H_2} = 0$.

$$\text{Therefore,} \quad 2\hat{h}_{C_2H_2} = 95,360$$
$$\hat{h}_{C_2H_2} = 47,680 \text{ cal/mole}$$
$$= 47.7 \text{ kcal/mole}$$

Answer is (C)

CHEMISTRY-57

A chemical reaction involving the collision of two molecules of A and B goes through the following energy profile:

$$A_2 + B_2 \longrightarrow 2AB$$

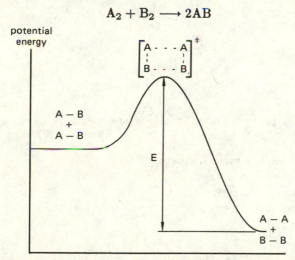

The energy, E, shown on the diagram represents which of the following?

(A) entropy of reaction
(B) enthalpy of reaction
(C) forward activation energy
(D) reverse activation energy
(E) free energy of reaction

PROFESSIONAL PUBLICATIONS, INC. • Belmont, CA

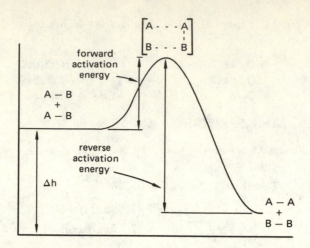

E is the energy required for the reverse reaction $(A-A + B-B \longrightarrow A-B + A-B)$ to proceed. Thus, E is the reverse activation energy.

Answer is (D)

CHEMISTRY–58

Reactions generally proceed faster at higher temperatures because of which of the following?

(A) The molecules collide more frequently.
(B) The activation energy is less.
(C) The molecules are less energetic.
(D) The bonds are easier to break.
(E) Both (A) and (B)

At higher temperatures, the molecules travel faster and, therefore, have a higher kinetic energy. This means that the molecules will collide more frequently and that the activation energy for a chemical reaction is smaller.

Answer is (E)

CHEMISTRY–59

Which of the following statements is _false_?

(A) In general, as reaction products are formed, they react with each other and re-form reactants.

(B) The net rate at which a reaction proceeds from left to right is equal to the forward rate minus the reverse rate.

(C) At equilibrium, the net reaction rate is zero.

(D) When a reaction mixture is far from its equilibrium composition, either the reverse or forward rate dominates, depending on whether the products or the reactants are in excess of the equilibrium value.

(E) The differential rate law is the mathematical expression that shows how the rate of a reaction depends on volume.

The differential rate law is the mathematical expression that shows how the rate of a reaction depends on concentration, not volume.

Answer is (E)

CHEMISTRY–60

For the reaction $3A + 2B \longrightarrow C + D$, the differential rate law is:

$$\frac{1}{3}\left(\frac{dA}{dr}\right) = \frac{dC}{dt} = k[A]^n[B]^m$$

Which of the following statements is _false_?

(A) The order of the reaction with respect to A is called n.

(B) The sum of $n + m$ is called the overall order of the reaction.

(C) The exponents of [A] and [B], n and m, are not necessarily equal to the stoichiometric coefficients of A and B in the net reaction.

(D) The overall order for the reaction can be predicted by or deduced from the equation for the reaction.

(E) The order with respect to each reagent must be found experimentally.

The order for the reaction must be found experimentally and cannot be determined from the equation for the reaction.

Answer is (D)

CHEMISTRY–61

Which of the following statements is false?

(A) When temperature is raised, the rate of any reaction is always increased.

(B) In general, when any two compounds are unmixed, a large number of reactions may be possible, but those which proceed the fastest are the ones observed.

(C) It is possible to influence the products of a chemical change by controlling the factors which affect reaction rates.

(D) Heterogeneous reactions are the reactions that take place at the boundary surface between two faces.

(E) As a chemical reaction progresses, the usual behavior of the reactant and product concentration is to change rapidly at first, then change slowly as the limiting concentrations are approached.

When temperature is increased, the rates of most reactions increase. However, the rates of some reactions do decrease.

Answer is (A)

CHEMISTRY–62

The following rate expression was found to accurately represent the kinetics of a chemical reaction.

$$r = kc_A^2 c_B$$

If c represents concentration in units of mole/l, what are the units of the rate constant, k?

(A) unitless

(B) s^{-1}

(C) $\dfrac{l}{mole \cdot s}$

(D) $\dfrac{l^2}{mole^2 \cdot s}$

(E) $\dfrac{l^3}{mole^3 \cdot s}$

The reaction rate always has units of mole/$l \cdot$ s. The units of k may be found as follows:

$$\frac{mole}{l \cdot s} = k \left(\frac{mole}{l} \right)^2 \left(\frac{mole}{l} \right)$$

$$k = \left(\frac{mole}{l \cdot s} \right) \left(\frac{l}{mole} \right)^3$$

$$= \frac{l^2}{mole^2 \cdot s}$$

Answer is (D)

CHEMISTRY–63

Let c represent the concentration of a reagent. For a first-order reaction, what would a plot of $\ln c$ versus t yield?

(A) a straight line whose slope is $-k$
(B) a straight line whose slope is k
(C) a horizontal line with an intercept of the $\ln c$ axis at $\ln c = k$
(D) a logarithmic curve approaching a value of k
(E) an exponential curve approaching a value of k

For a first order reaction:

$$-\frac{dc}{dt} = kc$$

where k = rate constant
          c = concentration

$$-\frac{dc}{c} = kdt$$

$$-\int_{c_0}^{c} \frac{dc}{c} = k \int_{0}^{t} dt$$

$$-\ln \frac{c}{c_0} = kt$$

$$\ln \frac{c}{c_0} = -kt$$

$$\ln c = -kt + \ln c_0$$

This is of the form $y = ax + y_0$. Therefore, the graph is a straight line with a slope of $-k$.

Answer is (A)

CHEMISTRY-64

The following kinetic data were collected for a specific chemical reaction. What is the rate constant for the reaction?

reaction: A + B $\longrightarrow$

experiment	c_A (mole/l)	c_B (mole/l)	initial rate (mole A/l · sec)
1	0.10	0.10	0.0010
2	0.20	0.10	0.0020
3	0.30	0.10	0.0030
4	0.10	0.20	0.0010
5	0.10	0.30	0.0010

(A) $0.010 \, \text{sec}^{-1}$ (B) $0.020 \, \text{sec}^{-1}$ (C) $0.020 \, l/\text{mole} \cdot \text{sec}$

(D) $0.030 \, l/\text{mole} \cdot \text{sec}$ (E) $0.040 \, l^2/\text{mole}^2$

First, determine the rate law. Experiments 4 and 5 show that the rate is not a function of c_B. Experiments 1, 2, and 3 show that the rate is directly proportional to c_A.

$$\text{Therefore,} \quad r = kc_A$$
$$k = \frac{r}{c_A}$$

Use the data from experiment 1 to determine k.

$$k = \frac{0.0010}{0.10}$$
$$= 0.01 \, \text{sec}^{-1}$$

Answer is (A)

CHEMISTRY-65

A certain temperature-dependent reaction proceeds 10 times faster at 500 K than it does at 300 K. How much faster will it react at 1000 K than it does at 300 K?

(A) 10   (B) 16   (C) 29   (D) 41   (E) 56

$$r = k \cdot f(c_i)$$

$$k = A \left[\exp\left(-\frac{E_A}{RT} \right) \right]$$

where E_A = activation energy

therefore, $r \propto \exp\left(-\frac{E_A}{RT} \right)$

$$\frac{r_1}{r_2} = \frac{\exp\left(\dfrac{-E_A}{RT_2} \right)}{\exp\left(\dfrac{-E_A}{RT_1} \right)}$$

$$= \exp\left[\frac{E_A}{R} \left(\frac{1}{T_1} - \frac{1}{T_2} \right) \right]$$

$$\ln 10 = \frac{E_A}{R} \left(\frac{1}{300} - \frac{1}{500} \right)$$

$$\frac{E_A}{R} = 1726.94$$

$$\frac{r_{1000}}{r_{300}} = \exp\left[1727 \left(\frac{1}{300} - \frac{1}{1000} \right) \right]$$

$$= 56$$

Answer is (E)

CHEMISTRY-66

A reaction rate is observed to triple as the result of raising the temperature from 0 °C to 20 °C. What is the activation energy of the reaction?

(A) $3900\,R$

(B) $4400\,R$

(C) $4700\,R$

(D) $5100\,R$

(E) $6200\,R$

$$r = C\exp\left(-\frac{E_A}{RT}\right)$$

where $C = $ constant

$E_A = $ activation energy

$$\frac{r_1}{r_2} = \exp\frac{E_A}{R}\left(\frac{1}{T_1} - \frac{1}{T_2}\right)$$

$$\frac{1}{3} = \exp\frac{E_A}{R}\left(\frac{1}{273} - \frac{1}{293}\right)$$

$$\ln\frac{1}{3} = -\frac{E_A}{R}\left(\frac{1}{273} - \frac{1}{293}\right)$$

$$E_A = -\frac{(R)\left(\ln\frac{1}{3}\right)}{\dfrac{1}{273} - \dfrac{1}{293}}$$

$$= 4393\,R$$

$$\approx 4400\,R$$

Answer is (B)

CHEMISTRY-67

In the troposphere, ozone is produced during the daylight and consumed during the darkness. Determine the half-life of ozone if it is depleted to 10% of its initial value after 10 hours of darkness.

(A) 3.0 hr   (B) 3.5 hr   (C) 4.0 hr

        (D) 4.5 hr   (E) 5.0 hr

$$C = C_0 e^{-kt}$$
$$C(10) = 0.1C_0$$
$$0.1 = e^{-10k}$$
$$-k = \frac{\ln 0.1}{10}$$
$$= -0.2303 \text{ hr}^{-1}$$
$$t_{\frac{1}{2}} = \text{half-life}$$
$$C(t_{\frac{1}{2}}) = 0.5C_0$$
$$0.5 = \exp\left[(-0.2303)t_{\frac{1}{2}}\right]$$
$$t_{\frac{1}{2}} = \frac{\ln 2}{0.2303}$$
$$= 3.0 \text{ hr}$$

Answer is (A)

CHEMISTRY–68

Which of the following statements is false?

(A) In considering chemical equilibrium, the relative stabilities of products and reactants are important.

(B) In considering chemical equilibrium, the pathway from the initial state to the final state is important.

(C) In treating reaction rates, the rate at which reactants are converted to products is important.

(D) In treating reaction rates, the sequence of physical processes by which reactants are converted to products is important.

(E) In treating reaction rates, the sequence of chemical processes by which reactants are converted to products is important.

Considerations of chemical equilibrium do not take into account the pathway from initial to final states.

Answer is (B)

CHEMISTRY–69

Consider the following reaction at equilibrium:

$$3H_2 \text{ (gas)} + N_2 \text{ (gas)} \longleftrightarrow 2NH_3 \text{ (gas)} + 92.0 \text{ kJ}$$

Which single change in conditions will cause a shift in equilibrium toward an increase in production of NH_3?

(A) addition of an inert gas
(B) removal of hydrogen gas
(C) increase in temperature
(D) increase in volume of the system
(E) increase in pressure on the system

The effects of each change in condition are as follows:

addition of inert gas	no effect on equilibrium
removal of hydrogen	shifts equilibrium to the reactants
increase in temperature	shifts equilibruim to the reactants
increase in volume	shifts equilibrium to the reactants
increase in pressure	shifts equilibrium to the products

Answer is (E)

CHEMISTRY–70

Consider the following reaction at equilibrium:

$$\tfrac{1}{2}N_2 + \tfrac{3}{2}H_2 \longleftrightarrow NH_3 \qquad \Delta\hat{H} = -11.0 \text{ kcal}$$

What would be the expected effect on the amount of NH_3 under each of the following conditions?

I.   raise the temperature
II.   compress the mixture
III. add additional H_2

(A) I: increase, II: increase, III: increase
(B) I: increase, II: increase, III: decrease
(C) I: increase, II: decrease, III: decrease
(D) I: decrease, II: increase, III: decrease
(E) I: decrease, II: decrease, III: increase

Each change has the following effects:

I.   raise the temperature: shifts equilibrium to the products
                                 because the reaction is exothermic

II. compress the mixture: shifts equilibrium to the products
                                 because products contain a
                                 smaller number of moles

III. add hydrogen gas: shifts equilibrium to the products
                                 because adding addtional reactants will
                                 force the formation of more products

Thus, all three changes result in an increase in the amount of NH_3.

Answer is (A)

CHEMISTRY–71

Consider the following reaction:

$$MgSO_4 \text{ (solid)} \longleftrightarrow MgO \text{ (solid)} + SO_3 \text{ (gas)}$$

What is the equilibrium constant for the given reaction?

(A) $k = \dfrac{[Mg][SO_3]}{2[MgSO_4]}$ (B) $k = \dfrac{[MgSO_4]}{[MgO][SO_3]}$ (C) $k = [MgO][SO_3]$

(D) $k = \dfrac{[SO_3]}{[MgSO_4]}$ (E) $k = [SO_3]$

Solids have a concentration of 1. Therefore, $k = [SO_3]$.

Answer is (E)

CHEMISTRY-72

The solubility of barium sulfate, $BaSO_4$, is 0.0091 g/l at 25 °C. What is the value of the solubility product constant k_{sp} for $BaSO_4$? The molecular weight of barium sulfate is 233 g/mole.

(A) 1.52×10^{-9} mole^2/l^2
(B) 4.24×10^{-8} mole^2/l^2
(C) 8.63×10^{-7} mole^2/l^2
(D) 2.98×10^{-6} mole^2/l^2
(E) 8.28×10^{-5} mole^2/l^2

$$BaSO_4(s) \longleftrightarrow Ba^{2+}(aq) + SO_4^{2-}(aq)$$
$$k_{sp} = [Ba^{2+}][SO_4^{2-}]$$
$$[BaSO_4] = \left(0.0091\,\frac{g}{l}\right)\left(\frac{mole}{233\,g}\right)$$
$$= 3.9 \times 10^{-5} \text{ mole}/l$$
$$[Ba^{2+}] = [SO_4^{2-}] = [BaSO_4]$$
$$k_{sp} = (3.9 \times 10^{-5})^2$$
$$= 1.52 \times 10^{-9} \text{ mole}^2/l^2$$

Answer is (A)

CHEMISTRY-73

Consider the reaction:

$$H_2 \text{ (gas)} + I_2 \text{ (gas)} \longleftrightarrow 2HI \text{ (gas)} \qquad k_{eq} = 25$$

Determine the number of moles of H_2 remaining when 1 mole of both H_2 and I_2 are equilibrated in a 1 liter box.

(A) $\frac{1}{6}$ mole

(B) $\frac{2}{7}$ mole

(C) $\frac{5}{7}$ mole

(D) $\frac{5}{6}$ mole

(E) $\frac{9}{8}$ mole

	H_2 (gas)	+	I_2 (gas)	$\longleftrightarrow$	2HI (gas)
initial moles	1		1		0
final moles	$1-x$		$1-x$		2x

$$k = \frac{[HI]^2}{[H_2][I_2]}$$

$$= \frac{(2x)^2}{(1-x)(1-x)}$$

$$= 25$$

$$\frac{4x^2}{(1-x)^2} = 25$$

$$4x^2 = 25(1-2x+x^2)$$

$$21x^2 - 50x + 25 = 0$$

$$x = \frac{+50 \pm \sqrt{(50)^2 - (4)(25)(21)}}{(2)(21)}$$

$$= \frac{5}{3} \text{ mole or } \frac{5}{7} \text{ mole}$$

Only the second value for x makes sense, because the first value is greater than the initial amounts of H_2 and I_2. Thus, the remaining number of moles of H_2 at equilibrium is $1 - \frac{5}{7} = \frac{2}{7}$ mole.

Answer is (B)

CHEMISTRY–74

Given the reaction:

$$A + B \longleftrightarrow 2C; \quad k_{eq} = 50$$

Determine the final concentration of C when 1 mole of both A and B are added to a 1 liter container containing 0.1 mole of C.

(A) 0.78 mole   (B) 0.88 mole   (C) 1.66 mole

       (D) 1.85 mole   (E) 2.00 mole

	A	+	B	$\longleftrightarrow$	2C
initial moles	1		1		0.1
final moles	$1 - x$		$1 - x$		$2x + 0.1$

$$k_{eq} = \frac{[C]^2}{[A][B]}$$

$$= \frac{(2x + 0.1)^2}{(1 - x)(1 - x)} = 50$$

$$4x^2 + 0.4x + 0.1 = 50x^2 - 100x + 50$$

$$46x^2 - 99.6x + 49.9 = 0$$

$$x = \frac{99.6 \pm \sqrt{(99.6)^2 - (4)(46)(49.9)}}{(2)(46)}$$

$$= 1.378 \text{ mole or } 0.78 \text{ mole}$$

However, x cannot be 1.378 mole because this is greater than the initial amounts of A and B. Therefore, $x = 0.78$ mole. The final number of moles of C is $2x = 1.66$ mole.

$2x + 0.1$

$2 \times .78 + 0.1 = 1.66\ Mole$

Answer is (C)

CHEMISTRY-75

The voltage of a galvanic cell does not depend on which of the following parameters?

(A) concentration of solutions
(B) temperature
(C) pressure
(D) volume
(E) chemical substances

The cell potential, $\mathcal{E}$, is dependent on all of the above except volume.

Answer is (D)

CHEMISTRY-76

Given the electrochemical cell below, what is the reaction at the anode?

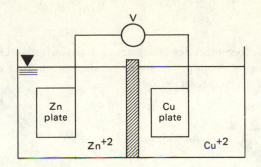

(A) $Cu \longrightarrow Cu^{2+} + 2e^-$
(B) $Cu^{2+} + 2e^- \longrightarrow Cu$
(C) $Zn \longrightarrow Zn^{2+} + 2e^-$
(D) $Zn^{2+} + 2e^- \longrightarrow Zn$
(E) $O_2 + 2H_2O + 4e^- \longrightarrow 4OH^-$

Zinc has a higher potential and will, therefore, act as the anode. By definition, the anode is where electrons are given off. Thus, the reaction at the anode of the electrochemical cell is $Zn \longrightarrow Zn^{2+} + 2e^-$.

Answer is (C)

CHEMISTRY-77

Which of the following statements regarding a galvanic cell is false?

(A) A negative value of cell voltage $\Delta\mathcal{E}$ means that the reaction in the cell proceeds spontaneously from right to left.

(B) If the standard potential of a cell is zero, a concentration difference alone is sufficient to generate a voltage.

(C) When a current I flows through the voltage difference $\Delta\mathcal{E}^0$ for a time, t, the electrical work performed is $\Delta\mathcal{E}^0 \times I \times t$.

(D) It is possible to have a gas electrode in a galvanic cell.

(E) The cell voltage, $\Delta\mathcal{E}$, is totally independent of the number of electrons transferred in a given reaction.

For the reaction $aA + bB \longrightarrow cC + dD$, the Nernst equation states:

$$\Delta \mathcal{E} = \Delta \mathcal{E}^0 - \frac{0.059}{n} \log \left(\frac{[C]^c [D]^d}{[A]^a [B]^b} \right)$$

Here, n = number of moles of electrons transferred in the reaction. Therefore, the cell voltage does depend on the number of electrons transferred in a given reaction.

Answer is (E)

CHEMISTRY-78

Consider the Nernst equation:

$$\Delta \mathcal{E} = \Delta \mathcal{E}^0 - \frac{0.059}{n} \log \left(\frac{[C]^c [D]^d}{[A]^a [B]^b} \right)$$

Which of the following statements is false?

(A) n = number of moles of electrons transferred in the reaction as written.
(B) The cell must be operating at a temperature of 25 °C.
(C) The equation holds for the reaction $aA + bB \longrightarrow cC + dD$.
(D) $\Delta \mathcal{E}^0$ is the standard cell potential.
(E) The factor of 0.059 is common to all cells, regardless of temperature.

The factor of 0.059 applies only to cells with an operating temperature of 25 °C.

Answer is (E)

CHEMISTRY-79

A zinc-copper standard cell with $\Delta\mathcal{E}^0 = 1.10$ volts is connected to an independent variable voltage supply such that the variable voltage opposes the cell voltage. Given the following reaction, what happens?

$$Zn + Cu^{2+} \longleftrightarrow Cu + Zn^{2+} \qquad \Delta\mathcal{E}^0 = 1.10 \text{ volt}$$

(A) When the variable voltage is below 1.10 volts, the cell reaction $Cu + Zn^{2+} \longrightarrow Cu^{2+} + Zn$ predominates.

(B) When the variable voltage is above 1.10 volts, the cell reaction $Cu + Zn^{2+} \longrightarrow Cu^{2+} + Zn$ predominates.

(C) When the variable voltage is above 1.10 volts, the cell reaction $Zn + Cu^{2+} \longrightarrow Cu + Zn^{2+}$ predominates.

(D) When the variable voltage is equal to 1.10 volts, the cell reaction $Cu + Zn^{2+} \longrightarrow Cu^{2+} + Zn$ predominates.

(E) When the variable voltage is equal to 1.10 volts, the cell reaction $Zn + Cu^{2+} \longrightarrow Cu + Zn^{2+}$ predominates.

When the variable voltage is below 1.10 volts, the reaction $Zn + Cu^{2+} \longrightarrow Cu + Zn^{2+}$ predominates. When it is equal to 1.10 volts, no net reaction occurs. When the variable voltage is above 1.10 volts, the reverse reaction, $Cu + Zn^{2+} \longrightarrow Zn + Cu^{2+}$ predominates.

Answer is (B)

CHEMISTRY-80

Given that $\Delta\mathcal{E}^0 = 0.03$ volt, $[Ni] = 1$ M, $[Co] = 0.1$ M, and $T = 25$ °C, calculate the cell voltage for the following equation:

$$Co + Ni^{2+} \longrightarrow Co^{2+} + Ni$$

(A) 0.01 V   (B) 0.03 V   (C) 0.06 V

(D) 0.09 V   (E) 0.12 V

Use the Nernst equation.

$$\Delta\mathcal{E} = \Delta\mathcal{E}^0 - \frac{0.059}{n} \log\left(\frac{[Co^{2+}]}{[Ni^{2+}]}\right)$$

$$n = 2$$

$$\Delta\mathcal{E} = 0.03 - \frac{0.059}{2} \log 0.1$$

$$= 0.03 + 0.03$$

$$= 0.06 \text{ volts}$$

Answer is (C) ✓

CHEMISTRY-81

In organic chemistry, which compound families are associated with the following bonds?

1.   C–C
2.   C=C
3.   C≡C

(A) 1: alkene, 2: alkyne, 3: alkane
(B) 1: alkyne, 2: alkane, 3: alkene
(C) 1: alkane, 2: alkene, 3: alkyne
(D) 1: alkane, 2: alkyne, 3: alkene
(E) 1: alkene, 2: alkane, 3: alkyne

An alkane is a saturated organic compound. Thus, the carbons may only have single bonds. In an alkene, the carbon atoms may have double bonds. In alkynes, the carbon atoms may have triple bonds.

Answer is (C)

CHEMISTRY-82

Which one of the following statements regarding organic substances is false?

(A) Organic matter is generally stable at very high temperatures.

(B) All organic matter contains carbon.

(C) Organic substances generally do not dissolve in water.

(D) Organic substances generally dissolve in high-concentration acids.

(E) The ionization of an organic substance is, in general, more difficult to achieve than the ionization of an inorganic substance.

Organic matter contains carbon, is generally insoluble in water, soluble in high concentration acids, not easily ionizable, and unstable at high temperatures.

Answer is (A)

CHEMISTRY-83

Which one of the following is most likely to prove that a substance is inorganic?

(A) The substance is heated together with copper oxide and the resulting gases are found to have no effect on limestone.

(B) The substance evaporates in room temperature and pressure.

(C) Analysis shows that the substance contains hydrogen.

(D) The substance floats in water.

(E) The substance is proved to be reproducible in the laboratory.

The carbon from organic matter generally reacts with copper oxide to produce carbon dixoide. This gas darkens limestone.

Answer is (A)

CHEMISTRY-84

Which of the following organic chemicals is most soluble in water?

(A) CH_3CH_3 (B) CH_3OH (C) CCl_4

(D) $CH_3-(CH_2)_n-CH_3$ (E) CH_4

Water is a polar molecule. Thus, a polar substance is more likely to dissolve in water than a nonpolar substance. Methanol (CH_3OH) is polar and, therefore, very miscible in water. All of the other molecules are nonpolar.

Answer is (B)

CHEMISTRY-85

Which statement describes all of the following three chemical formulas?

$$\begin{array}{ccc} H & H & CH \\ | & | & | \\ H-C-C-C-CHO \\ | & | & | \\ H & H & H \end{array} \qquad \begin{array}{ccc} H & H & H \\ | & | & | \\ H-C-C-C-CHO \\ | & | & | \\ H & OH & H \end{array} \qquad \begin{array}{ccc} H & H & H \\ | & | & | \\ OH-C-C-C-CHO \\ | & | & | \\ H & H & H \end{array}$$

(A) They are isotopes of a certain substance.
(B) They are the only possible forms of $C_4H_2O_2$.
(C) They are incorrectly written.
(D) They are isomers.
(E) They are polymers.

When a compound has one chemical formula, but different possible physical structures, the different structures are called isomers. The three formulas are all possible structures of $C_4H_8O_2$. Therefore, they are isomers.

Answer is (D)

CHEMISTRY-86

What structures do both aldehydes and ketones contain?

(A) the carboxyl group   (B) the carbonyl group   (C) the hydroxyl group

(D) the carbon-carbon double bond   (E) the amino group

Aldehydes and ketones both contain the carbonyl group.

Answer is (B)

CHEMISTRY-87

Identify the following acid structures:

I   II   III

(A) I: formic acid, II: oxalic acid, III: acetic acid
(B) I: oxalic acid, II: acetic acid, III: formic acid
(C) I: acetic acid, II: formic acid, III: oxalic acid
(D) I: formic acid, II: acetic acid, III: oxalic acid
(E) I: oxalic acid, II: formic acid, III: acetic acid

HCOOH is formic acid. CH_3COOH is acetic acid. $C_2H_2O_4$ is oxalic acid.

Answer is (D)

8 STATICS

STATICS–1

What is the length of the resultant of $\mathbf{A} + \mathbf{B} + \mathbf{C}$?

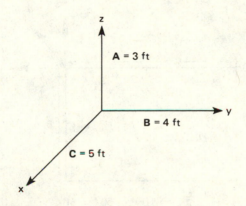

(A) 3.464 ft   (B) 4.268 ft   (C) 7.071 ft

      (D) 10.104 ft   (E) 12.707 ft

$$|\mathbf{A} + \mathbf{B} + \mathbf{C}| = \sqrt{A^2 + B^2 + C^2}$$
$$= \sqrt{3^2 + 4^2 + 5^2}$$
$$= 7.071 \text{ ft}$$

Answer is (C)

STATICS–2

Determine the magnitude of the moment of the force F about the corner A.

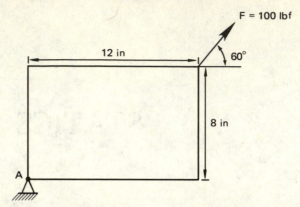

(A) 120 in-lbf   (B) 240 in-lbf   (C) 320 in-lbf
          (D) 560 in-lbf   (E) 640 in-lbf

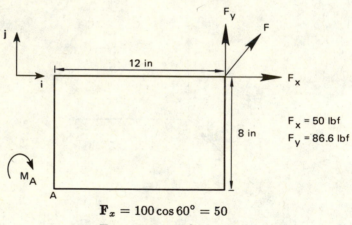

$$\mathbf{F}_x = 100 \cos 60° = 50$$
$$\mathbf{F}_y = 100 \sin 60° = 86.6$$
$$\mathbf{F} = 50\mathbf{i} + 86.6\mathbf{j}$$
$$\mathbf{r} = 12\mathbf{i} + 8\mathbf{j}$$
$$\sum M_A = 0$$
$$M_A - \mathbf{r}\mathbf{F} = 0$$
$$M_A = \mathbf{r}\mathbf{F}$$
$$= (12\mathbf{i} + 8\mathbf{j})(50\mathbf{i} + 86.6\mathbf{j})$$
$$M_A = 640 \text{ in-lbf}$$

Answer is (E)

STATICS–3

A cube of side a is acted upon by a force F as shown. Determine the magnitude of the moment of F about the diagonal AB.

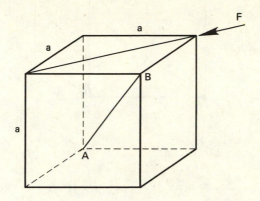

(A) $\dfrac{1}{\sqrt{8}}\,aF$ (B) $\dfrac{1}{\sqrt{6}}\,aF$ (C) $\dfrac{1}{\sqrt{4}}\,aF$

(D) $\dfrac{1}{\sqrt{3}}\,aF$ (E) $\dfrac{1}{\sqrt{2}}\,aF$

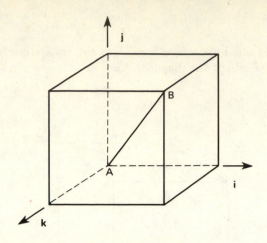

$$M_A = a(\mathbf{i} + \mathbf{j}) \times \frac{F}{\sqrt{2}}(-\mathbf{i} + \mathbf{j})$$

$$M_A = \frac{aF}{\sqrt{2}}(\mathbf{i} - \mathbf{j} + \mathbf{k})$$

$$|M_{AB}| = M_A \times \frac{\mathbf{AB}}{|\mathbf{AB}|}$$

$$= \left[\frac{aF}{\sqrt{2}}(\mathbf{i} - \mathbf{j} + \mathbf{k})\right] \times \left[\frac{1}{\sqrt{3}}(\mathbf{i} + \mathbf{j} + \mathbf{k})\right]$$

$$= \frac{aF}{\sqrt{6}}(1 - 1 + 1)$$

$$= \frac{1}{\sqrt{6}}aF$$

Answer is (B)

STATICS-4

The boom shown has negligible weight, but has sufficient strength to support the 1000 lbf load. Find the tension in the supporting cable between points A and B.

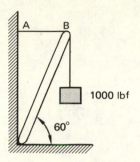

1000 lbf

60°

(A) 0 lbf   (B) 433 lbf   (C) 577.3 lbf
          (D) 866 lbf   (E) 1000 lbf

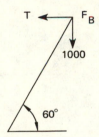

T ← F_B

1000

60°

Since $\sum F_y = 0$, and the cable is horizontal, the boom must support the vertical load of 1000 lbf.

$$F_B \sin 60° = 1000$$
$$F_B = 1154.7 \text{ lbf}$$
$$\sum F_x = 0$$
$$F_B \cos 60° - T = 0$$

where T = tension in the cable
$$= (1154.7)(\cos 60°)$$
$$= 577.3 \text{ lbf}$$

Answer is (C)

STATICS–5

Find the tensions, T_1 and T_2, in the ropes so that the system is in equilibrium.

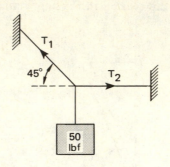

(A) $T_1 = 50$ lbf, $T_2 = 0$ lbf
(B) $T_1 = 50$ lbf, $T_2 = 50$ lbf
(C) $T_1 = 70.71$ lbf, $T_2 = 50$ lbf
(D) $T_1 = 70.70$ lbf, $T_2 = 70.71$ lbf
(E) $T_1 = 100$ lbf, $T_2 = 50$ lbf

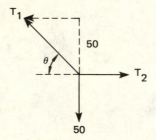

Since $\sum F_y = 0$, and T_2 is horizontal, T_1 must support 50 lbf vertically.

$$T_1 \sin \theta = 50$$
$$T_1 \sin 45° = 50$$
$$T_1 = 70.71 \text{ lbf}$$
$$\sum F_x = 0$$
$$T_1 \cos 45° - T_2 = 0$$
$$T_2 = T_1 \cos 45°$$
$$= 50 \text{ lbf}$$

Answer is (C)

STATICS–6

Determine the value of h which puts the system in the following figure in equilibrium. Given: $W = 100$ lbf, $P = 20$ lbf, and $d = 10''$.

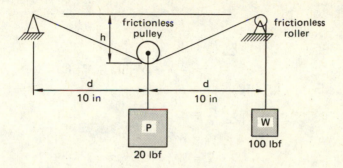

(A) 0.50 in   (B) 1.0 in   (C) 1.5 in

(D) 2.1 in   (E) 4.0 in

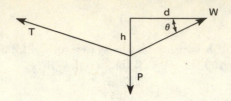

$$\sum F_y = 0$$
$$0 = -P + W \sin \theta + T \sin \theta$$
$$(T + W) \sin \theta = P$$
$$\sum F_x = 0$$
$$0 = W \cos \theta - T \cos \theta$$
$$T = W$$
$$= 100 \text{ lbf}$$
$$2W \sin \theta = P$$
$$\sin \theta = \frac{P}{2W}$$
$$\frac{h}{\sqrt{h^2 + d^2}} = \frac{P}{2W}$$
$$= 0.1$$
$$h^2 = 0.01(h^2 + d^2)$$
$$0.99h^2 = 0.01d^2$$
$$= d\sqrt{\frac{0.01}{0.99}}$$
$$h = 1 \text{ in}$$

Answer is (B)

STATICS–7

Hinges A and B support a 10,000 lbf bank vault door. Determine the horizontal force in the hinge pin at A.

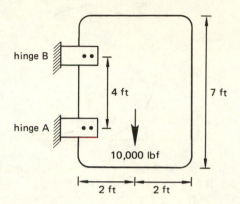

(A) 0 lbf   (B) 2500 lbf   (C) 5000 lbf

(D) 7500 lbf   (E) 10,000 lbf

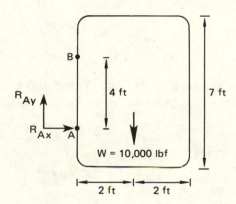

Sum the moments around joint B.

$$\sum M_B = 0$$

$$0 = (10,000)(2) - (4)(R_{Ax})$$

$$R_{Ax} = 5000 \text{ lbf}$$

Answer is (C)

STATICS-8

A cylindrical tank is set at rest as shown. What force, F, should be applied horizontally at point C to raise the tank?

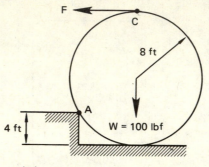

(A) 25.4 lbf   (B) 57.7 lbf   (C) 66.7 lbf

(D) 100 lbf   (E) 125 lbf

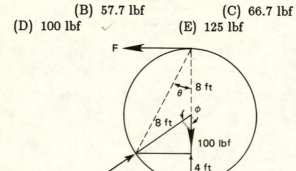

In order to raise the tank, $\sum M_A \geq 0$. Therefore, the minimum force that must be applied at point C can be found as follows:

$$\sum M_A = 0$$

$$F(12) - W(8)\sin\phi = 0$$

$$F = \frac{2W\sin\phi}{3}$$

$$\cos\phi = \frac{1}{2}$$

$$\phi = 60°$$

$$F = \frac{2W\sin 60°}{3}$$

$$= \frac{(2)(100)\sin 60°}{3}$$

$$= 57.7 \text{ lbf}$$

Answer is (B)

STATICS-9

A weight is attached to the lever as shown. Determine the expression for θ when the system is at equilibrium. The spring constant is K, the length of the lever is L, the radius of the wheel is r, and the magnitude of the weight is W.

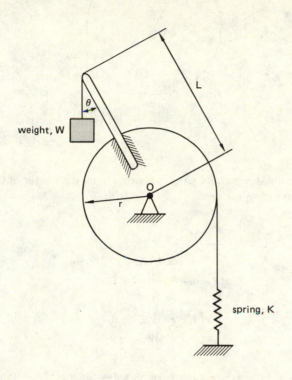

(A) $\theta = \dfrac{W L \sin \theta}{K r}$

(B) $\theta = \dfrac{W L \sin \theta}{K r^2}$

(C) $\theta = \dfrac{W r}{K L}$

(D) $\theta = \dfrac{W L \cos \theta}{K r}$

(E) $\theta = \dfrac{W L \cos \theta}{K r^2}$

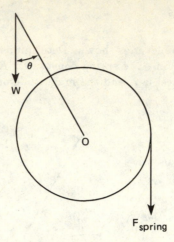

At equilibrium, the sum of the moments about the center of the wheel (point O) must equal zero.

$$\sum M_O = 0$$
$$= r F_{\text{spring}} - W L$$
$$r F_{\text{spring}} = W L \sin \theta$$
$$F_{\text{spring}} = K r \theta$$
$$r^2 K \theta = W L \sin \theta$$
$$\theta = \frac{W L \sin \theta}{K r^2}$$

Successive iterations are necessary to solve for θ.

Answer is (B)

STATICS–10

Two identical spheres weighing 100 lbf each are placed as shown. The line connecting the two centers of the spheres makes an angle of 30° to the horizontal surface. Determine the reaction force at A.

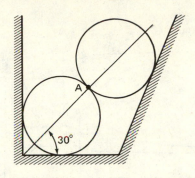

(A) 33.3 lbf   (B) 66.7 lbf   (C) 75.0 lbf

(D) 100 lbf   (E) 133 lbf

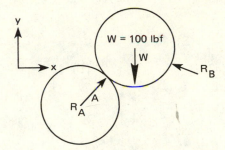

$$\sum F_x = 0$$

$$R_A \cos 30° - R_B \cos 30° = 0$$

$$R_A = R_B$$

$$\sum F_y = 0$$

$$-W + R_A \sin 30° + R_B \sin 30° = 0$$

$$R_A + R_B = \frac{W}{\sin 30°}$$

$$2R_A = \frac{W}{\sin 30°}$$

$$R_A = W$$

$$= 100 \text{ lbf}$$

Answer is (D)

STATICS–11

Determine the force, P, that must be exerted on the handles of the bolt cutter.

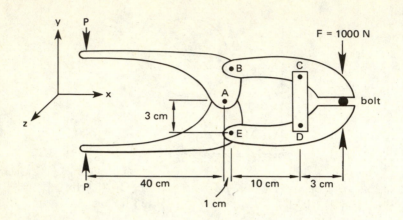

(A) 7.5 N   (B) 30.0 N   (C) 52.5 N

     (D) 250 N   (E) 325 N

First, consider the upper jaw.

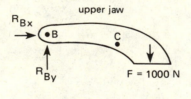

$$\sum F_x = 0$$
$$R_{Bx} = 0$$
$$\sum M_C = 0$$
$$-(R_{By})(10) + (F)(3) = 0$$
$$R_{By} = \frac{3}{10}F$$
$$= 300 \text{ N}$$

Now, consider the upper handle.

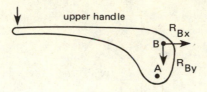

$$\sum M_A = 0$$
$$(R_{By})(1) + (P)(40) = 0$$
$$P = \frac{R_{By}}{40}$$
$$= \frac{300}{75}$$
$$= 7.5 \text{ N}$$

Answer is (A)

STATICS-12

Determine the force, F, required to keep the package from sliding down the plane.

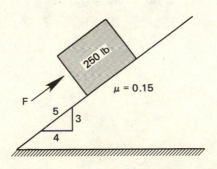

(A) 0 lbf   (B) 30 lbf   (C) 60 lbf
            (D) 90 lbf   (E) 120 lbf

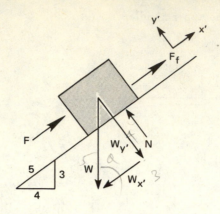

$$\sum F_{y'} = 0$$

$$W_{y'} - N = 0$$

$$W_{y'} = \frac{4}{5}W$$

$$N = \frac{4}{5}W$$

$$= 200 \text{ lbf}$$

$$F_f = \mu N$$

$$= (0.15)(200)$$

$$= 30 \text{ lbf}$$

$$\sum F_{x'} = 0$$

$$F - W_{x'} + F_f = 0$$

$$F = W_{x'} - F_f$$

$$W_{x'} = \frac{3}{5}W$$

$$= 150 \text{ lbf}$$

$$F = 150 - 30$$

$$= 120 \text{ lbf}$$

Answer is (E)

STATICS–13

Determine the minimum coefficient of friction at point B required for the man to use the ladder shown. Assume that there is no friction at point A.

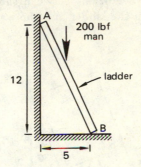

(A) $\mu = 0.20$ (B) $\mu = 0.28$ (C) $\mu = 0.42$

           (D) $\mu = 0.56$ (E) $\mu = 0.72$

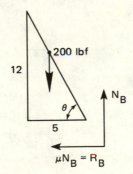

The total reaction force at B ($N_B + R_B$) must point along the ladder.

$$\text{Therefore,} \quad \frac{R_B}{N_B} = \cot\theta$$

$$\frac{\mu N_B}{N_B} = \cot\theta$$

$$\mu = \cot\theta$$

$$= \frac{5}{12}$$

$$= 0.42$$

Answer is (C)

STATICS-14

A 12 foot ladder weighing 40 lbf is placed as shown. When a 180 lbf man reaches a point 8 feet from the lower end (point A), the ladder is just about to slip. Determine the friction coefficient between the ladder and the floor. The coefficient of friction between the ladder and the wall is 0.20.

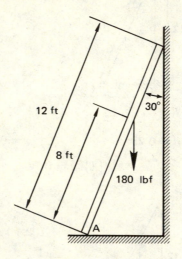

(A) $\mu = 0.20$ (B) $\mu = 0.25$ (C) $\mu = 0.30$
(D) $\mu = 0.35$ (E) $\mu = 0.40$

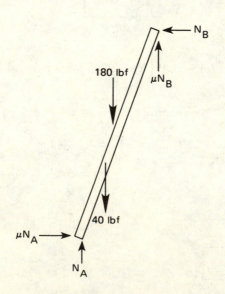

$$\sum M_A = 0$$

$$
\begin{aligned}
0 = \; & (40)(6)\sin 30° + (180)(8)\sin 30° \\
& - N_B(12)\cos 30° - (0.20)N_B(12)\sin 30°
\end{aligned}
$$

$$N_B = \frac{120 + 720}{11.6}$$

$$= 72.5 \text{ lbf}$$

$$\sum F_y = 0$$

$$0 = N_A + \mu_w N_B - 180 - 40$$

$$= N_A + (0.20)(72.5) - 180 - 40$$

$$N_A = 205 \text{ lbf}$$

$$\sum F_x = 0$$

$$0 = \mu N_A - N_B$$

$$\mu N_A = N_B$$

$$\mu = \frac{N_B}{N_A}$$

$$= \frac{72.5}{205}$$

$$= 0.35$$

Answer is (D)

STATICS–15

A block of mass, M, is released at position A on an inclined plane that is tangent to a circular arc. The plane is tilted 15° from horizontal, and the coefficient of friction is 0.4. Which point is the equilibrium position of the block?

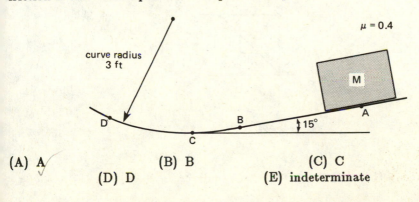

(A) A

(B) B

(C) C

(D) D

(E) indeterminate

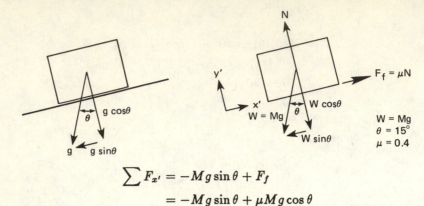

$$\sum F_{x'} = -Mg\sin\theta + F_f$$
$$= -Mg\sin\theta + \mu Mg\cos\theta$$

For the block to slide downhill, $\sum F_x < 0$.

$$-Mg\sin\theta + \mu Mg\cos\theta < 0$$
$$\mu Mg\cos\theta < Mg\sin\theta$$
$$\mu < \tan\theta$$
$$< \tan 15°$$
$$< 0.27$$

Thus, for the block to move, $\mu < 0.27$. However, $\mu = 0.4$. Therefore, the block never moves. It stays at point A.

Answer is (A)

STATICS–16

Determine the force in member AC in terms of P.

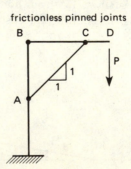

frictionless pinned joints

(A) P

(B) $\frac{4}{3}P$

(C) $\sqrt{2}P$

(D) $\frac{\sqrt{3}}{2}P$

(E) $2P$

$$AC = \text{force in member AC}$$
$$\sum F_y = 0$$
$$0 = (AC)\cos 45° - P$$
$$AC = \frac{P}{\cos 45°}$$
$$= \sqrt{2}\, P$$

Answer is (C)

STATICS–17

Determine the force in member CD.

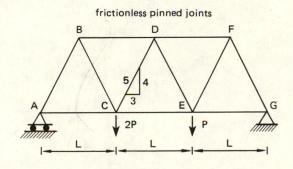

frictionless pinned joints

(A) $\frac{1}{12}P$

(B) $\frac{1}{3}P$

(C) $\frac{5}{12}P$

(D) P

(E) $\frac{5}{4}P$

Use the method of sections.

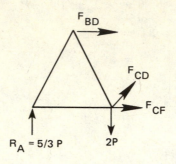

Only CD can support a vertical force.

$$\sum F_y = 0$$
$$0 = R_A - 2P + CD_y$$
$$CD_y = \frac{P}{3}$$
$$CD = \frac{5}{4} CD_y$$
$$= \left(\frac{5}{4}\right)\left(\frac{P}{3}\right)$$
$$= \frac{5P}{12}$$

Answer is (C)

STATICS–18

A truss is subjected to three loads. The truss is supported by a roller at A and by a pin joint at B. Determine the reaction force at A.

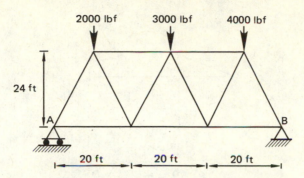

(A) 3830 lbf   (B) 4420 lbf   (C) 4860 lbf
(D) 4950 lbf   (E) 5170 lbf

The rolling support at A can only support a vertical reaction force. R_A is the reaction force at A.

$$\sum M_B = 0$$

$$0 = (R_A)(60) - (2000)(50) - (3000)(30) - (4000)(10)$$

$$R_A = 3830 \text{ lbf}$$

Answer is (A)

STATICS–19

A scissor jack is used to raise a car. If the jack supports a weight of 2000 lbf, determine the force in member BE.

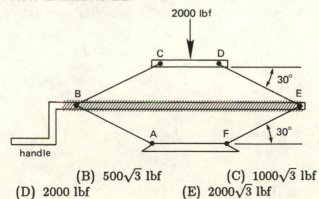

(A) 0 lbf   (B) $500\sqrt{3}$ lbf   (C) $1000\sqrt{3}$ lbf
(D) 2000 lbf   (E) $2000\sqrt{3}$ lbf

PROFESSIONAL PUBLICATIONS, INC. • Belmont, CA

CB and DE equally share the 2000 lbf load. Therefore, $BC_y = DE_y = 1000$ lbf, and $DE = 2000$ lbf. Use the method of joints at E.

$$\sum F_y = 0$$
$$0 = -FE_y + DE_y$$
$$FE_y = 1000 \text{ lbf}$$
$$FE = 2000 \text{ lbf}$$
$$\sum F_x = 0$$
Therefore, $$BE = FE_x + DE_x$$
$$= 1000\sqrt{3} + 1000\sqrt{3}$$
$$BE = 2000\sqrt{3} \text{ lbf}$$

Answer is (E)

STATICS–20

When loaded, which support has a reaction involving more than a single force?

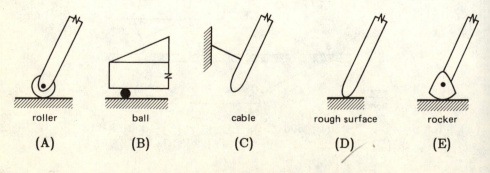

| roller | ball | cable | rough surface | rocker |
| (A) | (B) | (C) | (D) | (E) |

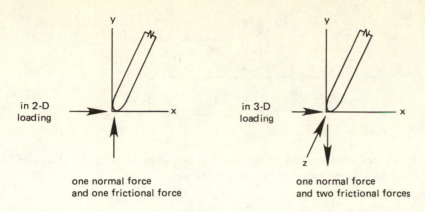

in 2-D
loading

one normal force
and one frictional force

in 3-D
loading

one normal force
and two frictional forces

(A), (B), and (E) have a normal force. Only (C) has a single force
along the cable. Only (D) can support two reaction forces.

Answer is (D)

STATICS–21

A beam, securely fastened to a wall, is subjected to three-dimensional loading.
How many components of reaction are possible?

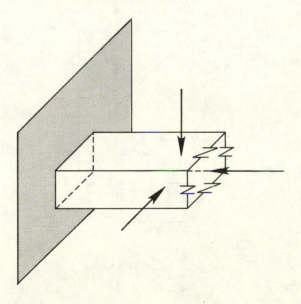

(A) two   (B) three   (C) four   (D) five   (E) six

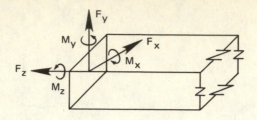

There are six components of reaction possible: three components of force and three components of momentum.

Answer is (E)

STATICS-22

Determine the vertical force per unit width at F necessary to resist the wind load of 0.5 psi on DEF.

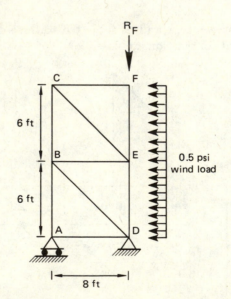

(A) 216 lbf/ft   (B) 432 lbf/ft   (C) 648 lbf/ft

          (D) 864 lbf/ft   (E) 1150 lbf/ft

$$\sum M_A = 0$$

$$0 = \int_0^{12} \left(0.5\,\frac{lbf}{in^2}\right)\left(144\,\frac{in^2}{ft^2}\right) y\,dy - 8R_F$$

$$8R_F = 5184$$

$$R_F = 648\ \text{lbf/ft}$$

Answer is (C)

0.5 × 12 ×12 × 6×12 - 8 RF

= 144 × 0·5 ×72 = 8 Ft

Ft = 144 × /12 × 72/8

STATICS–23

Compute the equilibrium displacement of the spring in the illustration shown.

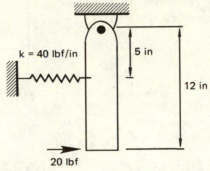

(A) 0 in   (B) 0.83 in   (C) 1.2 in

(D) 2 in   (E) 4.8 in

Sum the moments around the hinge.

$$\sum M_H = 0$$

$$0 = (F_{spring})(5) + (20)(12)$$

$$F_{spring} = -48\ \text{lbf}\quad \text{(opposite in direction to the 20 lbf load)}$$

$$= -k\delta_{spring}$$

$$\delta_{spring} = -\frac{F_{spring}}{k}$$

$$= \frac{48}{40}$$

$$= 1.2\ \text{in}$$

Answer is (C)

STATICS–24

The system shown is in equilibrium prior to the application of the 30 lbf force. After equilibrium is reestablished, what is the displacement at A?

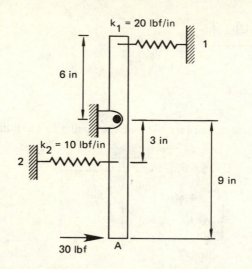

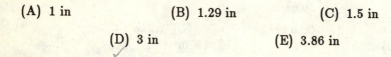

(A) 1 in   (B) 1.29 in   (C) 1.5 in

        (D) 3 in   (E) 3.86 in

Sum the moments around the hinge, and use Hooke's law ($F = -k\delta$) to find the spring forces.

$$\sum M_H = 0$$

$$0 = (-k_1\delta_1)(6) + (30)(9) - (k_2\delta_2)(3)$$

The ratio of the deflection of the bar at a point to the bar's distance from the hinge is equal to the angular displacement of the bar. Since the bar does not bend, this ratio is the same for any point on the bar.

Therefore, $\dfrac{\delta_1}{6} = \dfrac{\delta_2}{3}$

$$\delta_1 = 2\delta_2$$

Substitute for δ_1 in the equation for the moment.

$$0 = -(20)(2\delta_2)(6) + (30)(9) - (10)(\delta_2)(3)$$

$$\delta_2 = -\frac{(30)(9)}{-240 - 30}$$

$$= \frac{270}{270}$$

$$= 1 \text{ in}$$

$$\frac{\delta_A}{9} = \frac{\delta_2}{3}$$

$$\delta_A = 3\delta_2$$

$$= 3 \text{ in}$$

Answer is (D)

STATICS–25

What is the reaction force at support B on the simply supported beam with a linearly varying load?

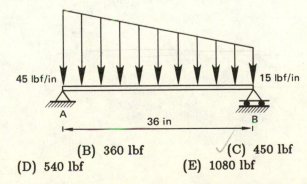

45 lbf/in   15 lbf/in   A   36 in   B

(A) 180 lbf   (B) 360 lbf   (C) 450 lbf
(D) 540 lbf   (E) 1080 lbf

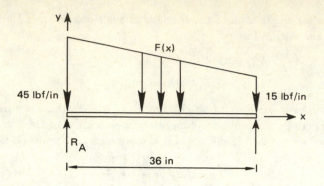

$$F(x) = 45\,\frac{\text{lbf}}{\text{in}} - \left(45\,\frac{\text{lbf}}{\text{in}} - 15\,\frac{\text{lbf}}{\text{in}}\right)\left(\frac{x}{36}\right)$$

$$= 45 - \frac{5x}{6}\ \text{lbf/in}$$

Sum the moments around support A.

$$M_A = 0$$

$$0 = -\int_0^{36} F(x) \cdot x\,dx + (R_B)(36)$$

$$R_B = \frac{1}{36}\int_0^{36}\left(45 - \frac{5x}{6}\right)x\,dx$$

$$= \frac{45}{36}\int_0^{36} x\,dx - \frac{5}{216}\int_0^{36} x^2\,dx$$

$$= \frac{1}{36}\left[\frac{45x^2}{2} - \frac{5x^3}{18}\right]_0^{36}$$

$$= \frac{1}{36}\left[\frac{(45)(36)^2}{2} - \frac{(5)(36)^3}{18}\right]$$

$$= (45)(18) - (10)(36)$$

$$= 450\ \text{lbf}$$

Answer is (C)

STATICS–26

For the simply supported beam with the linearly varying load shown, what is the sum of the reactions at the supports?

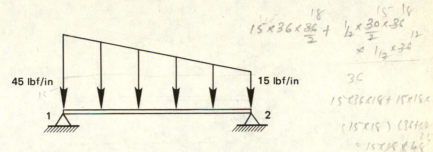

45 lbf/in   15 lbf/in

1   2

(A) 540 lbf   (B) 1080 lbf   (C) 1590 lbf
            (D) 1605 lbf   (E) 2160 lbf

$$\sum F_y = 0$$

$$0 = \int_0^{36} F(x)\,dx - R_1 - R_2$$

$$R_1 + R_2 = \int_0^{36} F(x)\,dx$$

$$F(x) = 45\,\frac{\text{lbf}}{\text{in}} - \left(45\,\frac{\text{lbf}}{\text{in}} - 15\,\frac{\text{lbf}}{\text{in}}\right)\frac{x}{36}$$

$$= \left(45 - \frac{5x}{6}\right)\frac{\text{lbf}}{\text{in}}$$

$$R_1 + R_2 = \int_0^{36} \left(45 - \frac{5x}{6}\right)dx$$

$$= (45)(36) - \frac{5x^2}{12}\Big|_0^{36}$$

$$= (45)(36) - \frac{(5)(36)^2}{12}$$

$$= (45)(36) - (15)(36)$$

$$= 1080\ \text{lbf}$$

Answer is (B)

STATICS-27

A beam is subjected to a distributed load as shown. Determine the reactions at the right support, A.

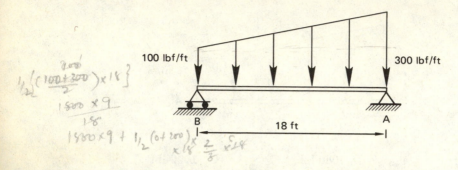

100 lbf/ft

300 lbf/ft

B   18 ft   A

(handwritten notes in left and lower margin:)

$\frac{1}{2}\left(\frac{100+300}{2}\right)\times 18$

1800×9

$18'$

$1800\times 9 + \frac{1}{2}(0+200)\times 18 \cdot \frac{2}{3}\times 24$

(A) 2100 lbf   (B) 2200 lbf   (C) 2300 lbf
          (D) 2500 lbf   (E) 2600 lbf

$$\sum F_x = 0$$

$$R_{Ax} = 0$$

$$\sum M_B = 0$$

$$0 = (R_{Ay})(18) - \int_0^{18} F(x)x\,dx$$

$$F(x) = 100 + \frac{300 - 100}{18}x$$

$$= 100 + \frac{100\,x}{9}$$

$$R_{Ay} = \frac{1}{18}\int_0^{18}\left(\frac{100x}{9} + 100\right)x\,dx$$

$$= \frac{1}{18}\int_0^{18}\left(\frac{100x^2}{9} + 100x\right)dx$$

$$= \frac{1}{18}\left[\frac{100x^3}{27} + 50x^2\right]_0^{18}$$

$$= \frac{(100)(18)^2}{27} + (50)(18)$$

$$= 2100 \text{ lbf}$$

Since $R_{Ax} = 0$, $R_{Ay} = R_A$. Thus, $R_A = 2100$ lbf.

Answer is (A)

STATICS-28

The beam in the figure is hinged at the wall and loaded as shown. What is the tension in the cable?

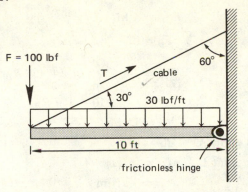

F = 100 lbf
T
cable
60°
30°
30 lbf/ft
10 ft
frictionless hinge

(A) 200 lbf   (B) 250 lbf   (C) 430 lbf
(D) 500 lbf   (E) 725 lbf

$$\sum M_{\text{hinge}} = 0$$

$$0 = (100)(10) + \int_0^{10} 30x\,dx - T(\sin 30°)(10)$$

$$= 1000 + 15x^2 \Big|_0^{10} - 5T$$

$$5T = 1000 + (30)(50)$$

$$T = 500 \text{ lbf}$$

Answer is (D)

(handwritten, right margin):
$100 \times 10 + T\sin 30 \times 10$
$- 30 \times 10 \times 5$
$T\sin 30 \times 10 = 1000 + 1500$
$T/2 = \dfrac{2500}{10}$
$T = 500$

STATICS-29

Determine the position of maximum moment in the beam ABC.

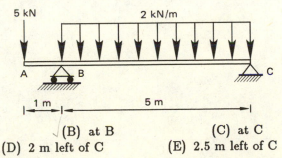

5 kN
2 kN/m
A
B
C
1 m
5 m

(A) at A   (B) at B   (C) at C
(D) 2 m left of C   (E) 2.5 m left of C

$$\sum M_B = 0$$

$$0 = (5)(1) - \int_0^5 2x\,dx + (R_{Cy})(5)$$

$$R_{Cy} = 4 \text{ kN}$$

$$\sum F_y = 0$$

$$0 = -5 - \int_0^5 2\,dx + 4 + R_B$$

$$= -5 - 10 + 4 + R_B$$

$$R_B = 11 \text{ kN}$$

at point A, $M_A = 0$

at point B, $M_B = (5)(1)$

$$= 5 \text{ kN·m}$$

from C to B, $M_x = (R_{Cy})x - \int_0^x 2x\,dx$

$$= 4x - \frac{2x^2}{2}$$

$$= 4x - 2x^2$$

$$\frac{dM_x}{dx} = 4 - 2x$$

$$= 0 \quad \text{where } M_x \text{ is a max}$$

$$0 = 4 - 2x$$

$$x = 2 \text{ m}$$

$$M_{x,\text{max}} = 4(2) - 4$$

$$= 4 \text{ kN·m} < M_B$$

Thus, the maximum moment is 5 kN · m, and occurs at point B.

Answer is (B)

STATICS–30

For the cantilever beam with the distributed load shown, what is the moment at the support?

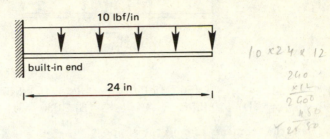

10 lbf/in

built-in end

24 in

$10 \times 24 \times 12$

240
$\times 12$
2600
480
2880

(A) 240 lbf   (B) 480 lbf-in   (C) 960 lbf-in

(D) 2880 lbf-in   (E) 3480 lbf-in

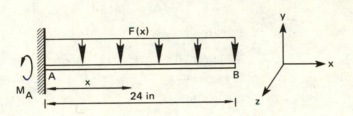

$F(x)$

A x B

M_A

24 in

$$\sum M_A = 0$$

$$0 = M_A - \int_0^{24} F(x)x\,dx$$

$$M_A = \int_0^{24} F(x)x\,dx$$

$$= \int_0^{24} 10x\,dx$$

$$= 5x^2 \Big]_0^{24}$$

$$= (5)(24)^2$$

$$= 2880 \text{ lbf-in}$$

Answer is (D)

STATICS-31

For the cantilever beam shown, what is the moment acting at the support?

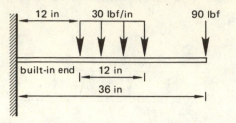

(A) 270 lbf-ft   (B) 315 lbf-ft   (C) 540 lbf-ft
          (D) 810 lbf-ft   (E) 1080 lbf-ft

Sum the moments around the support.

$$\sum M_S = M_S - (90)(36) - \int_{12}^{24} 30x\,dx = 0$$

$$M_S = (90)(36) + \int_{12}^{24} 30x\,dx$$

$$= (90)(36) + 15x^2 \Big]_{12}^{24}$$

$$= (90)(36) + 15(24^2 - 12^2)$$

$$= 3240 + 6480$$

$$= 9720 \text{ lbf-in}$$

$$= 810 \text{ lbf-ft}$$

Answer is (D)

STATICS-32

Determine the resultant moment at A for the cantilever beam shown.

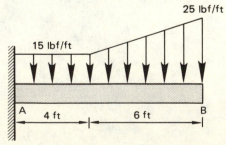

(A) 660 ft-lbf   (B) 990 ft-lbf   (C) 1080 ft-lbf
          (D) 1170 ft-lbf   (E) 1230 ft-lbf

Break the beam into two sections: from 0 to 4 feet where $F_1 = 15$ lbf/ft, and from 4 to 10 feet where $F_2 = 5(x+5)/3$ lbf/ft. Sum the moments around point A.

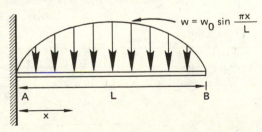

$$\sum M_A = 0$$

$$0 = M_A - \int_0^4 F_1(x)x\,dx - \int_4^{10} F_2(x)x\,dx$$

$$M_A = \int_0^4 15x\,dx + \int_4^{10} \frac{5(x+5)}{3}x\,dx$$

$$= \frac{15x^2}{2}\bigg]_0^4 + \left[\frac{5x^3}{9} + \frac{25x^2}{6}\right]_4^{10}$$

$$= (15)(8) + \frac{5(10^3 - 4^3)}{9} + \frac{25(10^2 - 4^2)}{6}$$

$$= 990 \text{ ft-lbf}$$

Answer is (B)

STATICS–33

Determine the moment at A for the beam shown.

$$w = w_0 \sin \frac{\pi x}{L}$$

(A) $\dfrac{w_0 L^2}{\pi}$

(B) $w_0 L$

(C) $\dfrac{2w_0 L^2}{\pi}$

(D) $\dfrac{w_0 L^2}{2\pi}$

(E) $\dfrac{w_0 L^2 \cos \dfrac{\pi}{4}}{\pi}$

$$\sum M_A = 0$$

$$0 = M_A - \int_0^L \left(w_0 \sin \frac{\pi x}{L}\right) x \, dx$$

$$M_A = \int_0^L \left(w_0 \sin \frac{\pi x}{L}\right) x \, dx$$

Integrate by parts.

$$u = x \qquad dv = \left(w_0 \sin \frac{\pi x}{L}\right) dx$$

$$du = dx \qquad v = -\frac{w_0 L}{\pi} \cos \frac{\pi x}{L}$$

$$M_A = \left[-\frac{x w_0 L}{\pi} \cos \frac{\pi x}{L}\right]_0^L + \frac{w_0 L}{\pi} \int_0^L \left(\cos \frac{\pi x}{L}\right) dx$$

$$= -\left(\frac{L^2 w_0}{\pi}\right)\left(\cos \pi - \cos 0\right) + \left[\left(\frac{w_0 L^2}{\pi^2}\right) \sin \frac{\pi x}{L}\right]_0^L$$

$$= -\left(\frac{L^2 w_0}{\pi}\right)\left(-1 - 1\right) + \left(\frac{w_0 L^2}{\pi^2}\right)\left(0 - 0\right)$$

$$= \frac{2 w_0 L^2}{\pi}$$

Answer is (C)

STATICS–34

The beam shown is statically indeterminate, but the reaction force at B is known to be 40 kips. What is the bending moment at point P?

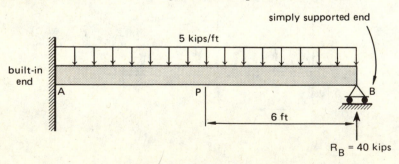

(A) 90 ft-kips   (B) 150 ft-kips   (C) 240 ft-kips
          (D) 300 ft-kips   (E) 330 ft-kips

Measure x from point B.

The moment at point P is:

$$M_P = R_B(6) - \int_0^6 F(x) \cdot x\,dx$$

$$= (40)(6) - \int_0^6 5x\,dx$$

$$= 240 - \frac{5x^2}{2}\bigg]_0^6$$

$$= 240 - 90$$

$$= 150 \text{ ft-kips}$$

Answer is (B)

STATICS–35

Determine the height of the centroid, $\bar{y}$, of the semi-circle with radius r.

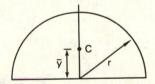

(A) $\bar{y} = \frac{2}{3}r$ (B) $\bar{y} = \frac{2\pi}{5}r$ (C) $\bar{y} = \frac{4}{3\pi}r$

(D) $\bar{y} = \frac{3}{4}r$ (E) $\bar{y} = \frac{3}{4\pi}r$

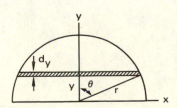

By definition, $\bar{y}A = \int ydA$

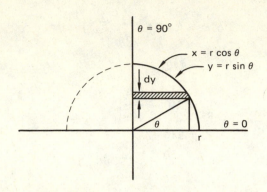

$A = \frac{1}{2}\pi r^2$ for the semi-circle

$x = r\cos\theta$

$y = r\sin\theta$

$dy = d(r\sin\theta) = r\cos\theta\,d\theta$

$dA = x\,dy = r\cos\theta\,r\cos\theta\,d\theta$

Since the area is symmetrical, area for $0 \le \theta \le 90°$ is half the total area.

$$\int ydA = 2\int_0^{\frac{\pi}{2}} (r\sin\theta)(r\cos\theta)(r\cos\theta)\,d\theta$$

$$= 2\int_0^{\frac{\pi}{2}} r^3\sin\theta\cos^2\theta\,d\theta$$

$$= \frac{-2r^3\cos^3\theta}{3}\Bigg]_0^{\frac{\pi}{2}}$$

$$= \frac{2r^3}{3}$$

$$\bar{y}A = \frac{2r^3}{3}$$

$$\bar{y} = \frac{\dfrac{2r^3}{3}}{\dfrac{\pi r^2}{2}}$$

$$= \frac{4r}{3\pi}$$

Answer is (C)

STATICS-36

What is the height of the center of mass of the cone-sphere system?

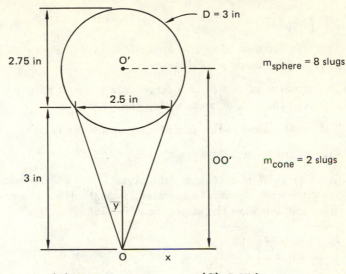

(A) 2.88 in   (B) 3.25 in   (C) 3.85 in

      (D) 4.05 in   (E) 4.25 in

By symmetry, the center of mass of the system is on the y-axis. First, find the height of the center of mass for each part of the system.

$$OO' = \text{height of sphere's center of mass}$$
$$= 3 + 2.75 - r_{\text{sphere}}$$
$$= 5.75 - 1.5$$
$$= 4.25 \text{ in}$$

$$h_c = \text{height of cone's center of mass}$$
$$= \left(\frac{3}{4}\right)(\text{height of the cone})$$
$$= \left(\frac{3}{4}\right)(3)$$
$$= 2.25 \text{ in}$$

$$h = \frac{(m_{\text{sphere}})(OO') + (m_{\text{cone}})(h_c)}{m_{\text{sphere}} + m_{\text{cone}}}$$
$$= \frac{(8)(4.25) + (2)(2.25)}{8 + 2}$$
$$= 3.85 \text{ in}$$

Answer is (C)

STATICS–37

Which statement about area moments of inertia is false?

(A) $I = \int d^2(dA)$

(B) The area moment of inertia arises whenever the magnitude of the surface force varies linearly with distance.

(C) The moment of inertia of a large area is equal to the summation of the inertia of the smaller areas within the large area.

(D) The areas closest to the axis of interest are the most significant.

(E) Area moments of inertia are always positive.

Area moment of inertia is defined as $I = \int d^2(dA)$ where d is the distance from the axis to the area element. Thus, the areas farthest from the axis have the largest contribution.

Answer is (D)

STATICS–38

Determine the moment of inertia around the centroidal axis of the following beam.

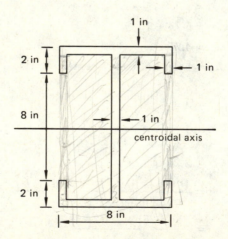

(A) 416 in⁴   (B) 650 in⁴   (C) 735 in⁴
         (D) 1067 in⁴   (E) 1152 in⁴

The moment of inertia of the beam is the moment of inertia of a solid beam with the same dimensions (a height of 12″ and a width of 8″) minus the moments of inertia around the central axis of the missing sections.

$$I = \left(\frac{1}{12}\right)(8)(12)^3 - 2\left[\left(\frac{1}{12}\right)(2.5)(10)^3\right]$$

$$-\frac{(8)(12)^3}{12} - 2\left[\frac{(2.5)(10)^3}{12}\right]$$

$$-2\left[\left(\frac{1}{12}\right)(1)(8)^3\right]$$

$$-2\left[\frac{(1)(8)^3}{12}\right]$$

$$= 1152 - 416.7 - 85.3$$

$$= 650 \text{ in}^4$$

Answer is (B)

9

MATERIALS SCIENCE

MATERIALS SCIENCE–1

Which of the following affects most of the engineering properties of materials, such as mechanical and strength characteristics?

(A) the atomic weight expressed in grams per gram-atom
(B) the electrons, particularly the outermost ones
(C) the magnitude of electrical charge of the protons
(D) the weight of the atoms
(E) the weight of the protons

> The outermost electrons are responsible for determining most of the material's properties.

> Answer is (B)

MATERIALS SCIENCE–2

The atomic weight of hydrogen is 1 gram per gram-atom. What is the mass of a hydrogen atom?

(A) 1.66×10^{-24} g/atom
(B) 6.02×10^{-23} g/atom
(C) 1.0×10^{-10} g/atom
(D) 1 g/atom
(E) The mass is too small to calculate.

By definition, the mass of an atom is its atomic weight divided by Avogadro's number.

$$W = \frac{1}{6.02 \times 10^{23}} = 1.66 \times 10^{-24} \text{ g/atom}$$

Answer is (A)

MATERIALS SCIENCE-3

What are valence electrons?

(A) the outershell electrons
(B) electrons with positive charge
(C) the electrons of complete quantum shells
(D) the K-quantum shell electrons
(E) the M-quantum shell electrons

By definition, the outermost electrons are the valence electrons.

Answer is (A)

MATERIALS SCIENCE-4

What is the strong bond between hydrogen atoms known as?

(A) the ionic bond
(B) the metallic bond
(C) ionic and metallic bonds
(D) the covalent bond
(E) There is no such bond.

Covalent bonds provide the strongest attractive forces between atoms.

Answer is (D)

MATERIALS SCIENCE–5

What are Van der Waals forces?

(A) weak secondary bonds between atoms
(B) primary bonds between atoms
(C) forces between electrons and protons
(D) forces not present in liquids
(E) forces present only in gases

By definition, Van der Waals forces are weak attractive forces between molecules.

Answer is (A)

MATERIALS SCIENCE–6

Which of the following curves best illustrates the relationship between inter-atomic forces and interatomic spacing?

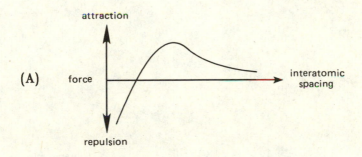

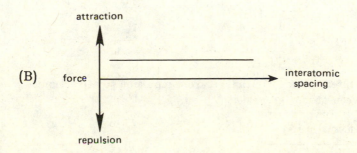

(C)

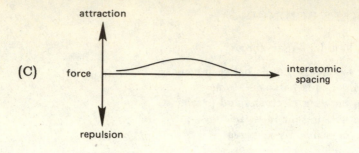

(D)

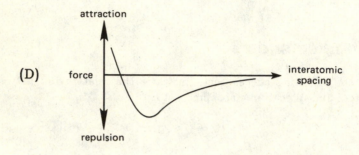

(E)

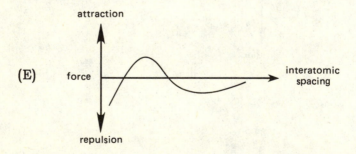

The interatomic force changes from repulsion to attraction as spacing between atoms increases.

Answer is (A)

MATERIALS SCIENCE–7

Compare the metallic iron atom Fe and the ferric ions Fe^{2+} and Fe^{3+}. Which has the smallest atomic radius? (Assume all are at the same temperature.)

(A) Fe
(B) Fe^{2+}
(C) Fe^{3+}
(D) They have the same radii.
(E) There is not enough information for comparison.

Ionizing removes valence electrons, causing the remaining electrons to be pulled in closer to the nucleus. Further reduction in spacing occurs with the removal of more electrons.

Answer is (C)

MATERIALS SCIENCE–8

Cesium (Cs) and sodium (Na) both have the same valence (+1), yet with chlorine (Cl), cesium has a coordination number of 8 in CsCl, while sodium has a coordination number of only 6 in NaCl. What is the main reason for this difference?

(A) The atomic weight of Cs is larger than the weight of Na.
(B) Cs forms covalent bonds in CsCl.
(C) Cs contains more electrons than Na.
(D) Cs is too large to be coordinated by only 6 chloride ions.
(E) All of the above are true.

Since the Cl atoms are of constant size, the larger coordination number for Cs means that more Cl atoms are needed to fit around a Cs atom than around a Na atom. Therefore, the Cs atom is larger than the Na atom.

Answer is (D)

MATERIALS SCIENCE-9

Which of the following statements is false?

(A) Ceramics are inorganic, nonmetallic solids that are processed or used at high temperatures.

(B) Metals are chemical elements that form substances that are opaque, lustrous, and good conductors of heat and electricity.

(C) Oxides, carbides, and nitrides are considered to be within the class of materials known as glasses.

(D) Most metals are strong, ductile, and malleable. In general, they are heavier than most other substances.

(E) Polymers are formed from many atoms with low molecular weight, bonded together by primary valence bonds.

The classes of materials are ceramics, metals, and polymers. Oxides, carbides, nitrides, and glasses are all ceramics.

Answer is (C)

MATERIALS SCIENCE-10

Which of the following materials is not a viscoelastic material?

(A) plastic   (B) metal   (C) rubber

(D) glass   (E) concrete

A material which is viscoelastic exhibits time-dependent elastic strain. Of the choices, only metal does not fit this description. Metal is considered to be an elastoplastic material.

Answer is (B)

MATERIALS SCIENCE-11

In molecules of the same composition, what are variations of atomic arrangements known as?

(A) polymers
(B) noncrystalline structures
(C) monomers
(D) crystal systems
(E) isomers

Isomers are molecules that have the same composition but different atomic arrangements.

Answer is (E)

MATERIALS SCIENCE-12

Which of the following accurately describes differences between crystalline polymers and simple crystals?

I. Crystalline polymers are made of folded chains of atoms unlike simple crystals.
II. Crystal size can be increased by raising the crystallization temperature only in polymers.
III. While a simple crystal may be totally crystallized, a polymer can reach only partial crystallization.

(A) I only   (B) II only   (C) III only

(D) I and II   (E) I and III

Only crystalline polymers are composed of folded chains and at best exhibit partial crystallization. The crystal size of both simple crystals and polymers can be increased by raising temperature.

Answer is (E)

MATERIALS SCIENCE-13

Polymers that favor crystallization are least likely to have which of the following?

(A) an atactic configuration of side groups
(B) small side groups
(C) only one repeating unit
(D) small chain lengths
(E) an isotactic arrangement of side groups

In order for crystallization to be favored, the molecules must be able to arrange themselves into an orderly structure. Atactic refers to a random configuration of side groups in the polymer; this would hinder crystallization.

Answer is (A)

MATERIALS SCIENCE-14

What is the atomic packing factor for a simple cubic crystal?

(A) 0.48   (B) 0.52   (C) 1.00   (D) 1.05   (E) 1.92

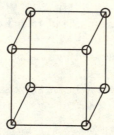

For a simple cubic crystal, there is one complete atom of radius r per cell. The cell has edges of length $2r$. By definition:

$$\text{atomic packing factor} = \frac{\text{volume of atoms}}{\text{volume of unit cell}}$$

$$= \frac{\frac{4\pi(r)^3}{3}}{(2r)^3}$$

$$= 0.52$$

Answer is (B)

MATERIALS SCIENCE–15

How many atoms are in the unit cell of a body-centered cubic structure?

(A) 1 (B) 2 (C) 3 (D) 4 (E) 9

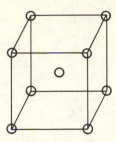

There is 1 atom at the center position and $\frac{1}{8}$ of an atom at each of the corners of the cube, since the atom present at each corner is shared by the adjoining unit cells. Therefore,

$$\text{total number of atoms} = 1 + \left(\tfrac{1}{8}\right)(8) = 2 \text{ atoms/cell}$$

Answer is (B)

MATERIALS SCIENCE–16

How many atoms are there per unit cell for a face-centered cubic structure?

(A) 1 (B) 2 (C) 3 (D) 4 (E) 8

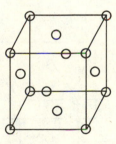

Like a body-centered cubic structure, there is $\frac{1}{8}$ of an atom at each corner of the cube. There is also $\frac{1}{2}$ of an atom at the center of each of the six faces, since each atom here is shared by the neighboring unit cell.

$$\text{total number of atoms} = \left(\tfrac{1}{8}\right)(8) + \left(\tfrac{1}{2}\right)(6) = 1 + 3 = 4 \text{ atoms/cell}$$

Answer is (D)

MATERIALS SCIENCE–17

What is the first coordination number of a body-centered cubic structure?

(A) 4   (B) 6   (C) 8   (D) 10   (E) 12

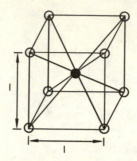

The first coordination number is the number of nearest neighbor atoms. In a body-centered cubic cell of edge length l, the minimum distance between atoms is $\frac{\sqrt{3}}{2}l$. By inspection of the figure, there are 8 neighboring atoms at this distance.

Answer is (C)

MATERIALS SCIENCE–18

What is the first coordination number of a face-centered cubic structure?

(A) 2   (B) 4   (C) 8   (D) 12   (E) 16

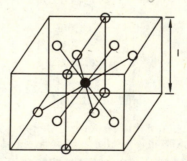

The closest atoms in a face-centered cubic cell of edge length l are $l/\sqrt{2}$ apart. Each atom in the center of a face has 12 such neighboring atoms. The atoms are: $(\pm\frac{1}{2}l, \pm\frac{1}{2}l, 0)$ $(\pm\frac{1}{2}l, 0, \pm\frac{1}{2}l)$ $(0, \pm\frac{1}{2}l, \pm\frac{1}{2}l)$.

Answer is (D)

MATERIALS SCIENCE–19

Which of the following statements is false?

A) Both copper and aluminum have a face-centered cubic crystal structure.

B) Both magnesium and zinc have a hexagonal close-packed crystal structure.

(C) Iron can have either a face-centered or a body-centered cubic crystal structure.

(D) All of the alkali metals have a body-centered cubic crystal structure.

(E) Both lead and cadmium have a hexagonal close-packed crystal structure.

Lead does not have a hexagonal close-packed structure. Its structure is face-centered cubic.

Answer is (E)

MATERIALS SCIENCE–20

Which of the following statements is false?

(A) The coordinates of the unique lattice points for a body-centered cubic unit cell are: 0,0,0 and $\frac{1}{2},\frac{1}{2},\frac{1}{2}$.

(B) The coordinates of the unique lattice points for a face-centered cubic unit cell are: 0,0,0; $\frac{1}{2},\frac{1}{2},0$; $\frac{1}{2},0,\frac{1}{2}$; and 0,$\frac{1}{2},\frac{1}{2}$.

(C) The coordinates of the unique lattice points for a simple cubic unit cell are: 0,0,0.

(D) The coordinates of the unique lattice points for a hexagonal unit cell are: 0,0,0.

(E) The coordinates of the unique lattice points for a rhombohedral unit cell are: $\frac{1}{2},\frac{1}{2},\frac{1}{2}$.

The rhombohedral Bravais lattice is a primitive cell and has only the point 0,0,0.

Answer is (E)

MATERIALS SCIENCE–21

How are the close-packed planes in a face-centered cubic metal designated?

(A) (100)   (B) (200)   (C) (110)   (D) (220)   (E) (111)

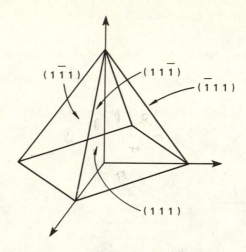

The close-packed planes are as shown.

Answer is (E)

MATERIALS SCIENCE–22

Which crystal structure possesses the highest number of close-packed planes and close-packed directions?

(A) simple cubic
(B) body-centered cubic
(C) face-centered cubic
(D) close-packed hexagonal
(E) rhombohedral

The face-centered cubic structure has four close-packed planes: (111), ($\bar{1}$11), (1$\bar{1}$1), and (11$\bar{1}$). Each plane has 3 close-packed directions.

Answer is (C)

PROFESSIONAL PUBLICATIONS, INC. • Belmont, CA

MATERIALS SCIENCE–23

What are the most common slip planes for face-centered cubic and body-centered cubic structures, respectively?

(A) face-centered: (111), body-centered: (110)
(B) face-centered: (100), body-centered: (110)
(C) face-centered: (110), body-centered: (111)
(D) face-centered: (111), body-centered: (100)
(E) face-centered: (100), body-centered: (111)

Slip planes are usually the most closely packed planes, since they have the largest spacing. The close-packed planes are (111) and (110) for the respective crystal structures.

Answer is (A)

MATERIALS SCIENCE–24

Comparing the face-centered cubic lattice with the hexagonal close-packed lattice, which of the following features describes the hexagonal close-packed structure only?

(A) It has the closest packed lattice structure.
(B) Its coordination number is 12.
(C) Its deformation properties are more directional.
(D) Its stacking order is ABCABC.
(E) It has four octahedral planes.

(A) and (B) are true for both face-centered cubic and hexagonal close-packed structures, while (D) and (E) are true for the face-centered cubic lattice. (C) applies to the hexagonal close-packed lattice only.

Answer is (C)

MATERIALS SCIENCE-25

In the following unit cell, what direction is indicated by the arrow?

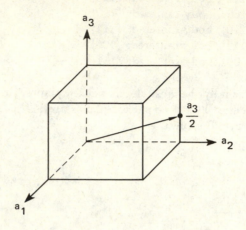

(A) [012]

(B) [01$\frac{1}{2}$]

(C) [210]

(D) [0$\frac{1}{2}$1]

(E) [121]

Direction is given by the Miller indices:

$$\left(\frac{x_1}{a_1}\right)^{-1}\left(\frac{x_2}{a_2}\right)^{-1}\left(\frac{x_3}{a_3}\right)^{-1}$$

x_1, x_2, and x_3 are the intercepts of a given direction with a_1, a_2, or a_3. If there is no intercept for an axis, its x-intercept goes to infinity. Therefore,

$$\left[\left(\frac{x_1}{a_1}\right)^{-1}\left(\frac{x_2}{a_2}\right)^{-1}\left(\frac{x_3}{a_3}\right)^{-1}\right] = \left[\left(\frac{\infty}{a_1}\right)^{-1}\left(\frac{1}{1}\right)^{-1}\left(\frac{1}{2}\right)^{-1}\right]$$

$$= [012]$$

Answer is (A)

MATERIALS SCIENCE–26

What are the Miller indices of this plane?

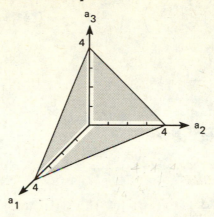

(A) (444)   (B) (111)   (C) $(\frac{1}{4}\frac{1}{4}\frac{1}{4})$ (D) (222)   (E) $(\frac{1}{2}\frac{1}{2}\frac{1}{2})$

The x, y, and z intercepts are a_1, a_2, and a_3, respectively. The Miller indices are:

$$\left(\left(\frac{a_1}{a_1}\right)^{-1} \left(\frac{a_2}{a_2}\right)^{-1} \left(\frac{a_3}{a_3}\right)^{-1} \right) = (111)$$

Answer is (B)

MATERIALS SCIENCE–27

What are the Miller indices of this plane?

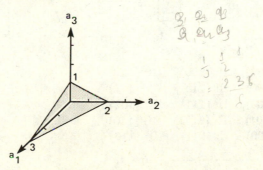

(A) (321)   (B) $(\frac{1}{3}\frac{1}{2}1)$ (C) $(\frac{2}{6}\frac{3}{6}\frac{6}{6})$ (D) (236)   (E) (123)

The intercepts for this plane are a_1, $(\frac{2}{3})a_2$, and $(\frac{1}{3})a_3$. The Miller indices are:

$$\left(\left(\frac{a_1}{a_1}\right)^{-1}\left(\frac{\frac{2a_2}{3}}{a_2}\right)^{-1}\left(\frac{\frac{a_3}{3}}{a_3}\right)^{-1}\right) = (1\tfrac{3}{2}3) = (236)$$

Answer is (D)

MATERIALS SCIENCE–28

A plane intercepts the coordinate axis at $x = 1$, $y = 3$, and $z = 2$. What are the Miller indices of the plane?

(A) (132)   (B) (123)   (C) (623)   (D) (326)   (E) (264)

The Miller indices are computed by taking the reciprocal of each intercept and converting to whole numbers of the same ratio:

$$(1\tfrac{1}{3}\tfrac{1}{2}) = (623)$$

Answer is (C)

MATERIALS SCIENCE–29

Which of the following gives the correct designations for the planes shown?

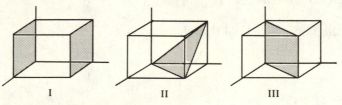

I   II   III

(A) I: (111),   II: (200),   III: (110)
(B) I: (100),   II: (110),   III: (111)
(C) I: (100),   II: (111),   III: (102)
(D) I: (101),   II: (213),   III: (110)
(E) I: (100),   II: (111),   III: (110)

Answer is (E)

MATERIALS SCIENCE-30

Using the 4-index scheme for a hexagonal crystal system, how would the directions **F** and **G** shown be defined?

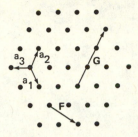

(A) $\mathbf{F} = [\bar{1}010]$, $\mathbf{G} = [1\bar{2}10]$
(B) $\mathbf{F} = [0110]$, $\mathbf{G} = [0300]$
(C) $\mathbf{F} = [0\bar{1}10]$, $\mathbf{G} = [\bar{1}3\bar{2}0]$
(D) $\mathbf{F} = [10\bar{1}0]$, $\mathbf{G} = [\bar{1}2\bar{1}0]$
(E) $\mathbf{F} = [1010]$, $\mathbf{G} = [1\bar{3}20]$

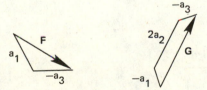

The **F** vector is the sum of one unit in the positive a_1 direction and one unit in the negative a_3 direction. The **G** vector is the sum of one unit in the negative a_1 direction, two units in the positive a_2 direction, and one unit in the negative a_3 direction.

Answer is (D)

MATERIALS SCIENCE-31

Given that a is a lattice constant and that h, k, and l are the Miller indices, which of the following equations describes the interplanar distance d in a cubic crystal?

(A) $d = \dfrac{2a}{\sqrt{\left(\dfrac{1}{h}\right)^2 + \left(\dfrac{1}{k}\right)^2 + \left(\dfrac{1}{l}\right)^2}}$

(B) $d = a\left(\dfrac{1}{h} + \dfrac{1}{k} + \dfrac{1}{l}\right)$

(C) $d = \left(\dfrac{a}{2}\right)\sqrt{h^2 + k^2 + l^2}$

(D) $d = \dfrac{a}{\sqrt{h^2 + k^2 + l^2}}$

(E) $d = 2a\sqrt{\left(\dfrac{1}{h}\right)^2 + \left(\dfrac{1}{k}\right)^2 + \left(\dfrac{1}{l}\right)^2}$

By geometry, it can be determined that $1/d^2 = (h^2 + k^2 + l^2)/a^2$. Therefore,

$$d = \frac{a}{\sqrt{h^2 + k^2 + l^2}}$$

Answer is (D)

MATERIALS SCIENCE-32

Calculate the theoretical density of copper given that the unit cell is face-centered cubic and the lattice parameter is 3.61 Å. The atomic weight of copper is 63.5 g/mole.

(A) 4.49 g/cm^3 (B) 7.86 g/cm^3 (C) 8.78 g/cm^3

(D) 8.97 g/cm^3 (E) 11.2 g/cm^3

There are 4 atoms per unit cell for a face-centered cubic structure. By definition,

$$\text{density} = \frac{\text{mass}}{\text{volume}}$$

$$= \frac{\left(4\dfrac{\text{atoms}}{\text{unit cell}}\right)\left(63.5\dfrac{\text{g}}{\text{mole}}\right)\left(\dfrac{1}{6.02 \times 10^{23}}\dfrac{\text{mole}}{\text{atom}}\right)}{\left[\left(3.61\ \text{Å}\right)\left(\dfrac{1 \times 10^{-8}\ \text{cm}}{\text{Å}}\right)\right]^3}$$

$$= 8.97\ \text{g/cm}^3$$

Answer is (D)

MATERIALS SCIENCE–33

Determine the planar density of copper atoms in the (100) plane given that the unit cell is face-centered cubic and the lattice parameter is 3.61 Å.

(A) 7.68×10^{18} atoms/m^2
(B) 1.53×10^{19} atoms/m^2
(C) 2.30×10^{19} atoms/m^2
(D) 3.84×10^{19} atoms/m^2
(E) 7.68×10^{19} atoms/m^2

There are two atoms total in the (100) plane. The planar density is therefore:

$$\text{planar density} = \frac{2}{(3.61 \times 10^{-10})^2} = 1.53 \times 10^{19}\ \text{atoms/m}^2$$

Answer is (B)

MATERIALS SCIENCE–34

Which of the following statements is not true regarding X-ray diffraction?

(A) The geometrical structure factor $F(hkl)$ is the ratio of the amplitude of the X-ray reflected from a plane in a crystal to the amplitude of the X-ray scattered from a single electron.

(B) X-ray diffraction is only useful for studying simpler crystals such as the body-centered cubic structure, rather than more complex crystals like the hexagonal close-packed structure.

(C) X-ray diffraction can be used to determine the grain size of a specimen.

(D) Bragg's law states that $\frac{n\lambda}{2d} = \sin\theta$ (n is an integer, λ is the wavelength of the X-ray, d is the interplanar spacing, and θ is the scattering angle).

(E) X-ray diffraction can be used to detect micro-stresses in a crystal.

X-ray diffraction is used to study all types of crystals. It is not limited to simple crystals.

Answer is (B)

MATERIALS SCIENCE–35

A sample of face-centered cubic nickel (Ni) was placed in an X-ray beam of wavelength $\lambda = 0.154$ nm. If the lattice parameter for Ni is $a_0 = 0.352$ nm, what is the first-order angle of diffraction?

(A) 5.68°   (B) 6.97°   (C) 12.6°   (D) 19.0°   (E) 22.9°

Using Bragg's law, with $n = 1$, $\lambda = 0.154$ nm, and $d = 0.352$ nm,

$$\lambda = 2d\sin\theta$$
$$\theta = \sin^{-1}\left[\frac{0.154}{(2)(0.352)}\right]$$
$$= 12.6°$$

Answer is (C)

MATERIALS SCIENCE–36

In a crystal structure, what is an interstitial atom?

(A) an extra atom sitting at a non-lattice point
(B) an atom missing at a lattice point
(C) a different element at a lattice point
(D) a line defect
(E) an atom that has identical surroundings with all other atoms in the crystal

An interstitial atom is an extra atom lodged within the crystal structure; it is a point defect.

Answer is (A)

MATERIALS SCIENCE–37

Which of the following is a line defect in a lattice crystal structure?

(A) tilt boundary
(B) screw dislocation
(C) vacancy
(D) Schottky imperfection
(E) Frenkel defect

The most common type of line defect is a dislocation.

Answer is (B)

MATERIALS SCIENCE–38

It is often desired to know the number of atoms, n, in a crystal structure that possess more than a specified amount of energy, E. Which of the following equations gives n, given that N is the total number of atoms present, M is a constant, k is the Boltzmann constant, and T is the temperature of the specimen?

(A) $n = \dfrac{M}{N}e^{-kE/T}$　　(B) $n = \dfrac{EM}{N}e^{-kT}$　　(C) $n = MNe^{-E/kT}$

(D) $n = MNe^{-kT/E}$　　(E) $n = \dfrac{N}{E}e^{-kMT}$

Equation (C) is the correct relationship for thermal energy distribution within a specimen.

Answer is (C)

MATERIALS SCIENCE-39

Which of the following describes diffusion in a crystal structure?

(A) It is not possible.
(B) It occurs only in alloys, never in pure crystals.
(C) It often uses an exchange or vacancy mechanism.
(D) It occurs primarily as a result of mechanical work.
(E) It does not affect material strength.

Diffusion is the movement of a defect from one point to another.

Answer is (C)

MATERIALS SCIENCE-40

What is Fick's first law for one-dimensional, steady state diffusion? C is the volume concentration of atoms, x is the distance along which diffusion occurs, D is the diffusion coefficient, and J is the flux or current density.

(A) $J = -D\dfrac{\partial C}{\partial x}$ (B) $J = C\dfrac{\partial D}{\partial x}$ (C) $J = -\dfrac{1}{D}\dfrac{\partial C}{\partial x}$

(D) $J = 2D\dfrac{\partial C}{\partial x}$ (E) $J = -\dfrac{C}{D}\dfrac{\partial C}{\partial x}$

The flux is proportional to the diffusion constant and the concentration gradient $\partial C/\partial x$.

Answer is (A)

MATERIALS SCIENCE-41

Which of the following are true about Fick's first law for diffusion?

I. It is only applicable to liquids, not solids.
II. The law states that the flux moves from high to low concentration.
III. J, the flux, may be in units of $cm^3/cm^2 \cdot s$.

(A) I only   (B) II only   (C) III only
(D) I and II   (E) II and III

Fick's law says that the flux of diffusion is proportional to the negative volume concentration gradient; the negative sign indicates that the flux is in the down-gradient direction. It applies to diffusion in a crystal. Flux, by definition, is the amount of volume moving across a unit surface area in unit time.

Answer is (E)

MATERIALS SCIENCE–42

What is the Arrhenius equation for the rate of a thermally activated process? (A = reaction constant, T = absolute temperature, R = gas constant, Q = activation energy)

(A) rate = $Ae^{-Q/RT}$
(B) rate = Ae^{-QRT}
(C) rate = $Ae^{Q/RT}$
(D) rate = Ae^{QRT}
(E) rate = $-Ae^{-RT/Q}$

The rate increases as the thermal energy increases.

Answer is (A)

MATERIALS SCIENCE–43

Which of the following statements is false?

(A) The surface energy of a liquid tends toward a minimum.

(B) The surface energy is the work required to create a unit area of additional space.

(C) The energy of an interior atom is greater than the energy of an atom on the surface of a liquid.

(D) Total surface energy is directly proportional to the surface area.

(E) For liquids, surface energy and surface tension have the same numerical value.

In a liquid, the energy of a surface atom is greater than the surface energy of an interior atom. Note: Although surface energy and tension have the same numerical value, they have different units.

Answer is (C)

MATERIALS SCIENCE-44

Which point on the stress-strain curve below gives the ultimate stress?

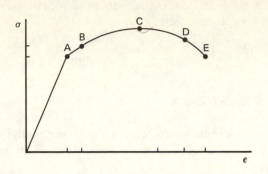

(A) A   (B) B   (C) C   (D) D   (E) E

The ultimate stress corresponds to the point of maximum load, beyond which further strain is accompanied by a reduction in load.

Answer is (C)

MATERIALS SCIENCE-45

A stress-strain diagram for a certain polymer is shown. Identify items A, B, and C.

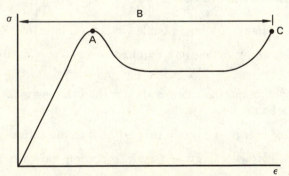

(A) A = lower yield point; B = plastic deformation; C = upper yield point
(B) A = lower yield point; B = proportional limit; C = upper yield point
(C) A = yield point; B = elastic deformation; C = elastic limit
(D) A = yield point; B = elongation at fracture; C = fracture
(E) A = upper yield point; B = yield zone; C = fracture

Beginning at the yield point, considerable elongation occurs with no noticeable increase in tensile stress. Eventually, fracture occurs. The total strain at fracture is known as the elongation.

Answer is (D)

MATERIALS SCIENCE–46

Which statement is true for the stress-strain relationship for the metal represented by the diagram below?

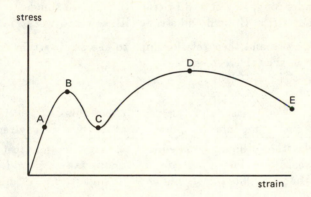

(A) Point A is the lower yield point.
(B) Point D is the fracture stress point.
(C) Point B is the upper yield point.
(D) The range from point C to point D is known as the elastic range.
(E) For the region up to point C, Hooke's law applies.

Point A is the elastic limit, and point D is the ultimate stress point. The region between points C and D is not the elastic, but the plastic region. Hooke's law does not apply in this plastic region. Only (C) is true.

Answer is (C)

MATERIALS SCIENCE–47

Identify the properties of the materials whose stress-strain diagrams are shown.

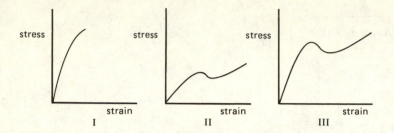

(A) I: soft and weak; II: soft and tough; III: hard and brittle
(B) I: hard and brittle; II: soft and weak; III: hard and tough
(C) I: soft and tough; II: hard and brittle; III: hard and strong
(D) I: hard and strong; II: soft and brittle; III: soft and tough
(E) I: hard and brittle; II: hard and strong; III: soft and weak

The properties and their relationships to the stress-strain diagrams are given in the table below:

	elastic modulus	yield point	elongation at fracture	ultimate strength
hard and brittle	high	undefined	low	moderate to high
soft and weak	low	low	moderate	low
hard and tough	high	high	high	high

Answer is (B)

MATERIALS SCIENCE–48

Which statement is most accurate regarding the two materials represented in the stress-strain diagrams below?

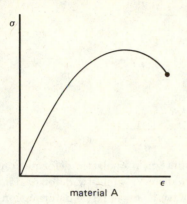

material A

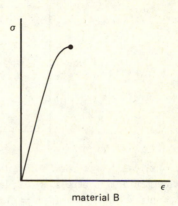

material B

(A) Material B is more ductile and has a lower modulus of elasticity than A.

(B) Material B would require more total energy to fracture than material A.

(C) Material A will withstand more stress before plastically deforming than material B.

(D) Material A is less stiff than material B but will withstand a higher load.

(E) Material B will withstand a higher load than material A but is more likely to fracture suddenly.

> From the graphs, the modulus of elasticity of material B is greater than that of material A. This means that material A is more ductile, that is, it can undergo more strain before fracturing. However, material B can withstand higher loads than material A. Only choice (E) is correct.

Answer is (E)

MATERIALS SCIENCE–49

If the diagram below represents deformation of rigid bodies, what do x, y, m, and n refer to?

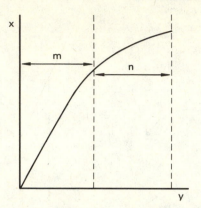

(A) x = stress, y = strain, m = plastic deformation, n = elastic deformation
(B) x = strain, y = stress, m = plastic deformation, n = elastic deformation
(C) x = stress, y = strain, m = elastic deformation, n = plastic deformation
(D) x = strain, y = stress, m = elastic deformation, n = plastic deformation
(E) x = stress, y = strain, m = plastic deformation, n = flow

(C) is the only choice that fits the graph.

Answer is (C)

MATERIALS SCIENCE–50

Which of the following best describes the 0.2% offset yield stress?

(A) It is the elastic limit after which a measurable plastic strain has occurred.
(B) It is the stress at which the material plastically strains 0.2%.
(C) It is the stress at which the material elastically strains 0.2%.
(D) It is 0.2% below the fracture point of the material.
(E) It is the proportional limit of $E = \sigma/\epsilon$.

By definition, the offset yield stress is where the material undergoes a 0.2% plastic strain.

Answer is (B)

MATERIALS SCIENCE–51

Which of the following is true regarding the ductile to brittle transition temperature?

I. It is important for structures used in cold environments.
II. It is the point at which the size of the shear lip or tearing rim goes to zero.
III. It is the temperature at which 20 joules of energy causes failure in a Charpy v-notch specimen of standard dimensions.

(A) I only   (B) I and II   (C) II and III
     (D) I and III   (E) I, II, and III

II is the only choice that is false. A test piece broken at 20 J of energy usually has a small shear lip.

Answer is (D)

MATERIALS SCIENCE–52

Which of the following are true regarding creep?

I. It is caused by the diffusion of vacancies to edge dislocations, permitting dislocation climb.
II. It involves the plastic deformation of materials at loads below the yield stress.
III. It may involve whole grain sliding.

(A) I only   (B) II only   (C) I and III
     (D) II and III   (E) I, II, and III

All are true.

Answer is (E)

MATERIALS SCIENCE–53

Under very slow deformation and at high temperature, it is possible to have some plastic flow in a crystal at a shear stress lower than the critical shear stress. What is this phenomenon called?

(A) slip   (B) twinning   (C) creep

        (D) bending   (E) shear

Creep involves the flow of material.

Answer is (C)

MATERIALS SCIENCE–54

What does the Charpy impact test measure?

I.   the energy required to break a test sample
II.   the strength of a test sample
III. the ductile to brittle transition point of metals

(A) I only   (B) II only   (C) III only

        (D) I and III   (E) I, II, and II

The Charpy test measures toughness, the energy required to break a sample. By conducting the test at different temperatures, the brittle transition temperature can be determined.

Answer is (D)

MATERIALS SCIENCE–55

A shaft made of good quality steel breaks in half due to fatigue. What would the surface of the fracture site look like?

(A) like a cup and cone

(B) quite smooth to the unaided eye, yet ripples are apparent under low-power magnification

(C) smooth over most of the surface although it appears torn at the location of fracture

(D) very jagged and rough

(E) pockmarked, with very small cavities

Typically, the surface is mostly smooth. Where final fracture took place however, the surface is torn.

Answer is (C)

MATERIALS SCIENCE-56

To which of the following can the large discrepancy between the actual and theoretical strengths of metals mainly be attributed?

(A) heat   (B) dislocations   (C) low density
        (D) stress direction   (E) large cracks

Although point defects do contribute to the discrepancy in strengths, the major reason for the difference is the presence of dislocations.

Answer is (B)

MATERIALS SCIENCE-57

The ease with which dislocations are able to move through a crystal under stress accounts for which of the following?

I.   ductility
II.   lower yield strength
III.   hardness

(A) I only   (B) II only   (C) III only
      (D) I and II   (E) II and III

The ease with which dislocations move through a crystal accounts for its ductility and lower yield strength.

Answer is (D)

MATERIALS SCIENCE-58

As the amount of slip increases, additional deformation becomes more difficult and decreases until the plastic flow finally stops. Slip may begin again only if a larger stress is applied. What is this phenomenon known as?

(A) cooling
(B) crowding
(C) strain hardening
(D) twinning
(E) elastic deformation

This is known as strain hardening.

Answer is (C)

MATERIALS SCIENCE-59

Which word combination best completes the following sentence?

"Plastic deformation of a single crystal occurs either by _____ or by _____, but _____ is the more common method."

(A) high pressure; high temperature; high pressure
(B) high temperature; high pressure; high temperature
(C) slip; twinning; slip
(D) twinning; slip; twinning
(E) compressive stress; tensile stress; compressive stress

Slip is a more common method of plastic deformation than twinning.

Answer is (C)

MATERIALS SCIENCE-60

Which one of these statements is true for twinning?

(A) It occurs at a lower shear stress than slip.
(B) It is the most significant form of plastic deformation.
(C) It cannot be caused by impact or thermal treatment.
(D) It rarely occurs in body-centered cubic structures.
(E) It frequently occurs in hexagonal close-packed structures.

Choices (A), (B), (C), and (D) are false. Twinning requires a relatively high shear stress, is much less common than slip, and can be caused by impact or thermal treatment. It occurs in both body-centered cubic and hexagonal close-packed crystal structures.

Answer is (E)

MATERIALS SCIENCE–61

Which of the following does not produce vacancies, interstitial defects, or impurity defects in a material?

(A) plastic deformation
(B) slow equilibrium cooling
(C) quenching
(D) increasing the temperature (which increases atomic energy)
(E) irradiation with high energy particles

Slow equilibrium cooling does not create variations in the material.

Answer is (B)

MATERIALS SCIENCE–62

Which of the following are true statements about the modulus of elasticity, E?

(A) It is the same as the rupture modulus.
(B) It is the slope of the stress-strain diagram in the linearly elastic region.
(C) It is the ratio of stress to volumetric strain.
(D) Its value depends only on the temperature of the material.
(E) It is dimensionless.

The modulus of elasticity is equal to the ratio of stress to strain for a particular material. It is the slope of the stress-strain diagram in the linearly elastic region.

Answer is (B)

MATERIALS SCIENCE-63

What is the modulus of elasticity, E, for a composite material in which the fibers take up 20% of the total volume, and the load is applied parallel to the fibers as shown?

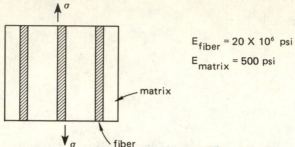

$E_{fiber} = 20 \times 10^6$ psi

$E_{matrix} = 500$ psi

— matrix

— fiber

(A) 4000 ksi   (B) 5630 ksi   (C) 7040 ksi
          (D) 7630 ksi   (E) 8600 ksi

The matrix and fibers experience the same strain, ϵ. The total stress, σ, is the sum of the stresses carried by the fibers and the matrix:

$$\sigma = E_f \epsilon V_f + E_m \epsilon (1 - V_f)$$

V_f is the fraction of the total volume taken up by the fibers. Thus,

$$E = \frac{\sigma}{\epsilon} = E_f V_f + E_m (1 - V_f)$$
$$= (20 \times 10^6)(0.2) + (500)(1 - 0.2)$$
$$= 4000 \text{ ksi}$$

Answer is (A)

MATERIALS SCIENCE-64

What is the proper relationship between the modulus of elasticity, E, the Poisson ratio, ν, and the bulk modulus of elasticity, K?

(A) $E = K(1 - 2\nu)$
(B) $E = K(1 - \nu)$
(C) $E = \dfrac{3K}{1 - 2\nu}$
(D) $E = 3K(1 - 2\nu)$
(E) $E = 3K(1 - \nu)$

For an element in triaxial stress, the unit volume change can be obtained from Hooke's law. The resultant equation is given by (D).

Answer is (D)

PROFESSIONAL PUBLICATIONS, INC. • Belmont, CA

MATERIALS SCIENCE–65

A crystal is subjected to a tensile load acting along its axis. α is the angle between the tensile axis and the slip plane as shown. At what value of α will the shear stress in the slip plane be a maximum?

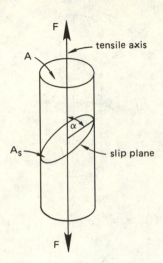

(A) $\alpha = 0°$ (B) $\alpha = 30°$ (C) $\alpha = 45°$
              (D) $\alpha = 60°$ (E) $\alpha = 90°$

The component of force along the shear surface is equal to $F \cos \alpha$. The area of the shear surface, A_s, is related to the cross-sectional area, A, by $A_s = A/\sin \alpha$. Using the equation for shear stress and maximizing with respect to α,

$$\tau = \frac{F \cos \alpha}{A/\sin \alpha}$$

$$= \left(\frac{F}{A}\right) \sin \alpha \cos \alpha$$

Taking the first derivative and setting it equal to zero,

$$\frac{\partial \tau}{\partial \alpha} = \left(\frac{F}{A}\right) (\cos^2 \alpha - \sin^2 \alpha) = 0$$

$$\cos^2 \alpha - \sin^2 \alpha = 0$$

$$\cos \alpha = \sin \alpha$$

$$\alpha = 45°$$

Answer is (C)

MATERIALS SCIENCE–66

An axial stress $\sigma = \frac{F}{A}$ is applied as shown. Calculate the resolved shear stress, τ_y, along the slip plane.

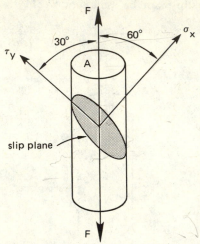

(A) $\tau_y = \frac{1}{4}\sigma$ (B) $\tau_y = \frac{1}{2}\sigma$ (C) $\tau_y = \frac{\sqrt{2}}{3}\sigma$

(D) $\tau_y = \frac{3}{4}\sigma$ (E) $\tau_y = \frac{\sqrt{3}}{2}\sigma$

$$\tau_y = \frac{F \sin 60°}{A / \cos 30°}$$

$$= \left(\frac{F}{A}\right) \sin 60° \cos 30°$$

$$= \sigma \left(\frac{\sqrt{3}}{2}\right)\left(\frac{\sqrt{3}}{2}\right)$$

$$= \frac{3}{4}\sigma$$

Answer is (D)

MATERIALS SCIENCE–67

If G is the shear modulus, b is the Burgers vector, and r is half the distance between particles, what is the local stress, τ, required to bend dislocations around a particle?

(A) $\tau = \frac{Gb}{r}$ (B) $\tau = Gbr$ (C) $\tau = \frac{br}{G}$

(D) $\tau = \frac{Gr}{b}$ (E) $\tau = \frac{G}{br}$

Line tension is given by $\tau = \frac{2T}{bl}$. $T = Gb^2$ and $l = 2r$. Therefore, $\tau = \frac{Gb}{r}$.

Answer is (A)

MATERIALS SCIENCE–68

Given that d is the distance between dislocations, and b is the Burgers vector, what is the expression for the misorientation angle θ of a tilt boundary?

(A) $\sin\theta = \frac{d}{b}$ (B) $\tan\theta = \frac{b}{d}$ (C) $\theta = \frac{b}{d}$

(D) $\theta = \frac{d}{b}$ (E) $\theta = db$

By definition, θ is given by $\tan\theta = \frac{b}{d}$.

Answer is (B)

MATERIALS SCIENCE–69

In general, what are the effects of cold-working a metal?

(A) increased strength and ductility
(B) increased strength, decreased ductility
(C) decreased strength and ductility
(D) decreased strength, increased ductility
(E) increased strength with no effect on ductility

The strength of the metal will increase at the expense of a loss in ductility.

Answer is (B)

MATERIALS SCIENCE-70

Which of the following does cold-working cause?

(A) elongation of grains in the flow direction, an increase in dislocation density, and an overall increase in energy of the metal

(B) elongation of grains in the flow direction, a decrease in dislocation density, and an overall decrease in energy of the metal

(C) elongation of grains in the flow direction, a decrease in dislocation density, and an overall increase in energy of the metal

(D) shortening of grains in the flow direction, a decrease in dislocation density, and an overall decrease in the energy of the metal

(E) shortening of grains in the flow direction, an increase in dislocation density, and an overall increase in the energy of the metal

There is an elongation of grains coupled with increases in both dislocation density and energy.

Answer is (A)

MATERIALS SCIENCE-71

Which of the following statements is false?

(A) The amount or percentage of cold work cannot be obtained from information about change in the area or thickness of a metal.

(B) The process of applying force to a metal at temperatures below the temperature of crystallization in order to plastically deform the metal is called cold-working.

(C) Annealing eliminates most of the defects caused by the cold-working of a metal.

(D) Annealing reduces the hardness of the metal.

(E) Internal stresses are released during the annealing of a metal that has been cold-worked.

The percentage of cold work can be calculated directly from the reduction in thickness or area of the metal.

Answer is (A)

MATERIALS SCIENCE–72

Which of the following statements is false?

(A) There is a considerable increase in the hardness and the strength of a cold-worked metal.

(B) Cold-working a metal significantly reduces its ductility.

(C) Cold-working causes a slight decrease in the density and electrical conductivity of a metal.

(D) Cold work distorts the equiaxial microstructure of metals, causes the formation of crystal defects, and frequently causes elongation of the grains.

(E) Cold work decreases the yield point of the metal.

Cold-working increases the yield point as well as the strength and hardness of the metal.

Answer is (E)

MATERIALS SCIENCE–73

Which of the following statements is false?

(A) Hot-working can be regarded as the simultaneous combination of cold-working and annealing.

(B) Hot-working increases the density of the metal.

(C) One of the primary goals of hot-working is to produce a fine-grained product.

(D) In hot-working, no noticeable changes in mechanical properties occur.

(E) Hot-working causes much strain hardening of the metal.

In hot-working, the high temperature immediately releases any strain hardening which could occur in the deformation of the metal.

Answer is (E)

MATERIALS SCIENCE–74

Which of the following is false?

(A) Grain size is of minor importance in considering the properties of poly-crystalline materials.

(B) Fine-grained materials usually exhibit greater yield stresses than coarse-grained materials at low temperatures.

(C) At high temperatures, grain boundaries become weak and sliding occurs.

(D) Grain boundary sliding is the relative movement of two grains by a shear movement parallel to the grain boundary between them.

(E) Grain sliding can cause the formation of voids along the boundary.

Grain size is an important factor to consider in understanding the properties of polycrystalline materials because it affects the area and length of the grain boundaries.

Answer is (A)

MATERIALS SCIENCE–75

Which of the following correctly describes atoms located at grain boundaries?

(A) They are subjected to the same type of interatomic forces that are present in the interior atoms of the crystal.

(B) They are located primarily in highly strained and distorted positions.

(C) They have a higher free energy than atoms in the undisturbed part of the crystal lattice.

(D) They are oriented so as to attain a minimum energy state.

(E) All of the above are correct.

All are correct statements regarding atoms at the grain boundary.

Answer is (E)

MATERIALS SCIENCE–76

What causes the vinyl interiors of automobiles to crack when subjected to prolonged direct sunlight?

(A) the volatilization (evaporation) of plasticizers
(B) repetitive expansion and contraction of the plastic
(C) oxidation of the plastic by sunlight and oxygen
(D) additional polymerization
(E) all of the above

All of the statements are true.

Answer is (E)

MATERIALS SCIENCE–77

Low density polyethylene undergoes extensive elongation (over 100%) prior to rupture, while polystyrene undergoes only 1 to 2% elongation. What is the main reason for this difference?

(A) The polyethylene is less dense.
(B) The large styrene groups in the polystyrene prevent slippage.
(C) More cross linking occurs in the polystyrene.
(D) Polyethylene is less crystalline.
(E) None of the above are true.

Polystyrene has large styrene groups on the side of its carbon chain which prevent slippage, making the polystyrene brittle.

Answer is (B)

MATERIALS SCIENCE–78

Which of the following describe the modulus of elasticity of an elastomer?

I.   It is directly proportional to the number of cross links in the elastomer.
II. Its value increases with temperature.
III. It is directly proportional to the number of double bonds in the chemical structure.

(A) I only   (B) II only   (C) III only
        (D) I and II   (E) I, II, and III

Choice III is false since a double bond prevents rotation along the bond, inhibiting elasticity.

Answer is (D)

MATERIALS SCIENCE–79

Which of these statements describe the glass transition temperature?

I.   It is the temperature at which the rate of volume contraction increases abruptly.
II. It is the temperature at which residual stresses in the glass can be relieved.
III. It is the point where the material behaves more like a solid than a viscous liquid.

(A) I only   (B) I and II   (C) II and III
        (D) I and III   (E) I, II, and III

The glass transition temperature is the point at which the free movement of the glass molecules past each other becomes difficult. The glass begins to act like a solid, increasing in specific volume.

Answer is (D)

MATERIALS SCIENCE-80

f the diagram below represents the sintering of the ceramic MgO, what could he curves x and y refer to?

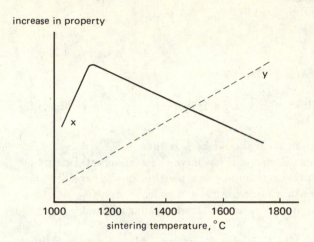

A) x = grain size; y = porosity
B) x = grain size; y = strength
C) x = porosity; y = grain size
D) x = porosity; y = strength
E) x = strength; y = grain size

As the sintering temperature increases, the strength of the ceramic will increase first and then drop abruptly. The grain size will increase linearly with rising temperature.

Answer is (E)

MATERIALS SCIENCE-81

Of the following inorganic glasses, which have tetrahedral lattice structures?

$$SiO_2, B_2O_3, BeF_2, GeO_2$$

(A) SiO_2 and B_2O_3
(B) SiO_2 and BeF_2
(C) SiO_2, B_2O_3, and BeF_2
(D) SiO_2 and GeO_2
(E) SiO_2, BeF_2, and GeO_2

SiO_2, BeF_2, and GeO_2 have tetrahedral structures. B_2O_3 has an almost triangular structure.

Answer is (E)

MATERIALS SCIENCE-82

Which of the following is not an important criterion for forming a complete binary solid solution?

(A) The difference in radii should be less than 15%.
(B) The constituent elements must have the same crystal structure.
(C) The atoms should be close to one another in the periodic table.
(D) The difference in atomic numbers should be small.
(E) The elements should have similar electronegativities.

All choices except (D) are criteria for a binary solid solution.

Answer is (D)

MATERIALS SCIENCE-83

How can an ordered solid solution be distinguished from a compound?

(A) In an ordered solid solution, the solute atoms occupy interstitial positions within the lattice.

(B) The solute atoms in an ordered solid solution substitute for atoms in the parent lattice.

(C) The atoms in an ordered solid solution form layers in the lattice structure.

(D) When heated, an ordered solid solution becomes disordered before melting.

(E) An ordered solid solution appears glassy.

Unlike a compound, an ordered solid solution becomes disordered when heated.

Answer is (D)

MATERIALS SCIENCE–84

What is transformed in a eutectoid reaction?

(A) One solid is transformed into two solids of different composition.
(B) A solid becomes a liquid at the eutectic temperature.
(C) A liquid becomes a solid at the solidus temperature.
(D) A solid becomes a liquid at the liquidus temperature.
(E) Two different metals form an alloy.

One solid is transformed into two different solids.

Answer is (A)

MATERIALS SCIENCE–85

Which of the following is the correct representation of a eutectic cooling reaction?
(The subscripts denote different compositions.)

(A) $(\text{liquid}) \longrightarrow (\text{solid})_1 + (\text{solid})_2$
(B) $(\text{solid})_1 + (\text{liquid}) \longrightarrow (\text{solid})_2$
(C) $(\text{solid})_1 \longrightarrow (\text{solid})_2 + (\text{solid})_3$
(D) $(\text{solid})_1 + (\text{solid})_2 \longrightarrow (\text{solid})_3$
(E) $(\text{liquid})_1 \longrightarrow (\text{solid}) + (\text{liquid})_2$

A eutectic reaction is the transformation from one liquid phase to two solid phases.

Answer is (A)

MATERIALS SCIENCE–86

Two pieces of copper are brazed together using a eutectic alloy of copper and silver. The braze material melts at 780 °C. If a second braze is attempted in order to attach another piece of copper, which of the following is true?

(A) The first braze will melt if the braze temperature is again 780 °C.
(B) The braze temperature must be lowered below 780 °C.
(C) The first braze will partially melt, causing the parts to slide.
(D) The braze temperature must be raised above 780 °C.
(E) The first braze will not melt at 780 °C, but the second braze will.

All compositions of copper and silver other than the eutectic will have a melting point higher than the eutectic temperature. The alloy of the first braze will dissolve somewhat into the copper pieces, changing its composition. It will not melt again at the second braze temperature of 780 °C.

Answer is (E)

MATERIALS SCIENCE–87

On an alloy phase diagram, what is the solidus temperature?

(A) The point at which all solids completely reach the liquid stage.

(B) The temperature of the liquid phase at which the first solid forms for a given overall composition.

(C) The temperature of the solid phase at which the first liquid forms for a given overall composition.

(D) The temperature at which the solid is at equilibrium.

(E) The temperature at which the components of the alloy separate.

The solidus temperature is the temperature at which liquid first forms.

Answer is (C)

MATERIALS SCIENCE–88

n this phase diagram, what can be said about the phases present in regions I, I, and III?

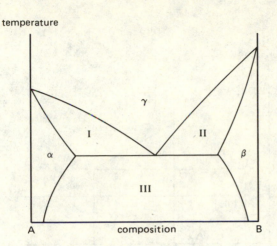

A) α, β, and γ are present in region I.
B) β and γ are present in region II.
C) α, β, and γ are present in region III.
D) α and γ are present in region III.
E) β and γ are present in region III.

β and γ are present in region II. γ is not present in region III, nor is β present in region I.

Answer is (B)

MATERIALS SCIENCE–89

Given the following phase diagram, determine the percentage of liquid remaining at 600 °C that results from the equilibrium cooling of an alloy containing 5% silicon and 95% aluminum.

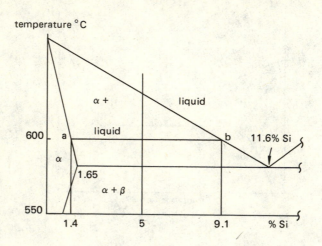

(A) 0%   (B) 47%   (C) 53%   (D) 67%   (E) 100%

At a there is 1.4% Si, while at b there is 9.1% Si. Therefore,

$$\text{percent liquid} = \left(\frac{5 - 1.4}{9.1 - 1.4}\right)(100) = 47\%$$

Answer is (B)

MATERIALS SCIENCE–90

Consider the Ag-Cu phase diagram. Calculate the equilibrium amount of β in an alloy of 30% Ag, 70% Cu at 850 °C.

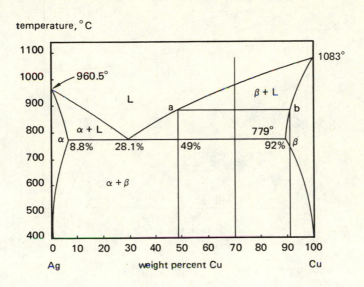

(A) 0%   (B) 22%   (C) 49%   (D) 52%   (E) 100%

At 70% Cu, $a = 49\%$ Cu and $b = 92\%$ Cu.

$$\text{percent } \beta = \left(\frac{70 - 49}{92 - 49}\right)(100) = 49\%$$

Answer is (C)

MATERIALS SCIENCE–91

Using the given phase diagram, what are the relative weights of phase α_1 and phase α_2, for an alloy of 70% B at temperature T_1?

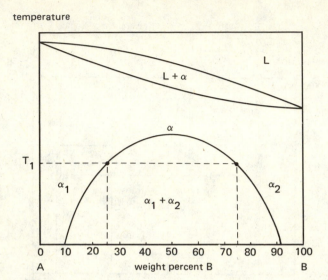

(A) 10% α_1, 90% α_2
(B) 30% α_1, 70% α_2
(C) 50% α_1, 50% α_2
(D) 70% α_1, 30% α_2
(E) 90% α_1, 10% α_2

Let W_{α_1} denote the weight fraction of α_1 and W_{α_2} denote the weight fraction of α_2. From the diagram, $C_{\alpha_1} = 25\%$ and $C_{\alpha_2} = 75\%$. Then,

$$W_{\alpha_1} + W_{\alpha_2} = 1 \tag{1}$$
$$W_{\alpha_1} C_{\alpha_1} + W_{\alpha_2} C_{\alpha_2} = C_0 \tag{2}$$

Solving the two equations using $C_0 = 70\%$,

$$W_{\alpha_1} = \frac{C_{\alpha_2} - C_0}{C_{\alpha_2} - C_{\alpha_1}} = \frac{75 - 70}{75 - 25} = 10\%$$

Answer is (A)

MATERIALS SCIENCE-92

For 50% B at 1275 °C as shown in the diagram, what is the relative amount of each phase present?

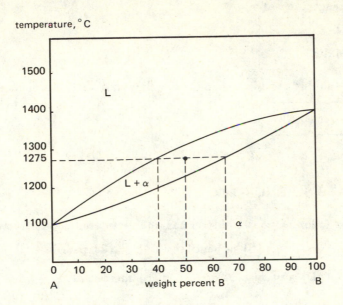

(A) 40% liquid, 60% solid
(B) 45% liquid, 55% solid
(C) 50% liquid, 50% solid
(D) 55% liquid, 45% solid
(E) 60% liquid, 40% solid

From the phase diagram, $C_\alpha = 65\%$ and $C_l = 40\%$. With C_0 given as 50%, and denoting the weight fraction of liquid and solid by W_l and W_α, respectively:

$$W_l + W_\alpha = 1$$
$$W_l C_l + W_\alpha C_\alpha = C_0$$

$$W_l = \frac{C_\alpha - C_0}{C_\alpha - C_l} = \frac{65 - 50}{65 - 40} = 60\%$$

Answer is (E)

MATERIALS SCIENCE–93

Which of the following is not a structural class of steels?

(A) carbon
(B) high-strength, low-alloy
(C) low-alloy
(D) stainless
(E) tool and die

"Tool and die" steel is an application class, not a structural class.

Answer is (E)

MATERIALS SCIENCE–94

Which of the following phases of steel has a face-centered cubic structure?

(A) ferrite   (B) cementite   (C) pearlite

(D) austenite   (E) martensite

Only austenite has a face-centered cubic structure.

Answer is (D)

MATERIALS SCIENCE–95

Low-carbon steels are generally used in the "as rolled" or "as fabricated" state. What is the reason for this?

(A) They come in many different shapes and thicknesses.
(B) Their strength generally cannot be increased by heat treatment.
(C) They degrade severely under heat treatment.
(D) Their chromium content is so low.
(E) Their density is so high.

Since their strength cannot be increased by heat treatment, low-carbon steels are used as fabricated.

Answer is (B)

MATERIALS SCIENCE–96

The equilibrium cooling of a steel containing 0.8% carbon results in a product with little use because it is extremely brittle. Which of the following is the primary reason for this poor characteristic?

A) The material has not been cold-worked.
B) The austenite grains are too small, and the carbide grains are too large.
C) Thick layers of iron carbide surround the coarse ferrite grains.
D) The carbide forms thin plates that are brittle.
E) The carbon content is too high.

When hypereutectoid steels are slow-cooled, brittle carbide plates are formed.

Answer is (D)

MATERIALS SCIENCE–97

Ductile cast iron and gray cast iron both contain 4% carbon. Ductile cast iron, however, has a higher tensile strength and is considerably more ductile. Which of the following is the major difference that accounts for the superior properties of the ductile iron?

(A) The gray cast iron contains iron carbide, whereas the ductile iron contains graphite.

(B) The gray cast iron contains flakes of graphite, whereas the ductile iron contains spheroids of graphite.

(C) The ductile iron is tempered to give better properties.

(D) The ferrite grains in the gray cast iron are excessively large.

(E) The graphite particles in the gray cast iron are excessively large.

Gray cast iron contains flakes of graphite while ductile cast iron contains spheroids. The difference in the shape of the graphite gives the ductile iron approximately twice the tensile strength and 20 times the ductility of the gray cast iron.

Answer is (B)

MATERIALS SCIENCE-98

In preparing a metallographic iron specimen, the grain boundaries are made most visible by which of the following steps?

(A) grinding the sample with silicon carbide abrasive
(B) polishing the sample with Al_2O_3
(C) mounting the sample in an epoxy resin mold
(D) etching the sample in a 2% solution of nitric acid in alcohol
(E) polishing the sample with powdered diamond dust in oil

Etching the specimen with nitric acid in alcohol dissolves metal from the surface and preferentially attacks the grain boundaries. It is the last step in the sample preparation process.

Answer is (D)

MATERIALS SCIENCE-99

Which of the following statements is false?

(A) Low-alloy steels are a minor group and are rarely used.

(B) Low-alloy steels are used in the heat-treated condition.

(C) Low-alloy steels contain small amounts of nickel and chromium.

(D) The addition of small amounts of molybdenum to low-alloy steels makes it possible to harden and strengthen thick pieces of the metal by heat treatment.

(E) In the AISI and the SAE steel specifications, the last two digits define the carbon content of the steel in weight percent.

Low-alloy steels are one of the most commonly used classes of structural steels.

Answer is (A)

MATERIALS SCIENCE–100

Which of the following statements is false?

(A) High-strength low-alloy steels are not as strong as non-alloy low-carbon steels.

(B) Small amounts of copper increase the tensile strength of steels.

(C) Small amounts of silicon in steels have little influence on toughness or fabricability.

(D) Small amounts of phosphorus in steels can provide an increase in the yield strength of the steel.

(E) Additions of small amounts of silicon to steel can cause a marked decrease in yield strength of the steel.

Additions of small amounts of silicon to steel increases both the yield strength and the tensile strength.

Answer is (E)

MATERIALS SCIENCE–101

Which of the following statements is false?

(A) Stainless steels contain large amounts of chromium.

(B) There are three basic types of stainless steels: martensitic, austenitic, and ferritic.

(C) The non-magnetic stainless steels contain large amounts of nickel.

(D) Stabilization of the face-centered cubic crystal structure of stainless steels imparts a non-magnetic characteristic to the alloy.

(E) The non-magnetic stainless steels are called austenitic.

There are only two basic types of stainless steels, magnetic (martensitic or ferritic) and non-magnetic (austenitic).

Answer is (B)

MATERIALS SCIENCE-102

For a completely corrosion-resistant stainless steel, what minimum percentage of chromium in the alloy is required?

(A) 1.1%   (B) 3.2%   (C) 8.3%   (D) 11%   (E) 15%

For complete corrosion resistance, the chromium content must be at least 11%.

Answer is (D)

MATERIALS SCIENCE-103

Which of the following would most likely require a steel containing 0.6% carbon that has been spheroidized, cold-drawn, and slightly tempered?

(A) a bridge beam
(B) a water pipe
(C) a cutting tool
(D) a ball bearing
(E) a car fender

A hypoeutectoid steel that has been worked using the above process has good strength and excellent toughness. A cutting tool undergoes tremendous stress loads due to the relatively small contact area. It requires a stronger material than the other objects.

Answer is (C)

MECHANICS OF MATERIALS

MECHANICS OF MATERIALS–1

Where do stress concentrations occur?

I. near the points of application of concentrated loads
II. along the entire length of high distributed loads
III. at discontinuities

(A) I and II   (B) I and III   (C) II and III
       (D) I, II, and III   (E) none of the above

Stress concentrations occur under concentrated loads and at discontinuities, not under distributed loads.

Answer is (B)

MECHANICS OF MATERIALS–2

What is the definition of normal strain, ϵ? (δ is elongation, and L is the length of the specimen.)

(A) $\epsilon = \dfrac{L + \delta}{L}$ (B) $\epsilon = \dfrac{L + \delta}{\delta}$ (C) $\epsilon = \dfrac{\delta}{L + \delta}$

       (D) $\epsilon = \dfrac{\delta}{L}$ (E) $\epsilon = \dfrac{L}{\delta}$

Strain is defined as elongation per unit length.

Answer is (D)

MECHANICS OF MATERIALS–3

What can the maximum load be on the column, if the cross-sectional area is 144 ft^2 and the compressive stress cannot exceed 200 lbf/ft^2?

(A) 20 kips   (B) 22 kips   (C) 28.8 kips

    (D) 30 kips   (E) 32 kips

The equation for axial stress is:

$$\sigma = \frac{F}{A}$$
$$F = \sigma A$$
$$= \left(200 \frac{\text{lbf}}{\text{ft}^2}\right)(144 \text{ ft}^2)$$
$$= 28,800 \text{ lbf} = 28.8 \text{ kips}$$

Answer is (C)

MECHANICS OF MATERIALS–4

A copper column of annular cross section has an outer diameter, d_2, of 15 feet, and is subjected to a force of 45 kips. The allowable compressive stress is 300 lbf/ft². What should be the wall thickness, t?

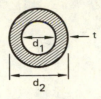

(A) 3 ft   (B) 3.52 ft   (C) 4.59 ft   (D) 5.03 ft   (E) 5.83 ft

For axial stress,

$$\sigma = \frac{F}{A}$$

Then,

$$A = \frac{F}{\sigma} = \frac{\pi}{4}(d_2^2 - d_1^2)$$

$$d_1 = \sqrt{d_2^2 - \frac{4F}{\pi\sigma}}$$

$$= \sqrt{(15)^2 - \frac{(4)(45)}{(\pi)(0.3)}}$$

$$= 5.83 \text{ ft}$$

Therefore,

$$t = \frac{d_2 - d_1}{2} = \frac{15 - 5.83}{2} = 4.59 \text{ ft}$$

Answer is (C)

MECHANICS OF MATERIALS–5

What is the stress at surface S of the cylindrical object if $F = 3500$ pounds and the specific weight of the material is $\gamma = 490$ lbf/ft^3?

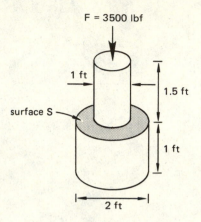

F = 3500 lbf

1 ft

1.5 ft

surface S

1 ft

2 ft

(A) 500 lbf/ft^2 (B) 1110 lbf/ft^2 (C) 1300 lbf/ft^2
        (D) 4460 lbf/ft^2 (E) 5190 lbf/ft^2

The stress at surface S is due to the weight of the material above it in addition to the force F. The total load is:

$$F_{total} = W + F = \gamma(\text{volume}) + F$$
$$= (490)\left(\frac{\pi}{4}\right)(1)^2(1.5) + 3500$$
$$= 4080 \text{ lbf}$$

$$\sigma = \frac{F_{total}}{A} = \frac{4080}{\pi(1)^2}$$
$$= 1300 \text{ lbf/ft}^2$$

Answer is (C)

MECHANICS OF MATERIALS–6

Considering the stress-strain diagram for aluminum, which point is the fracture point?

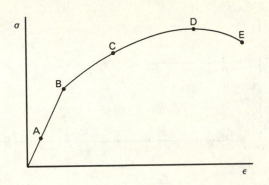

(A) A   (B) B   (C) C   (D) D   (E) E

Point E is where fracture occurs.

Answer is (E)

MECHANICS OF MATERIALS–7

In a stress-strain diagram, what is the correct term for the stress level at $\epsilon = 0.2\%$ offset?

(A) the elastic limit
(B) the plastic limit
(C) the offset rupture stress
(D) the offset yield stress
(E) the offset fatigue strength

This is known as the offset yield stress.

Answer is (D)

MECHANICS OF MATERIALS-8

Consider this stress-strain diagram for a typical steel in tension. Determine the region of perfect plasticity or yielding.

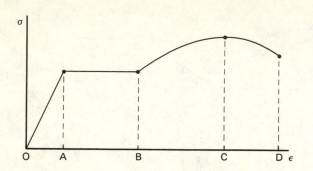

(A) O to A   (B) A to B   (C) B to C
    (D) C to D   (E) A to C

The plastic region is between points A and B. O to A is known as the linear region, B to C is where strain hardening occurs, and C to D is where reduction in area occurs.

Answer is (B)

MECHANICS OF MATERIALS-9

Under which type of loading does fatigue occur?

(A) static load
(B) plane load
(C) high load
(D) repeated load
(E) none of the above

Fatigue occurs under repeated loading cycles.

Answer is (D)

MECHANICS OF MATERIALS–10

A specimen is subjected to a load. When the load is removed, the strain disappears. From this information, which of the following can be deduced about this material?

(A) It is elastic.
(B) It is plastic.
(C) It has a high modulus of elasticity.
(D) It does not obey Hooke's law.
(E) It is ductile.

By definition, elasticity is the property of a material in which it returns to its original dimensions during unloading.

Answer is (A)

MECHANICS OF MATERIALS–11

Which of the following may be the Poisson ratio of a material?

(A) 0.45   (B) 0.50   (C) 0.55   (D) 0.60   (E) 0.85

The Poisson ratio must be in the range $0 < \nu < 0.5$. Choice (A) is the only answer that satisfies this condition.

Answer is (A)

MECHANICS OF MATERIALS–12

A 100″ long aluminum bar is subjected to a tensile stress, σ, of 25,000 psi. Find the elongation, δ. ($E = 10 \times 10^6$ psi)

(A) 0.025 in   (B) 0.25 in   (C) 0.45 in

          (D) 0.65 in   (E) 2.50 in

From Hooke's law,

$$\epsilon = \frac{\sigma}{E} = \frac{\delta}{L}$$

$$\delta = \frac{\sigma L}{E} = \frac{(25,000)(100)}{10,000,000}$$
$$= 0.25 \text{ in}$$

Answer is (B)

MECHANICS OF MATERIALS–13

A 22″ wide thin plate is in tension as shown. What is the width of the plate after the tension is removed? The modulus of elasticity, E, is 29×10^6 psi, and the Poisson ratio, ν, is 0.3.

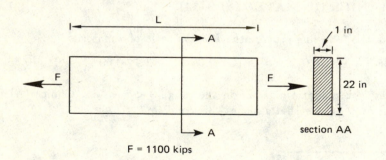

F = 1100 kips

(A) 21.995 in   (B) 22.00 in   (C) 22.001 in

(D) 22.0005 in   (E) 22.005 in

The Poisson ratio is defined as the negative ratio of lateral strain, ϵ_y, to axial strain, ϵ_x. Using this and the equation for axial stress and strain,

$$\nu = -\frac{\epsilon_y}{\epsilon_x}$$

$$\epsilon_y = -\nu\epsilon_x \qquad\qquad 1$$

$$\epsilon_x = \frac{\sigma}{E} = \frac{F}{EA} \qquad\qquad 2$$

Combining equations 1 and 2,

$$\epsilon_y = -\frac{\nu F}{EA} = -\frac{(0.3)(-1100 \times 10^3)}{(29 \times 10^6)(22)}$$

$$= 5.17 \times 10^{-4}$$

Therefore, the width after removing the load is

$$\text{width} = 22(1 + 5.17 \times 10^{-4}) = 22.001 \text{ inches}$$

Answer is (C)

MECHANICS OF MATERIALS–14

Find the lateral strain, ϵ_y, of the steel specimen shown if $F_x = 600$ kips, $E = 2.8 \times 10^7$ psi, and $\nu = 0.29$.

(A) -4×10^{-4} in   (B) -1.04×10^{-4} in   (C) 1.04×10^{-4} in
(D) 4×10^{-4} in   (E) 6×10^{-4} in

From Hooke's law and the equation for axial stress,

$$\epsilon_x = \frac{\sigma}{E} = \frac{F}{EA}$$

$$= \frac{600,000}{(2.8 \times 10^7)(60)}$$

$$= 3.57 \times 10^{-4} \text{ in}$$

From the Poisson ratio,

$$\epsilon_y = -\nu\epsilon_x = -(0.29)(3.57 \times 10^{-4})$$

$$= -1.04 \times 10^{-4} \text{ in}$$

Answer is (B)

MECHANICS OF MATERIALS–15

A steel specimen is subjected to a tensile force, F, of 400 kips. If the Poisson ratio, ν, is 0.29, and the modulus of elasticity, E, is 2.8×10^7 psi, find the dilatation, e.

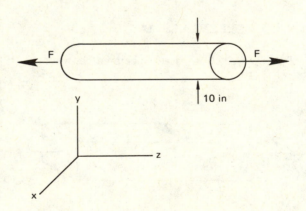

(A) 6.5×10^{-5} in   (B) 7.65×10^{-5} in   (C) 7.65×10^{-4} in

(D) 65 in   (E) 76.5 in

Dilatation is defined as the sum of the strain in all three coordinate directions. In the axial z direction,

$$\epsilon_z = \frac{F}{EA}$$

$$= \frac{400,000}{(2.8 \times 10^7)(78.5)}$$

$$= 1.82 \times 10^{-4}$$

From the Poisson ratio,

$$\nu = -\frac{\epsilon_x}{\epsilon_z} = -\frac{\epsilon_y}{\epsilon_z}$$

$$\epsilon_x = \epsilon_y = -\nu\epsilon_z$$

$$= -(0.29)(1.82 \times 10^{-4})$$

$$= -5.275 \times 10^{-5}$$

Therefore,

$$e = \epsilon_x + \epsilon_y + \epsilon_z$$

$$= (1.82 \times 10^{-4}) + 2(-5.275 \times 10^{-5})$$

$$= 7.65 \times 10^{-5} \text{ in}$$

Answer is (B)

MECHANICS OF MATERIALS–16

Given a shear stress of $\tau_{xy} = 5000$ psi and a shear modulus of $G = 1.15 \times 10^7$ psi, find the shear strain, ϵ_{xy}.

(A) 2.5×10^{-5} psi (B) 4.35×10^{-4} psi (C) 4.5×10^{-4} psi
(D) 8.25×10^{-4} psi (E) 435 psi

Hooke's law for shear gives:

$$\epsilon_{xy} = \frac{\tau_{xy}}{G} = \frac{5000}{1.15 \times 10^7}$$

$$= 4.35 \times 10^{-4} \text{ psi}$$

Answer is (B)

MECHANICS OF MATERIALS–17

A 6″ diameter rivet undergoes a shear force of $V = 1750$ pounds. Find the average shear stress in the rivet.

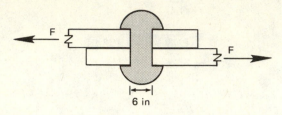

6 in

(A) 15.5 psi   (B) 37.0 psi   (C) 42.7 psi
          (D) 52.0 psi   (E) 61.9 psi

$$\tau = \frac{V}{A} = \frac{1750}{\left(\frac{\pi}{4}\right)(6)^2} = 61.9 \text{ psi}$$

Answer is (E)

MECHANICS OF MATERIALS–18

A steel bar carrying a 3000 kN load, F, is attached to a support by a round pin 0.3 meters in diameter. Calculate the average shear stress in the pin.

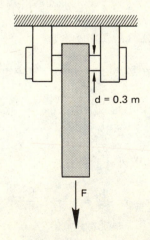

d = 0.3 m

F

(A) 9.97 MPa   (B) 12.2 MPa   (C) 21.2 MPa
          (D) 24.6 MPa   (E) 42.4 MPa

The pin will shear on two cross sections, hence:

$$\tau = \frac{F}{2A} = \frac{3000}{(2)\left(\frac{\pi}{4}\right)(0.3)^2} = 21.2 \text{ MPa}$$

Answer is (C)

MECHANICS OF MATERIALS–19

What is the maximum allowable load, F, if the factor of safety is 1.5 and the yield stress, σ_{yield}, is 3000 psi?

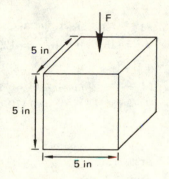

5 in

5 in

5 in

F

$\sigma_{allowable} = \frac{\sigma_{yield}}{2.5 \text{ SF}}$
$= \frac{3000 \times 10}{1.5}$
$= \text{grade}$

$\sigma = \frac{F}{A}$

$F = \sigma A$
$= 2000 \times 15^2$
$= 50 \text{ kip}$

(A) 50 kips   (B) 55 kips   (C) 60 kips

(D) 64 kips   (E) 900 kips

$$\sigma_{\text{allowable}} = \frac{\sigma_{\text{yield}}}{\text{SF}}$$
$$= \frac{3000}{1.5} = 2000 \text{ psi}$$
$$F = \sigma_{\text{allowable}} \times A$$
$$= (2000)(25)$$
$$= 50 \text{ kips}$$

Answer is (A)

MECHANICS OF MATERIALS-20

The allowable tensile stress for a $\frac{1}{4}$-20 bolt of thread length 7/32 inches is 30,000 psi. The allowable shear stress of the material is 15,000 psi. Where and how will such a bolt be most likely to fail?

(A) at the root diameter due to tension
(B) at the threads due to shear
(C) at the root diameter due to shear
(D) at the threads due to tension
(E) at the root diameter due to buckling

The bolt will most likely fail due to shearing of the threads or due to tensile failure of the bolt diameter.

$$F_{\text{allowable, thread}} = \left(\tau_{\text{allowable}}\right)(\text{average shear area})$$
$$= (15)(\tfrac{1}{2})(\tfrac{\pi}{4})(\tfrac{7}{32}) = 1290 \text{ lbf}$$

$$F_{\text{allowable, root}} = \left(\sigma_{\text{allowable}}\right)(\text{root area})$$
$$= (30)(\tfrac{\pi}{4})(\tfrac{1}{4})^2 = 1470 \text{ lbf}$$

The threads will shear first.

Answer is (B)

MECHANICS OF MATERIALS-21

Hexagonal nuts for $\frac{1}{4}$-20 bolts have a height of 7/32 inches. If the ultimate strength of the nut material in shear is 15,000 psi, find the maximum allowable shear force on the nut threads using a safety factor of 5.

(A) 41 lbf   (B) 82 lbf   (C) 129 lbf

            (D) 258 lbf   (E) 1290 lbf

$$\tau_{\text{allowable}} = \frac{\tau}{\text{SF}}$$
$$= \frac{15,000}{5}$$
$$= 3000 \text{ psi}$$
$$V = \tau_{\text{allowable}} A = \tau_{\text{allowable}}\left(\tfrac{1}{2}\pi dh\right)$$
$$= (3000)\left(\tfrac{1}{2}\right)(\pi)\left(\tfrac{1}{4}\right)\left(\tfrac{7}{32}\right)$$
$$= 258 \text{ lbf}$$

Answer is (D)

MECHANICS OF MATERIALS-22

Determine the length, L, of the fillet weld for the lap joint shown in the figure below. The weld has to resist a tension, F, of 115 kips. The effective throat for the weld, h, is 0.5 inches, and the allowable stress is 21 ksi.

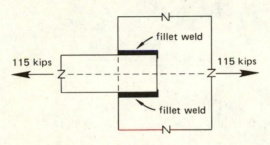

(A) 5.5 in   (B) 7.7 in   (C) 11.0 in

(D) 15.5 in   (E) 22.0 in

For a fillet weld, the average normal stress is:

$$\sigma = \frac{F}{hL}$$

$$L = \frac{F}{\sigma h} = \frac{115}{(21)(0.5)}$$
$$= 11.0 \text{ in}$$

Answer is (C)

MECHANICS OF MATERIALS-23

Find the elongation of the aluminum specimen shown in the figure when loaded to its yield point. $E = 10 \times 10^6$ psi, and $\sigma_{\text{yield}} = 37$ ksi. Neglect the weight of the bar.

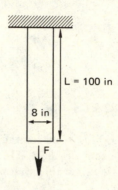

L = 100 in

8 in

F

(A) 0.25 in

(B) 0.37 in

(C) 0.43 in

(D) 0.65 in

(E) 0.83 in

From Hooke's law, the axial strain is:

$$\epsilon = \frac{\sigma}{E} = \frac{37,000}{10 \times 10^6} = 0.0037$$

The total elongation is:

$$\delta = \epsilon L = (0.0037)(100) = 0.37 \text{ in}$$

Answer is (B)

MECHANICS OF MATERIALS–24

What is the total elongation of this composite body under a force of 6000 pounds?
$E_1 = 10 \times 10^6$ psi, and $E_2 = 1.5 \times 10^7$ psi.

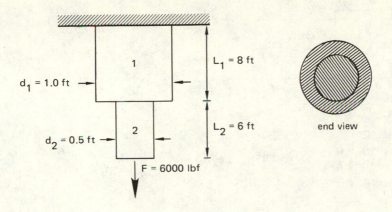

(A) 0.0015 in   (B) 0.0020 in   (C) 0.0023 in

(D) 0.0048 in   (E) 0.0063 in

Total elongation is the elongation of section 1 plus the elongation of
section 2.

$$\delta_{\text{total}} = \delta_1 + \delta_2 = \frac{F L_1}{A_1 E_1} + \frac{F L_2}{A_2 E_2}$$
$$= \frac{(6000)(8)}{(\frac{\pi}{4})(1)^2(10 \times 10^6)(144)} + \frac{(6000)(6)}{(\frac{\pi}{4})(0.5)^2(1.5 \times 10^7)(144)}$$
$$= 1.27 \times 10^{-4} \text{ ft} = 0.0015 \text{ in}$$

Answer is (A)

MECHANICS OF MATERIALS–25

Find the total elongation of the rod shown if $E = 1.43 \times 10^9$ lbf/ft^2.

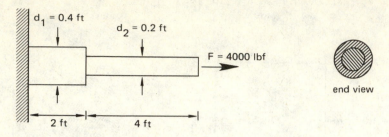

(A) 2.67×10^{-3} ft
(B) 4.01×10^{-3} ft
(C) 4.33×10^{-3} ft
(D) 4.83×10^{-3} ft
(E) 5.33×10^{-3} ft

$$\delta_{\text{total}} = \frac{F L_1}{E A_1} + \frac{F L_2}{E A_2} = \frac{F}{E}\left(\frac{L_1}{A_1} + \frac{L_2}{A_2}\right)$$

$$= \frac{4F}{\pi E}\left(\frac{L_1}{d_1^2} + \frac{L_2}{d_2^2}\right)$$

$$= \frac{(4)(40,000)}{(\pi)(1.43 \times 10^9)}\left[\frac{2}{(0.4)^2} + \frac{4}{(0.2)^2}\right]$$

$$= 4.01 \times 10^{-3} \text{ ft}$$

Answer is (B)

MECHANICS OF MATERIALS–26

A 200 meter cable is suspended vertically. At any point along the cable, the strain is proportional to the length of the cable below that point. If the strain at the top of the cable is 0.001, determine the total elongation of the cable.

(A) 0.05 m   (B) 0.10 m   (C) 0.15 m
          (D) 0.20 m   (E) 0.25 m

Since the strain is proportional to the length of the cable supported, it is necessary to integrate over the entire length to determine total elongation. Since the strain at $L = 200$ m is 0.001,

$$\epsilon = \left(\frac{0.001}{200}\right)L = (5 \times 10^{-6})L$$

By definition, $\epsilon = \dfrac{\delta}{L}$. Therefore,

$$\delta = (5 \times 10^{-6})L^2$$
$$d\delta = (1 \times 10^{-5})LdL$$

$$\delta_{total} = \int_0^{200} (1 \times 10^{-5})LdL = (1 \times 10^{-5})\left[\frac{L^2}{2}\right]_0^{200}$$

$$= 0.2 \text{ m}$$

Answer is (D)

MECHANICS OF MATERIALS–27

The figure below shows a two-member truss with a load $F = 11,500$ kips applied statically. Given that $L_1 = 4$ feet, $L_2 = 5$ feet, and that the beams have cross-sectional area $A = 6$ in^2, find the elongation of member AB after F is applied. Use $E = 29,000,000$ psi.

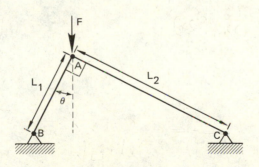

(A) −2.48 in   (B) −1.98 in   (C) −1.48 in
          (D) 1.48 in   (E) 2.48 in

A free-body diagram of joint A gives:

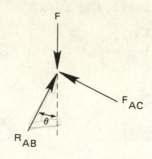

$$\cos\theta = \frac{R_{AB}}{F}$$
$$R_{AB} = F\cos\theta$$

$$R_{AB} = F\cos\theta = F\frac{L_2}{\sqrt{L_1^2 + L_2^2}}$$

$$= \frac{(11,500)(5)}{\sqrt{(4)^2 + (5)^2}} = 8980 \text{ kips}$$

$$F_{AB} = -R_{AB} = -8980 \text{ kips}$$

$$\delta_{AB} = \frac{F_{AB}L_1}{EA} = \frac{(-8980 \times 10^3)(4)}{(29,000,000)(6)}$$

$$= -0.206 \text{ ft}$$

$$= -2.48 \text{ in}$$

Member AB is compressed 2.48 inches.

Answer is (A)

MECHANICS OF MATERIALS–28

The two bars shown are attached on a surface. The bars have moduli of elasticity and areas as given. If a force of $F = 300$ kips compresses the assembly, what is the reduction in length?

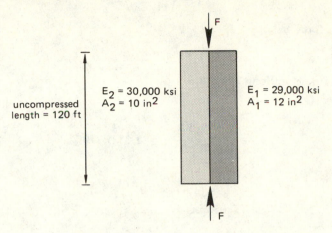

uncompressed length = 120 ft

E_2 = 30,000 ksi
A_2 = 10 in^2

E_1 = 29,000 ksi
A_1 = 12 in^2

(A) 0.052 in   (B) 0.056 in   (C) 0.060 in
(D) 0.103 in   (E) 0.120 in

From the principle of compatability, both bars are compressed the same length.

$$\epsilon_1 = \frac{\sigma_1}{E_1} \qquad \epsilon_2 = \frac{\sigma_2}{E_2}$$

$$= \frac{F_1}{A_1 E_1} \qquad = \frac{F_2}{A_2 E_2}$$

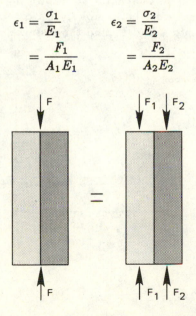

Since $\epsilon_1 = \epsilon_2$,

$$\frac{F_1}{A_1 E_1} = \frac{F_2}{A_2 E_2}$$

$$F_1 = \left(\frac{A_1 E_1}{A_2 E_2}\right) F_2 \tag{1}$$

From a force balance,

$$F_1 + F_2 = 300 \tag{2}$$

Combining equations 1 and 2,

$$F_2 = 139 \text{ kips}$$

$$\epsilon_2 = \frac{139}{(10)(30,000)} = 4.63 \times 10^{-4}$$

$$\delta = \epsilon L = (4.63 \times 10^{-4})(120)$$
$$= 0.056 \text{ in}$$

Answer is (B)

MECHANICS OF MATERIALS-29

A rigid weightless bar is suspended horizontally by cables 1 and 2 as shown. The cross-sectional areas of the cables are shown in the figure. The modulus of elasticity, ϵ, is the same for both cables. If a concentrated load of 300 kips is applied between points A and B, find the distance, x, for the bar to remain horizontal.

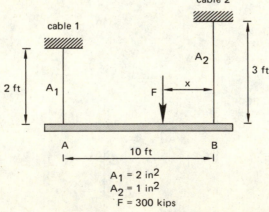

cable 2

cable 1

A_2

3 ft

2 ft A_1

F   x

A

10 ft

B

$A_1 = 2 \text{ in}^2$
$A_2 = 1 \text{ in}^2$
F = 300 kips

(A) 4.3 ft   (B) 5.0 ft   (C) 5.7 ft
          (D) 6.0 ft   (E) 7.5 ft

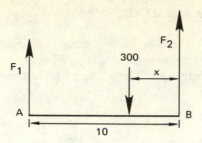

From the free-body diagram, taking moments about point B gives:

$$\sum M_B = 0 : 300x = 10F_1$$

$$x = \frac{F_1}{30} \qquad\qquad 1$$

Also, from a force balance,

$$F_1 + F_2 = 300 \qquad\qquad 2$$

For the bar to remain horizontal, the deflection of cable 1 must equal the deflection of cable 2:

$$\delta_1 = \delta_2$$

$$\frac{F_1 L_1}{E A_1} = \frac{F_2 L_2}{E A_2}$$

$$F_1 = \left(\frac{L_2 A_1}{L_1 A_2}\right) F_2 \qquad\qquad 3$$

Equations 2 and 3 give $F_1 = 129$ kips. Substituting into equation 1,

$$x = \frac{129}{30} = 4.3 \text{ ft}$$

Answer is (A)

MECHANICS OF MATERIALS–30

A prismatic bar at 50 °F is imbedded in a rigid concrete wall. The bar is 40″ long, and has a cross-sectional area of 4 in². What is the axial force in the bar if its temperature is raised to 100 °F? The coefficient of thermal expansion, α, is 5×10^{-6} °F^{-1}.

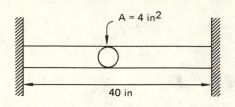

$A = 4$ in^2

40 in

E = modulus of elasticity
  = 32×10^6 psi
α = coefficient of thermal
  expansion
  = 5×10^{-6} in/in²-°F

(A) 5000 lbf   (B) 8000 lbf   (C) 24,000 lbf
          (D) 32,000 lbf   (E) 40,000 lbf

Elongation due to temperature change is given by:

$$\delta = \alpha L(T_2 - T_1)$$
$$= (5 \times 10^{-6})(40)(100 - 50)$$
$$= 0.01 \text{ in}$$

Elongation is related to stress:

$$\delta = \frac{FL}{EA}$$

$$F = \frac{\delta EA}{L} = \frac{(0.01)(32 \times 10^6)(4)}{40}$$
$$= 32,000 \text{ lbf}$$

Answer is (D)

MECHANICS OF MATERIALS-31

Determine the maximum axial load, F, that can be applied to the wood post shown without exceeding a maximum shear stress of 240 psi parallel to the grain.

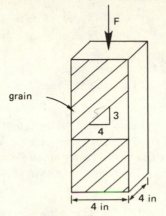

(A) 2300 lbf   (B) 6400 lbf   (C) 8580 lbf
(D) 11,000 lbf   (E) 11,300 lbf

The stresses acting on the inclined surface consist of shear and normal components. Transforming these to the x and y coordinate axes in the free-body diagram,

$$\sum F_x = 0 : \tau_x - \sigma_x = 0$$

$$240 \left(\frac{4}{5}\right) - \sigma \left(\frac{3}{5}\right) = 0$$

$$\sigma = 320 \text{ psi}$$

$$\sum F_y = 0 : \tau_y + \sigma_y - \frac{F}{A} = 0$$

$$(240)\left(\frac{3}{5}\right) + (320)\left(\frac{4}{5}\right) - \frac{F}{16} = 0$$

$$F = 6400 \text{ lbf}$$

Answer is (B)

MECHANICS OF MATERIALS-32

The shear strain, ϵ, along a shaft is:

$$\epsilon = r\frac{d\phi}{dx}$$

r is the shaft radius, and $\frac{d\phi}{dx}$ is the change of the angle of twist with respect to the axis of the shaft. Which condition is not necessary for the above equation to be valid?

(A) The area of interest must be free of connections and other load applications.
(B) The material must be isotropic and homogeneous.
(C) The loading must result in the stress being a couple acting along the axis.
(D) The cross-sectional area must be constant within the area of interest.
(E) r must be the full radius of the shaft.

The equation may be evaluated for any value of r, giving the stress distribution over the shaft cross section.

Answer is (E)

MECHANICS OF MATERIALS-33

A 6 foot diameter bar experiences a torque of 200 ft-lbf. What is the maximum shear stress in the bar?

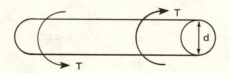

(A) 2.2 lbf/ft^2 (B) 2.5 lbf/ft^2 (C) 3.9 lbf/ft^2
            (D) 4.2 lbf/ft^2 (E) 4.7 lbf/ft^2

The equation for shear gives:

$$\tau = \frac{Tr}{J} = \frac{(200)(3)}{\left(\frac{\pi}{32}\right)(6)^4} = 4.7 \text{ lbf/ft}^2$$

Answer is (E)

MECHANICS OF MATERIALS–34

What is the angle of twist, ϕ, for the aluminum bar shown? The shear modulus of elasticity, G, is 3.7×10^6 psi.

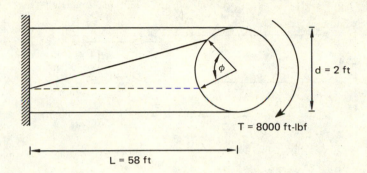

(A) 0.000554°   (B) 0.00554°   (C) 0.032°

(D) 0.08°   (E) 0.0875°

The angle of twist is given by:

$$\phi = \frac{TL}{GJ} = \frac{(8000)(58)}{(3.7 \times 10^6)(\frac{\pi}{32})(2)^4(144)}$$

$$= 0.000554 \text{ rad}$$

$$= 0.032°$$

Answer is (C)

MECHANICS OF MATERIALS-35

What torque (T) should be applied to the end of the shaft shown in order to produce a twist of 1.5 degrees? Use $G = 80$ GPa for the shear modulus of steel.

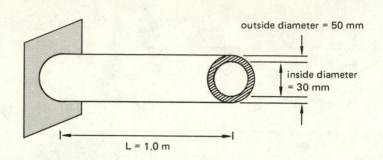

outside diameter = 50 mm

inside diameter = 30 mm

L = 1.0 m

(A) 420 N·m   (B) 560 N·m   (C) 830 N·m

(D) 920 N·m   (E) 1110 N·m

Converting the twist angle to radians and calculating the polar moment of inertia J,

$$\phi = 1.5\left(\frac{2\pi}{360}\right) = 0.026 \text{ rad}$$

$$r_1 = 0.015 \text{ m} \qquad r_2 = 0.025 \text{ m}$$

$$J = \frac{\pi}{2}(r_2^4 - r_1^4) = \frac{\pi}{2}\left[(0.025)^4 - (0.015)^4\right]$$

$$= 5.34 \times 10^{-7} \text{ m}^4$$

$$T = \frac{GJ}{L}\phi$$

$$= \frac{(80 \times 10^9)(5.34 \times 10^{-7})}{1}(0.026)$$

$$= 1110 \text{ N·m}$$

Answer is (E)

MECHANICS OF MATERIALS-36

Determine the maximum torque that can be applied to the shaft given that the maximum angle of twist is 0.0225 radians.

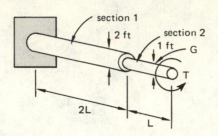

(A) $0.01\dfrac{\pi G}{L}$ (B) $0.05\dfrac{\pi G}{L}$ (C) $0.25\dfrac{\pi G}{L}$

(D) $0.5\dfrac{\pi G}{L}$ (E) $\dfrac{\pi G}{L}$

The angle of twist is:

$$\phi = \frac{TL}{GJ}$$

J for a circular bar of diameter d is $\frac{1}{2}\pi d^4$. The total angle of twist, ϕ_{total}, is equal to the sum of the angles of twist for the two different sections.

$$
\begin{aligned}
\phi_{\text{total}} &= \phi_1 + \phi_2 \\
&= \frac{T(2L)}{GJ_1} + \frac{T(L)}{GJ_2} \\
&= \frac{2TL}{\pi G}\left(\frac{2}{d_1^4} + \frac{1}{d_2^4}\right) \\
&= \frac{2TL}{\pi G}\left(\frac{2}{(2)^4} + \frac{1}{(1)^4}\right) \\
&= \frac{2.25TL}{\pi G} \\
T &= \frac{\pi G \phi_{\text{total}}}{2.25L} \\
&= 0.01\frac{\pi G}{L}
\end{aligned}
$$

Answer is (A)

MECHANICS OF MATERIALS-37

For the shaft below, what is the largest torque that can be applied if the shear stress is not to exceed 110 MPa?

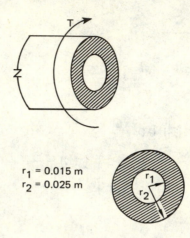

$r_1 = 0.015$ m
$r_2 = 0.025$ m

(A) 1720 N·m   (B) 1870 N·m   (C) 2350 N·m
          (D) 3360 N·m   (E) 5420 N·m

Since the shear stress is largest at the outer diameter, the maximum torque is found using this radius.

$$T_{max} = \frac{\tau J}{r_2}$$

For an annular region,

$$J = \frac{\pi}{2}(r_2^4 - r_1^4) = \frac{\pi}{2}\left[(0.025)^4 - (0.015)^4\right]$$
$$= 5.34 \times 10^{-7} \text{ m}^4$$
$$T_{max} = \frac{(5.34 \times 10^{-7})(110 \times 10^6)}{0.025}$$
$$= 2350 \text{ N·m}$$

Answer is (C)

MECHANICS OF MATERIALS–38

A hollow circular bar has an inner radius r_1 and an outer radius r_2. If $r_1 = \frac{1}{2}r_2$, what percentage of torque can the shaft carry in comparison with a solid shaft?

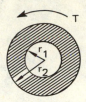

(A) 25%   (B) 50%   (C) 73%   (D) 94%   (E) 100%

The equation for torsional stress gives:

$$\tau = \frac{Tr}{J}$$

$$T = \frac{\tau J}{r}$$

For the hollow shaft,

$$T_h = \frac{\tau(\frac{\pi}{2})(r_2^4 - r_1^4)}{r_2}$$

For the solid shaft,

$$T_s = \frac{\tau(\frac{\pi}{2})(r_2^4)}{r_2}$$

Therefore,

$$\frac{T_h}{T_s} = \frac{\tau\left(\dfrac{\pi}{2}\right)\dfrac{(r_2^4 - r_1^4)}{r_2}}{\tau\left(\dfrac{\pi}{2}\right)\left(\dfrac{(r_2^4)}{r_2}\right)}$$

$$= \frac{r_2^4 - r_1^4}{r_2^4}$$

$$= \frac{r_2^4 - \left(\dfrac{r_2}{2}\right)^4}{r_2^4} = 0.94$$

$$= 94\%$$

Answer is (D)

MECHANICS OF MATERIALS–39

Of the following solid shaft diameters, which is the smallest that can be used for the rotor of a 6 horsepower motor operating at 3500 rpm, if the maximum shear stress for the shaft is 8500 psi?

(A) $\frac{5}{16}$ in   (B) $\frac{3}{8}$ in   (C) $\frac{1}{2}$ in

          (D) 1 in   (E) 2 in

The relationship between the horsepower, H, transmitted by a shaft and the torque, T, is:

$$H = \frac{2\pi nT}{33,000}$$

$n =$ rpm, $T =$ ft-lbf, and $H =$ hp

$$T = \frac{(33,000)(6)}{(2)(\pi)(3500)} = 9.0 \text{ ft-lbf}$$

$$\tau_{max} = \frac{Tr}{J} = \frac{Td}{2J}$$

$$J = \frac{Td}{2\tau_{max}} = \frac{\pi d^4}{32}$$

Therefore, $d = \left[\frac{32T}{2\pi\tau_{max}}\right]^{1/3} = \left[\frac{(16)(9)(12)}{(\pi)(8500)}\right]^{1/3}$

$$= 0.401 \text{ in}$$

The $\frac{1}{2}''$ shaft should be chosen.

Answer is (C)

MECHANICS OF MATERIALS–40

A beam supports a distributed load, w, as shown. Find the shear force at $x = 2$ feet.

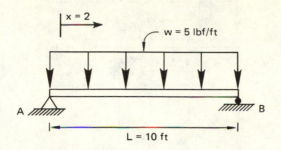

(A) 10 lbf   (B) 12 lbf   (C) 13 lbf

(D) 14 lbf   (E) 15 lbf

The reactions at A and B are found by symmetry to be $R_A = R_B = 25$ lbf. Sectioning the beam at $x = 2$ feet, the free-body diagram with shear force is:

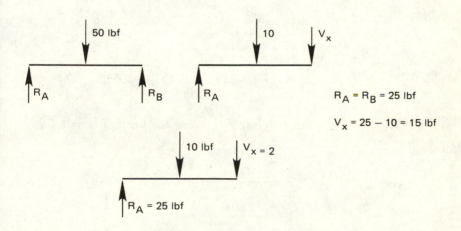

$$\sum F_y = 0 : 25 - 10 - V_{x=2} = 0$$

$$V_{x=2} = 25 - 10 = 15 \text{ lbf}$$

Answer is (E)

MECHANICS OF MATERIALS–41

For the beam shown, find the bending moment, M, at $x = 3$ feet.

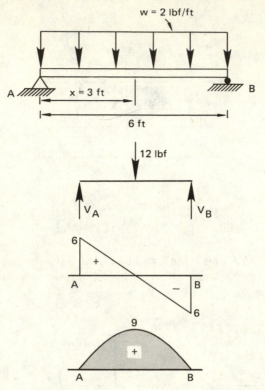

(A) 4.5 ft-lbf   (B) 6 ft-lbf   (C) 8 ft-lbf
          (D) 9 ft-lbf   (E) 12 ft-lbf

By symmetry, $R_A = R_B = 6$ lbf. Sectioning the beam at $x = 3$ feet gives:

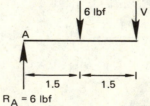

$$\sum M_{x=3} = 0 : -(6)(3) + (6)(1.5) + M = 0$$

$$M = 18 - 9$$

$$= 9 \text{ ft-lbf}$$

Answer is (D)

MECHANICS OF MATERIALS-42

Find the expression for the bending moment as a function of x for the following beam.

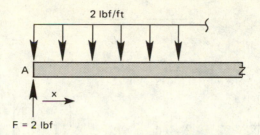

(A) $M(x) = -x^3 + 2x$ (B) $M(x) = -x^2 + 1$ (C) $M(x) = -x^2 + 2x$

(D) $M(x) = x^3 - 2x^2$ (E) $M(x) = x^2 - 2x$

$$\sum M = 0 : -2x + \left(2x\right)\left(\frac{1}{2}x\right) + M(x) = 0$$

$$M(x) = -x^2 + 2x$$

Answer is (C)

MECHANICS OF MATERIALS–43

Which of the following is the shear force diagram for this beam?

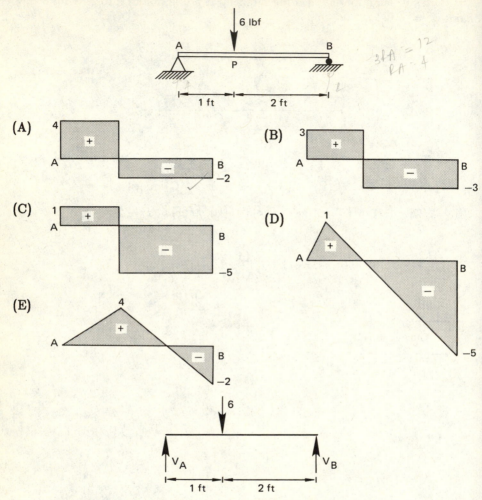

The reactions at points A and B are $R_A = 4$ lbf and $R_B = 2$ lbf. Drawing free-body diagrams of the left and right sections of the beam,

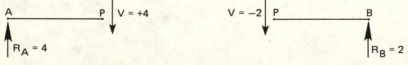

Thus, $V = +4$ lbf between points A and P, and $V = -2$ lbf between points P and B.

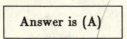

Answer is (A)

MECHANICS OF MATERIALS–44

Which of the following is the bending moment diagram for this beam?

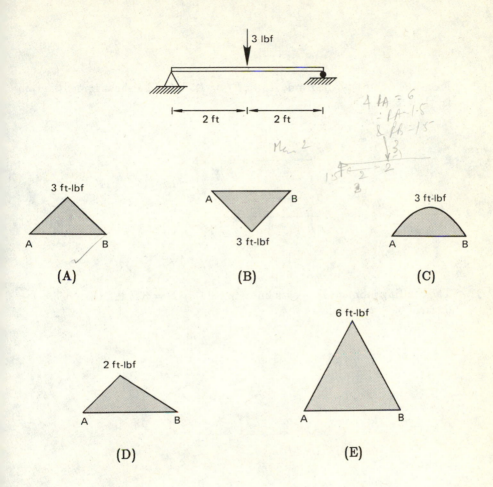

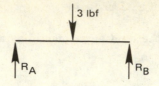

From the free-body diagram, $R_A = R_B = 1.5$ lbf. The shear force diagram is, therefore,

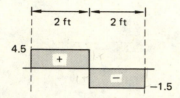

The bending moment increases linearly to $(1.5)(2) = 3$ ft-lbf, then decreases linearly back to 0 ft-lbf.

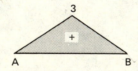

Answer is (A)

MECHANICS OF MATERIALS–45

What is the maximum shear force in the cantilever beam below? The beam is loaded by three concentrated forces as shown.

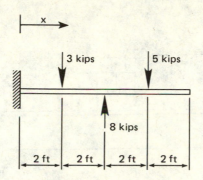

(A) 0 kips   (B) 2 kips   (C) 3 kips

(D) 5 kips   (E) 8 kips

Examining the shear force along the beam from left to right,

$$
\begin{aligned}
\text{for} \quad 0 < x < 2, \quad & V = 0 \text{ kips} \\
2 < x < 4, \quad & V = -3 \text{ kips} \\
4 < x < 6, \quad & V = 5 \text{ kips} \\
6 < x < 8, \quad & V = 0 \text{ kips}
\end{aligned}
$$

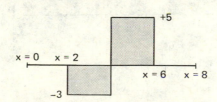

The maximum shear force is, therefore, 5 kips.

Answer is (D)

MECHANICS OF MATERIALS–46

Which of the following bending moment diagrams corresponds to the simply supported beam shown? The beam is subjected to a distributed load, w, between points B and C.

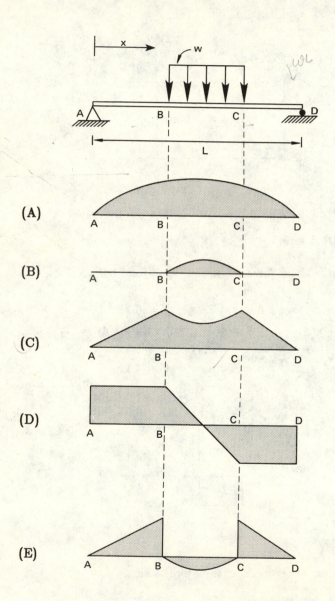

For sections AB and CD, the beam may be modeled as:

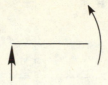

$M(x)$ is linear with respect to x. For section BC, the beam is modeled as:

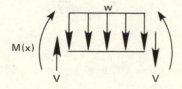

$M(x)$ is parabolic, reaching a maximum at the center.

Answer is (C)

MECHANICS OF MATERIALS–47

Determine the maximum allowable load, F, on the cantilever if the maximum compressive stress is 1000 psi and the maximum tensile stress is 800 psi.

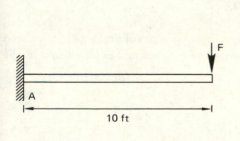

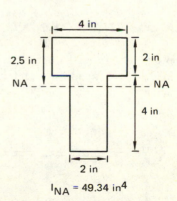

$I_{NA} = 49.34$ in^4

(A) 117 lbf   (B) 132 lbf   (C) 144 lbf
          (D) 152 lbf   (E) 168 lbf

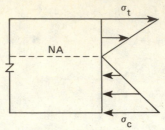

The maximum bending moment occurs at A, where $M = 10F$.

$$\sigma_{max} = \frac{Mc}{I} = \frac{10Fc}{I}$$

$$F = \frac{\sigma_{max}I}{10c}$$

For compression, $\sigma_{allowable} = 1000$ psi and $c = 3.5$ inches.

$$F_{\substack{allowable \\ compression}} = \frac{(1000)(49.34)}{(10)(12)(3.5)} = 117 \text{ lbf}$$

For tension, $\sigma_{allowable} = 800$ psi and $c = 2.5$ inches.

$$F_{\substack{allowable \\ tension}} = \frac{(800)(49.34)}{(10)(12)(2.5)} = 132 \text{ lbf}$$

The maximum allowable load is 117 lbf.

$$\boxed{\text{Answer is (A)}}$$

MECHANICS OF MATERIALS–48

A simply supported beam with the cross section shown supports a concentrated load, $F = 3$ kips, at its center, C. Calculate the maximum bending stress in the beam.

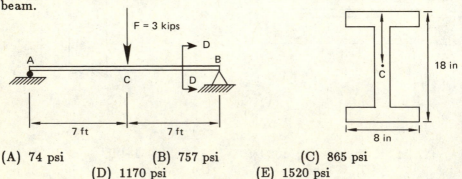

(A) 74 psi   (B) 757 psi   (C) 865 psi
            (D) 1170 psi   (E) 1520 psi

The reactions at A and B are $R_A = R_B = 1.5$ kips by symmetry. Since the maximum bending moment occurs at C,

$$M_{max} = (1500)(7) = 10,500 \text{ ft-lbf}$$

The moment of inertia is the difference between the moments of inertia of an area measuring $8'' \times 18''$ and two areas measuring $3.5'' \times 16''$:

$$I = \left(\tfrac{1}{12}\right)(8)(18)^3 - (2)\left(\tfrac{1}{12}\right)(3.5)(16)^3 = 1499 \text{ in}^4$$

Since $c = \left(\tfrac{1}{2}\right)(18) = 9$ in,

$$\sigma_{max} = \frac{Mc}{I} = \frac{(10,500)(9)(12)}{1499}$$
$$= 757 \text{ psi}$$

Answer is (B)

MECHANICS OF MATERIALS–49

For the cantilever beam shown, what is the maximum bending stress in the tension fiber?

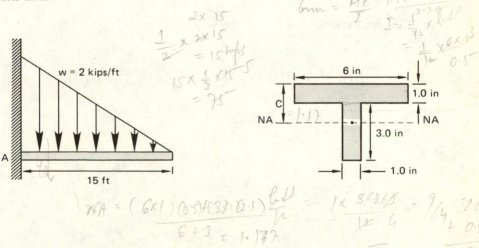

(A) 2.28 ksi   (B) 44.1 ksi   (C) 62.5 ksi

(D) 97.9 ksi   (E) 132 ksi

The maximum moment occurs at A and is a result of the distributed load w. (w is equivalent to a concentrated load, $W = (\frac{1}{2})(15)(2) = 15$ kips, acting at a point $(\frac{1}{3})(15) = 5$ feet from A.) The free-body diagram for the cantilever is:

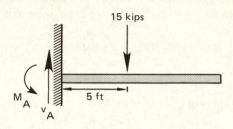

$$M_A = (15,000)(5) = 75,000 \text{ ft-lbf}$$

The upper part of the beam will be under tension, with c equal to the distance between the neutral axis, NA, and the top edge of the beam.

$$c = \frac{\sum A\bar{y}}{\sum A} = \frac{(1)(6)(0.5) + (1)(3)(2.5)}{6 + 3} = 1.17 \text{ in}$$

To find I, the parallel axis theorem is used:

$$I_{NA} = (\tfrac{1}{12})(6)(1)^3 + (6)(1)(1.2 - 0.5)^2 + (\tfrac{1}{12})(1)(3)^3 + (3)(1)(2.5 - 1.2)^2$$
$$= 10.76 \text{ in}^4$$

$$\sigma = \frac{Mc}{I} = \frac{(75,000)(1.17)(12)}{10.76}$$
$$= 97.9 \text{ ksi}$$

Answer is (D)

MECHANICS OF MATERIALS-50

A composite beam made of steel and wood is subjected to a uniform distributed load, w. Determine the maximum compressive stress in the steel.

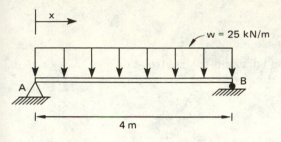

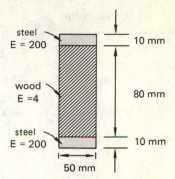

(A) 600 MPa
(B) 800 MPa
(C) 1000 MPa
(D) 1200 MPa
(E) 1400 MPa

The maximum moment is at the center of the beam, where $x = 2$ meters.

$$R_A = R_B = 50 \text{ kN}$$
$$M_{max} = (50)(2) - (25)(2)(1)$$
$$= 50 \text{ kN·m}$$

Since $E_{wood}/E_{steel} = 4/200 = 1/50$, the wood is equivalent to a steel web 1 mm thick.

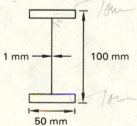

$$I = \left(\frac{1}{12}\right)(50)(100)^3 - (2)\left(\frac{1}{12}\right)(24.5)(80)^3$$
$$= 2.076 \times 10^6 \text{ mm}^4 = 2.076 \times 10^{-6} \text{ m}^4$$
$$\sigma = \frac{Mc}{I} = \frac{(50)(0.05)}{2.076 \times 10^{-6}}$$
$$= 1200 \text{ MPa}$$

Answer is (D)

MECHANICS OF MATERIALS-51

In bending of a rectangular beam under axial loading, where is the location of maximum shear stress?

(A) at the top edge
(B) at the bottom edge
(C) at the neutral axis
(D) at a location between the top edge and the neutral axis
(E) at a location between the bottom edge and the neutral axis

The shear distribution is:

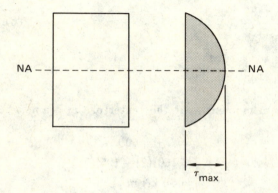

The maximum shear stress is at the neutral axis.

Answer is (C)

MECHANICS OF MATERIALS–52

An I-beam is loaded as shown. Given that the thickness of the web, t, is 20 millimeters, what is the maximum shear stress, τ, in the web at point C along the beam?

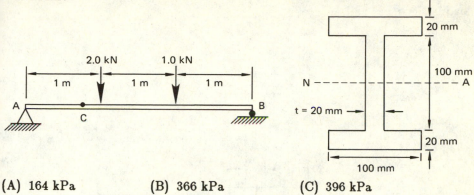

(A) 164 kPa   (B) 366 kPa   (C) 396 kPa
          (D) 747 kPa   (E) 984 kPa

The reaction at A is found by taking the moment about B:

$$\sum M_B = 0: -R_A(3) + (2)(2) + (1)(1) = 0$$
$$R_A = 1.67 \text{ kN}$$
$$V_C = R_A = 1.67 \text{ kN}$$

The shear stress is given by $\tau = VQ/It$, where Q is the first moment of either the upper half or the lower half of the cross-sectional area with respect to the neutral axis.

$$Q = (50)(20)(25) + (100)(20)(60) = 145,000 \text{ mm}^3$$
$$I = (\frac{1}{12})(100)(140)^3 - (\frac{1}{12})(80)(100)^3$$
$$= 16.2 \times 10^6 \text{ mm}^4$$

τ_{max} occurs at the neutral axis.

$$\text{Thus,} \quad \tau_{\text{max}} = \frac{VQ}{It} = \frac{(1670)(145,000)}{(16.2 \times 10^6)(20)}$$
$$= 0.747 \text{ N/mm}^2 = 747 \text{ kPa}$$

Answer is (D)

MECHANICS OF MATERIALS–53

An I-beam is made of three planks, each 20 mm × 100 mm in cross section, nailed together as shown. If the longitudinal spacing, x, between the nails is 25 mm, and the shear force, V, acting on the cross section is 600 N, what is the load in shear per nail, F?

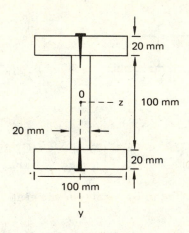

(A) 55.5 N   (B) 75 N   (C) 111 N   (D) 150 N   (E) 444 N

The shear force per unit distance along the beam's axis is given by:

$$f = \frac{VQ}{I}$$

For an I-beam, Q is the first moment of the upper flange area with respect to the z-axis.

$$Q = A\bar{y} = (60)(100)(20) = 120{,}000 \text{ mm}^3$$
$$I = (\tfrac{1}{12})(100)(140)^3 - (2)(\tfrac{1}{12})(40)(100)^3 = 16.2 \times 10^6 \text{ mm}^4$$
$$f = \frac{VQ}{I} = \frac{(600)(120{,}000)}{16.2 \times 10^6} = 4.44 \text{ N/mm}$$

The load capacity of the nails per unit length is F/x. Therefore,

$$F = xf = (25)(4.44) = 111 \text{ N}$$

Answer is (C)

MECHANICS OF MATERIALS–54

Considering the orientation of τ_{yx} in the figure, find the direction of the shear stress on the other three sides of the stress element.

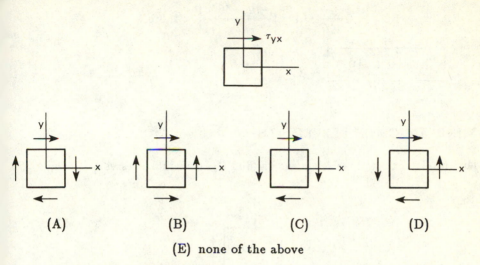

(A)   (B)   (C)   (D)

(E) none of the above

For static equilibrium, the shear stresses on opposite faces of an element must be equal in magnitude and opposite in direction. Also, the shear stresses on adjoining faces must not produce rotation of the element.

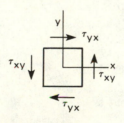

Answer is (D)

MECHANICS OF MATERIALS–55

If the principal stresses on a body are $\sigma_1 = 400$ psi, $\sigma_2 = -700$ psi, and $\sigma_3 = 600$ psi, what is the maximum shear stress?

(A) 100 psi   (B) 200 psi   (C) 550 psi

(D) 650 psi   (E) 1300 psi

The maximum shear stress is equal to one-half of the difference between the principal stresses. Comparing the three combinations, the maximum shear stress is given by the difference between σ_2 and σ_3.

$$\left|\frac{\sigma_2 - \sigma_3}{2}\right| = \left|\frac{-700 - 600}{2}\right| = 650 \text{ psi}$$

Answer is (D)

MECHANICS OF MATERIALS–56

For the element of plane stress shown, find the principal stresses $\sigma_{\max}$ and $\sigma_{\min}$.

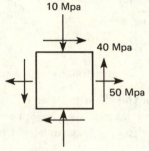

(A) $\sigma_{\max} = 30$ MPa, $\sigma_{\min} = 20$ MPa
(B) $\sigma_{\max} = 40$ MPa, $\sigma_{\min} = 50$ MPa
(C) $\sigma_{\max} = 70$ MPa, $\sigma_{\min} = -30$ MPa
(D) $\sigma_{\max} = 80$ MPa, $\sigma_{\min} = 10$ MPa
(E) $\sigma_{\max} = 90$ MPa, $\sigma_{\min} = -20$ MPa

The stresses on the element are:

$$\sigma_x = 50 \text{ MPa} \qquad \sigma_y = -10 \text{ MPa} \qquad \tau_{xy} = 40 \text{ MPa}$$

$$\sigma_{\max,\min} = \frac{\sigma_x + \sigma_y}{2} \pm \sqrt{\left(\frac{\sigma_x - \sigma_y}{2}\right)^2 + \tau_{xy}^2}$$

$$= \frac{50 - 10}{2} \pm \sqrt{\left(\frac{50 - 10}{2}\right)^2 + (40)^2}$$

$$= 70 \text{ MPa}, \ -30 \text{ MPa}$$

Answer is (C)

MECHANICS OF MATERIALS-57

What are the principal stresses of this stress element?

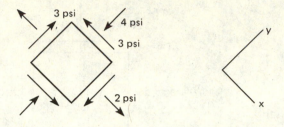

(A) $\sigma_{max} = 0.16$ psi, $\sigma_{min} = -6.16$ psi
(B) $\sigma_{max} = 2.00$ psi, $\sigma_{min} = -4.00$ psi
(C) $\sigma_{max} = 3.24$ psi, $\sigma_{min} = -5.24$ psi
(D) $\sigma_{max} = 5.24$ psi, $\sigma_{min} = -3.24$ psi
(E) $\sigma_{max} = 6.16$ psi, $\sigma_{min} = -0.16$ psi

The stresses on the element are:

$$\sigma_x = 2 \text{ psi} \qquad \sigma_y = -4 \text{ psi} \qquad \tau_{xy} = -3 \text{ psi}$$

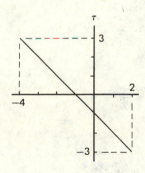

$$\sigma_{max,min} = \frac{\sigma_x + \sigma_y}{2} \pm \sqrt{\left(\frac{\sigma_x - \sigma_y}{2}\right)^2 + \tau_{xy}^2}$$

$$= \frac{2 - 4}{2} \pm \sqrt{\left(\frac{2 + 4}{2}\right)^2 + (-3)^2}$$

$$= 3.24 \text{ psi}, \ -5.24 \text{ psi}$$

Answer is (C)

MECHANICS OF MATERIALS-58

Find the maximum principal stress of the element.

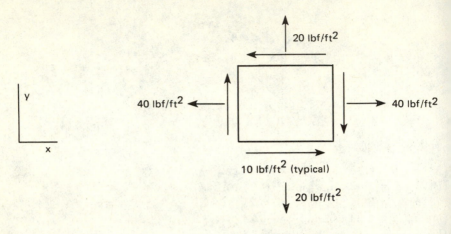

(A) 30.1 lbf/ft^2 (B) 34.4 lbf/ft^2 (C) 40.1 lbf/ft^2

(D) 44.1 lbf/ft^2 (E) 46.1 lbf/ft^2

The stresses on the element are:

$\sigma_x = 40 \text{ lbf/ft}^2$ $\sigma_y = 20 \text{ lbf/ft}^2$ $\tau_{xy} = -10 \text{ lbf/ft}^2$

$$\sigma_{\max} = \frac{\sigma_x + \sigma_y}{2} + \sqrt{\left(\frac{\sigma_x - \sigma_y}{2}\right)^2 + \tau_{xy}^2}$$

$$= \frac{40 + 20}{2} + \sqrt{\left(\frac{40 - 20}{2}\right)^2 (-10)^2}$$

$$= 44.1 \text{ lbf/ft}^2$$

Answer is (D)

MECHANICS OF MATERIALS–59

For the following stress element, determine the maximum shear stress.

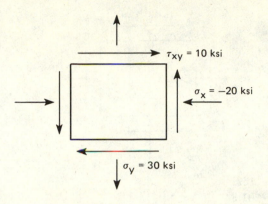

(A) 10 ksi   (B) 11.2 ksi   (C) 14.1 ksi

(D) 20 ksi   (E) 26.9 ksi

The maximum shear stress is:

$$\tau_{max} = \sqrt{\left(\frac{\sigma_x - \sigma_y}{2}\right)^2 + \tau_{xy}^2}$$

$$= \sqrt{\left(\frac{-20 - 30}{2}\right)^2 + (10)^2}$$

$$= 26.9 \text{ ksi}$$

Answer is (E)

MECHANICS OF MATERIALS–60

For the state of plane stress shown, what are the principal planes?

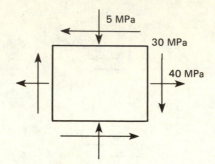

(A) 32° and 122°
(B) 25° and 115°
(C) −26.5° and −116.5°
(D) −11.5° and −101.5°
(E) −9.2° and −99.2°

The stresses on the element are:

$$\sigma_x = 40 \text{ MPa} \qquad \sigma_y = -5 \text{ MPa} \qquad \tau_{xy} = -30 \text{ MPa}$$

$$\tan 2\theta_p = \frac{2\tau_{xy}}{\sigma_x - \sigma_y} = \frac{(2)(-30)}{40 - (-5)}$$
$$= -1.33$$
$$2\theta_p = -53.0° \text{ or } -233°$$
$$\theta_p = -26.5° \text{ or } -116.5°$$

Answer is (C)

MECHANICS OF MATERIALS–61

A steel ($\sigma_{\text{yield}} = 30$ ksi) pressure tank is designed to hold pressures up to 1000 psi. The tank is cylindrical with a diameter of 3 feet. If the longitudinal stress must be less than 20% of the yield stress of the steel, what is the necessary wall thickness, t?

(A) 0.75 in   (B) 1.50 in   (C) 3 in

(D) 3.75 in   (E) 7.50 in

For a thin-walled cylinder of diameter d containing a pressure, p,

$$\sigma_{long} = \frac{pd}{4t} = 0.2\,\sigma_{yield}$$

$$d = (3)\left(12\,\frac{in}{ft}\right) = 36 \text{ in}$$

$$t = \frac{pd}{0.8\,\sigma_{yield}} = \frac{(1000)(36)}{(0.8)(30,000)}$$

$$= 1.50 \text{ in}$$

Answer is (B)

MECHANICS OF MATERIALS–62

In designing a cylindrical pressure tank 3 feet in diameter, a factor of safety of 2.5 is used. The cylinder is made of steel ($\sigma_{yield} = 30$ ksi), and will contain pressures up to 1000 psi. What is the required wall thickness, t, based on circumferential stress considerations?

(A) 0.75 in   (B) 1.50 in   (C) 3 in
         (D) 3.75 in   (E) 7.50 in

For a thin-walled cylinder of diameter d and containing a pressure, p,

$$\sigma_{circumferential} = \frac{pd}{2t} = \frac{\sigma_{yield}}{2.5}$$

$$d = (3)\left(12\,\frac{in}{ft}\right) = 36 \text{ in}$$

$$t = \frac{1.25pd}{\sigma_{yield}} = \frac{(1.25)(1000)(36)}{30,000}$$

$$= 1.50 \text{ in}$$

Answer is (B)

MECHANICS OF MATERIALS–63

Determine the maximum principal strain at a point where $\epsilon_x = 1500\ \mu$m, $\epsilon_y = -750\ \mu$m, and $\tau_{xy} = 1000\ \mu$m.

(A) 1160 μm   (B) 1485 μm   (C) 1606 μm
         (D) 1825 μm   (E) 2320 μm

The equation for principal strain gives:

$$\epsilon_{max,min} = \frac{\epsilon_x + \epsilon_y}{2} \pm \sqrt{\left(\frac{\epsilon_x - \epsilon_y}{2}\right)^2 + \left(\frac{\tau_{xy}}{2}\right)^2}$$

$$= \frac{1500 - 750}{2} \pm \sqrt{\left(\frac{1500 + 750}{2}\right)^2 + \left(\frac{1000}{2}\right)^2}$$

$$= 375 \pm 1231$$

$$\epsilon_{max} = 1606\ \mu m$$

Answer is (C)

MECHANICS OF MATERIALS–64

A beam of length L carries a concentrated load, F, at point C. Determine the deflection at point C in terms of F, L, E, and I, where E is the modulus of elasticity, and I is the moment of inertia.

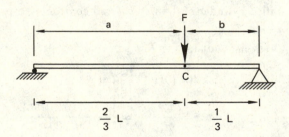

(A) $\dfrac{2FL^3}{243EI}$ (B) $\dfrac{4FL^3}{243EI}$ (C) $\dfrac{FL^3}{27EI}$

(D) $\dfrac{FL^3}{9EI}$ (E) $\dfrac{4FL^3}{9EI}$

The equation for bending moment in the beam is:

$$EI\delta'' = -M$$

Computing M for the different beam sections,

$$EI\delta'' = -\frac{Fbx}{L} \qquad (0 \le x \le a)$$

$$EI\delta'' = -\frac{Fbx}{L} + F(x-a) \qquad (a \le x \le L)$$

Integrating each equation twice gives:

$$EI\delta = -\frac{Fbx^3}{6L} + C_1 x + C_3 \qquad (0 \le x \le a)$$

$$EI\delta = -\frac{Fbx^3}{6L} + \frac{F(x-a)^3}{6} + C_2 x + C_4 \qquad (a \le x \le L)$$

The constants are determined by the following conditions: (1) at $x = a$, the slopes δ' and deflections δ are equal; (2) at $x = 0$ and $x = L$, the deflection $\delta = 0$. These conditions give:

$$C_1 = C_2 = \frac{Fb(L^2 - b^2)}{6L}, \quad C_3 = C_4 = 0$$

Evaluating the equation for $(0 \le x \le a)$ at $x = a = \frac{2}{3}L$ and $b = \frac{1}{3}L$:

$$EI\delta = \frac{F(\frac{1}{3}L)(\frac{2}{3}L)}{6L}\left(L^2 - \frac{L^2}{9} - \frac{4L^2}{9}\right)$$

$$\delta = \frac{4FL^3}{243EI}$$

Answer is (B)

MECHANICS OF MATERIALS–65

What is the deflection at point B of the beam below?

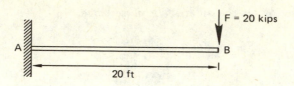

(A) 0.25 in   (B) 0.784 in   (C) 0.993 in

(D) 1.12 in   (E) 1.50 in

Using the moment-area method, the M/EI diagram is:

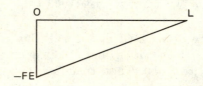

$$A = (\tfrac{1}{2})(L)(-FL)\left(\frac{1}{EI}\right) = -\frac{FL^2}{2EI}$$

The first moment with respect to point A is:

$$Q = A(\tfrac{2}{3}L) = -\left(\frac{FL^2}{2EI}\right)(\tfrac{2}{3}L) = -\frac{FL^3}{3EI}$$

The displacement is:

$$y = -Q = \frac{FL^3}{3EI} = \frac{(20)(240)^3}{(3)(29 \times 10^6)(3200)}$$
$$= 0.993 \text{ in}$$

Answer is (C)

MECHANICS OF MATERIALS-66

The propped cantilever shown is loaded by forces F_1 and F_2. The magnitudes of F_1 and F_2 are increased gradually. How many failure mechanisms are possible for this system?

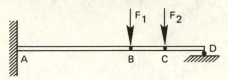

(A) one
(B) two
(C) three
(D) four
(E) none

The peak values of the bending moment will occur where loads or reactions are acting, that is, at points A, B, or C. Failure will occur if the beam hinges at any two of these cross sections, hence, there are three possible failure mechanisms.

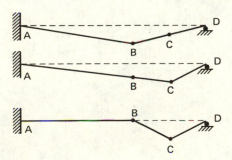

Answer is (C)

11 DYNAMICS

DYNAMICS-1

How many degrees of freedom does a coin rolling on the ground have?

(A)   one   (B) two   (C) three

          (D) five   (E) six

A coin has two translational degrees of freedom and one rotational degree of freedom.

Answer is (C)

DYNAMICS-2

What is the definition of instantaneous velocity?

(A) $v = dx\, dt$ (B) $v = \int x\, dt$ (C) $v = \dfrac{dx}{dt}$

    (D) $v = \lim\limits_{\Delta t \to 0} \dfrac{\Delta t}{\Delta x}$ (E) $v = xt$

By definition,

$$v = \lim_{\Delta t \to 0} \frac{\Delta x}{\Delta t} = \frac{dx}{dt}$$

Answer is (C)

DYNAMICS-3

A car travels 100 miles to city A in 2 hours, then travels 200 miles to city B in 3 hours. What is the average speed of the car for the trip?

(A) 45 mph   (B) 58.3 mph   (C) 60 mph
          (D) 66.7 mph   (E) 70 mph

Average velocity is defined as total distance traveled over total time:

$$v_{average} = \frac{\Delta x}{\Delta t} = \frac{100 + 200}{2 + 3}$$
$$= 60 \text{ mph}$$

Answer is (C)

DYNAMICS-4

The position of a particle moving along the x-axis is given by $x(t) = t^2 - t + 8$, where x is in units of feet, and t is in seconds. Find the velocity of the particle when $t = 5$ seconds.

(A) 9 ft/sec   (B) 10 ft/sec   (C) 11 ft/sec
          (D) 12 ft/sec   (E) 13 ft/sec

The velocity equation is the first derivative of the position equation with respect to time. Therefore,

$$v(t) = \frac{dx}{dt}$$
$$= \frac{d}{dt}(t^2 - t + 8) = 2t - 1$$
$$v(5) = (2)(5) - 1 = 9 \text{ ft/sec}$$

Answer is (A)

DYNAMICS-5

If a particle's position is given by the expression $x(t) = 3.4t^3 - 5.4t$ meters, what is the acceleration of the particle after $t = 5$ seconds?

(A) 1.02 m/s^2 (B) 3.40 m/s^2 (C) 18.1 m/s^2
          (D) 25.5 m/s^2 (E) 102 m/s^2

The acceleration is found from the second derivative of the position equation. Therefore,

$$a(t) = \frac{d^2 x}{dt^2}$$
$$= \frac{d^2}{dt^2}(3.4t^3 - 5.4t)$$
$$= \frac{d}{dt}(10.2t^2 - 5.4)$$
$$= 20.4t$$
$$a(5) = (20.4)(5) = 102 \text{ m/s}^2$$

Answer is (E)

DYNAMICS-6

A car starts from rest and moves with a constant acceleration of 6 m/s². What is the speed of the car after 4 seconds?

(A) 18 m/s   (B) 24 m/s   (C) 35 m/s

(D) 55 m/s   (E) 64 m/s

For uniformly accelerated motion,

$$v = v_0 + at = 0 + (6)(4)$$
$$= 24 \text{ m/s}$$

Answer is (B)

DYNAMICS-7

A car starts from rest and has a constant acceleration of 3 ft/sec². What is the average velocity during the first 10 seconds of motion?

(A) 12 ft/sec   (B) 13 ft/sec   (C) 14 ft/sec

(D) 15 ft/sec   (E) 16 ft/sec

The distance traveled by the car is:

$$x = x_0 + v_0 t + \tfrac{1}{2}at^2 = 0 + 0 + \tfrac{1}{2}(3)(10)^2$$
$$= 150 \text{ ft}$$

$$V_{average} = \frac{\Delta x}{\Delta t} = \frac{150}{10}$$
$$= 15 \text{ ft/sec}$$

Answer is (D)

DYNAMICS-8

A truck increases its speed uniformly from 13 km/hr to 50 km/hr in 25 seconds. What is the acceleration of the truck?

(A) 0.216 m/s²   (B) 0.411 m/s²   (C) 0.622 m/s²

(D) 0.924 m/s²   (E) 1.87 m/s²

For uniformly accelerated rectilinear motion,

$$v = v_0 + at$$
$$at = v - v_0$$
$$a = \frac{v - v_0}{t} = \left(\frac{50 - 13}{25}\right)\left(\frac{1000}{3600}\right)$$
$$= 0.411 \text{ m/s}^2$$

Answer is (B)

DYNAMICS-9

A bicycle moves with a constant deceleration of -2 ft/sec². If the initial velocity of the bike was 10 ft/sec, how far does it travel in 3 seconds?

(A) 19 ft   (B) 20 ft   (C) 21 ft   (D) 22 ft   (E) 23 ft

For constant acceleration,

$$x = x_0 + v_0 t + \tfrac{1}{2}at^2$$
$$= 0 + (10)(3) + (\tfrac{1}{2})(-2)(3)^2$$
$$= 21 \text{ ft}$$

Answer is (C)

DYNAMICS–10

A ball is dropped from a height of 60 meters above ground. How long does it take to hit the ground?

(A) 1.3 s   (B) 2.1 s   (C) 3.5 s
            (D) 5.5 s   (E) 8 s

The positive y direction is downward, and $y = 0$ at 60 m above ground. For uniformly accelerated motion,

$$y = y_0 + v_0 t + \tfrac{1}{2}at^2$$
$$60 = 0 + 0 + (\tfrac{1}{2})(9.81)t^2$$
$$t = 3.5 \text{ s}$$

Answer is (C)

DYNAMICS–11

A man driving a car at 45 mph suddenly sees an object in the road 60 feet ahead. What constant deceleration is required to stop the car in this distance?

(A) −42.6 ft/sec²   (B) −41.8 ft/sec²   (C) −39.8 ft/sec²
            (D) −36.3 ft/sec²   (E) −35.3 ft/sec²

For uniform deceleration, the velocity equation that is not a function of time is:

$$v^2 = v_0^2 + 2a(x - x_0)$$

Using $v = 0$, $v_0 = 45$ mph $= 66$ ft/sec, and $(x - x_0) = 60$ feet,

$$0 = (66)^2 + 2a(60)$$

$$a = -\frac{66}{(2)(60)} = -36.3 \text{ ft/sec}^2$$

Answer is (D)

DYNAMICS-12

A ball is thrown vertically upward with an initial speed of 80 ft/sec. How long will it take for the ball to return to the thrower?

(A) 2.25 sec   (B) 2.62 sec   (C) 4.06 sec
        (D) 4.97 sec   (E) 6 sec

At the apex of its flight, the ball has zero velocity and is at the midpoint of its flight time. If the total flight time is t_{total}, then the time elapsed at this point is $\frac{1}{2}t_{total}$.

$$v = v_0 + at$$
$$0 = 80 + (-32.2)(\tfrac{1}{2}t_{total})$$
$$t_{total} = \frac{80}{16.1} = 4.97 \text{ sec}$$

Answer is (D)

DYNAMICS-13

A projectile is launched upward from level ground at an angle of 60° with the horizontal. It has an initial velocity of 45 m/s. How long will it take before the projectile hits the ground?

(A) 4.1 s   (B) 5.8 s   (C) 7.94 s
        (D) 9.53 s   (E) 12 s

The projectile will experience acceleration only in the y direction due to gravity. The y component of velocity is:

$$v_{0y} = (45)(\sin 60°) = 39.0 \text{ m/s}$$

For uniform rectilinear motion with constant acceleration,

$$y = y_0 + v_{0y}t + \tfrac{1}{2}at^2$$
$$= 0 + 39t + \tfrac{1}{2}(-9.81)t^2$$
$$= 39t - 4.91t^2$$

When the body is on the ground, $y = 0$.

$$0 = 39t - 4.91t^2 = t(39 - 4.91t)$$
$$t = 7.94 \text{ s}$$

Answer is (C)

DYNAMICS–14

A man standing at a window 5 meters tall watches a falling ball pass by the window in 0.3 seconds. From how high above the top of the window was the ball released?

(A) 8.18 m   (B) 9.63 m   (C) 11.8 m
         (D) 21.3 m   (E) 46.7 m

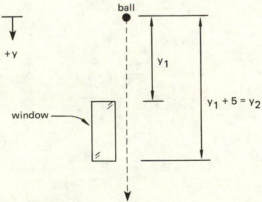

The positive y direction is taken as downward, and the initial release point is y_0. Then,

$$y = y_0 + v_o t + \tfrac{1}{2}at^2 = 0 + 0 + \tfrac{1}{2}at^2$$
$$y_1 = \tfrac{1}{2}at_1^2$$
$$y_2 = \tfrac{1}{2}at_2^2$$

However, $y_2 = y_1 + 5$, and $t_2 = t_1 + 0.3$. Therefore,

$$\tfrac{1}{2}at_1^2 + 5 = \tfrac{1}{2}a(t_1 + 0.3)^2$$
$$\tfrac{1}{2}at_1^2 + 5 = \tfrac{1}{2}at_1^2 + 0.3at_1 + 0.045a$$

$$t_1 = \frac{5 - (0.045)(9.81)}{(0.3)(9.81)} = 1.55 \text{ s}$$

Solving for y_1,

$$y_1 = \tfrac{1}{2}(9.81)(1.55)^2 = 11.8 \text{ m}$$

Answer is (C)

DYNAMICS-15

A block with a spring attached to one end slides along a rough surface with an initial velocity of 7 m/s. After it slides 4 meters, it impacts a wall for 0.1 seconds, and then slides 10 meters in the opposite direction before coming to a stop. If the block's deceleration is assumed constant and the contraction of the spring is negligible, what is the average acceleration of the block during impact with the wall?

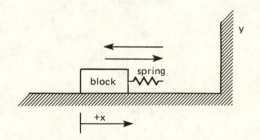

(A) -49 m/s^2 (B) -50 m/s^2 (C) -59 m/s^2
(D) -100 m/s^2 (E) -118 m/s^2

Let v_1 = initial velocity
v_2 = velocity just before impact
v_2' = velocity after impact
v_3 = final velocity
v_1 = 7 m/s
$v_2 = v_2'$ because Δx is small and energy is conserved.
$v_3 = 0$

$$v_2 = \sqrt{v_1^2 - 2a_{1-2}\,s_{1-2}}$$

$$= \sqrt{\left(7\frac{m}{s}\right)^2 - (2)(a_{1-2})(4\,m)} = \sqrt{49 - 8\,a_{1-2}}$$

or $\quad a_{1-2} = \dfrac{49 - v_2{}^2}{8}$

$$v_3 = \sqrt{v_2'{}^2 - 2\,a_{2-3}\,s_{2-3}}$$

$$0 = \sqrt{v_2'{}^2 - (2)a_{2-3}(10)}$$

or $\quad a_{2-3} = \dfrac{v_2'{}^2}{20}$

But $\quad a_{1-2} = a_{2-3}$, and $v_2 = v_2'$.

$$\frac{49 - v_2{}^2}{8} = \frac{v_2{}^2}{20}$$

$$980 - 20v_2{}^2 = 8v_2{}^2$$

$$980 = 28v_2{}^2$$

$$v_2 = 5.916 \text{ m/s}$$

Then, because of the direction change,

$$a_{2-2'} = \frac{v_2' - v_2}{\Delta t} = \frac{-5.916 - 5.916\,\frac{m}{s}}{0.10\,s}$$

$$= -118.3 \text{ m/s}^2$$

Answer is (E)

DYNAMICS-16

A car starting from rest moves with a constant acceleration of 10 mi/hr^2 for 1 hour, then decelerates at a constant -5 mi/hr^2 until it comes to a stop. How far has it traveled?

(A) 10 mi (B) 15 mi (C) 20 mi (D) 25 mi (E) 30 mi

For constant acceleration,

$$x = x_0 + v_0 t + \tfrac{1}{2}at^2$$

During acceleration, $x_0 = 0$, $v_0 = 0$, $a = 10$ mi/hr^2, and $t = 1$ hr. Then, the distance over which the car accelerates is:

$$x_{\text{acc}} = \tfrac{1}{2}(10)(1)^2 = 5 \text{ mi}$$

At the end of the hour, the car's velocity is:

$$v = v_0 + at = (10)(1)$$
$$= 10 \text{ mph}$$

During deceleration, $x_0 = 0$, $v_0 = 10$ mph, and $a = -5$ mi/hr^2. The car has velocity $v = 0$ when it stops. Therefore,

$$v = v_0 + at$$
$$0 = 10 + (-5)t$$
$$t = 2 \text{ hr}$$

The distance over which the car decelerates is:

$$x_{\text{dec}} = (10)(2) + \tfrac{1}{2}(-5)(2)^2 = 10 \text{ mi}$$
$$x_{\text{total}} = x_{\text{acc}} + x_{\text{dec}} = 5 + 10 = 15 \text{ mi}$$

Answer is (B)

DYNAMICS-17

A train with a top speed of 50 mph cannot accelerate or decelerate faster than 4 ft/sec^2. What is the minimum distance between two train stops in order for the train to be able to reach its top speed?

(A) 587 ft   (B) 1250 ft   (C) 1344 ft
          (D) 2500 ft   (E) 2689 ft

To travel the minimum distance, the train must accelerate from $v_0 = 0$ to $v = 50$ mph at a constant 4 ft/sec^2 and then decelerate at a constant -4 ft/sec^2 to $v = 0$. The train travels the same distance during acceleration as during deceleration, since the initial and final speeds are identical, as well as the magnitude of acceleration or deceleration. The following two equations apply for constant acceleration:

$$x = x_0 + v_0 t = \tfrac{1}{2}at^2 \qquad\qquad v = v_0 + at$$

During acceleration, $x_0 = 0$, $v_0 = 0$, $v = 50$ mph $= 73.3$ ft/sec^2, and $a = 4$ ft/sec^2. Then,

$$t = \frac{v - v_0}{a} = \frac{73.3 - 0}{4} = 18.3 \text{ sec}$$

$$x = \tfrac{1}{2}(4)(18.3)^2 = 672 \text{ ft}$$

The minimum total distance is:

$$x = (2)(672) = 1344 \text{ ft}$$

Answer is (C)

DYNAMICS–18

A block with a mass of 115 slugs slides down a frictionless wedge with a slope of 40°. The wedge is moving horizontally in the opposite direction at a constant velocity of 16 ft/sec. What is the absolute speed of the block 2 seconds after it is released from rest?

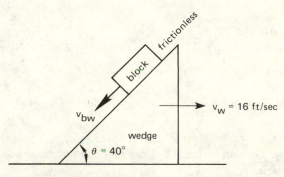

(A) 11.4 ft/sec   (B) 30.9 ft/sec   (C) 31.9 ft/sec

(D) 41.4 ft/sec   (E) 53.2 ft/sec

Let v_{bw} equal the velocity of the block relative to the wedge's slope. The component of gravitational force in this direction, F_{slope}, is $W \sin \theta$. Down the slope, relative to the wedge,

$$F_{slope} = W \sin \theta = m a_{slope}$$

$$a_{slope} = \frac{W \sin \theta}{m} = g \sin \theta$$

$$v_{bw} = v_0 + a_{slope} t = 0 + gt \sin \theta$$
$$= (32.2)(2)(\sin 40°)$$
$$= 41.4 \text{ ft/sec}$$

PROFESSIONAL PUBLICATIONS, INC. • Belmont, CA

The absolute velocity, v_b, can be found from a velocity triangle:

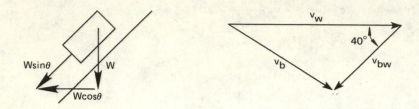

The law of cosines gives:

$$v_b^2 = v_w^2 + v_{bw}^2 - 2v_w v_{bw} \cos \theta$$
$$= (16)^2 + (41.4)^2 - 2(16)(41.4)(\cos 40°)$$
$$= 955$$
$$v_b = 30.9 \text{ ft/sec}$$

Answer is (B)

DYNAMICS–19

A stream flows at $v_s = 3$ mph. At what angle, θ, upstream should a boat travelling at $v_b = 8$ mph be launched in order to reach the shore directly opposite to the launch point?

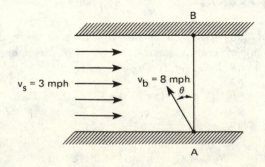

(A) 22°   (B) 24°   (C) 26°   (D) 28°   (E) 30°

Draw a velocity triangle:

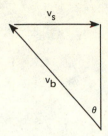

Thus,

$$\sin \theta = \frac{v_s}{v_b} = \frac{3}{8}$$

$$\theta = \sin^{-1}\left(\frac{3}{8}\right)$$

$$= 22°$$

Answer is (A)

DYNAMICS-20

An object is launched at 45° to the horizontal on level ground as shown. What is the range of the projectile if its initial velocity is 180 ft/sec? Neglect air resistance.

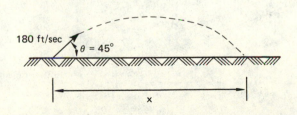

(A) 719 ft   (B) 1000 ft   (C) 1440 ft
          (D) 2050 ft   (E) 4100 ft

Choosing the launch point as the origin of the x and y axes, $x_0 = y_0 = 0$. For uniform acceleration,

$$x = x_0 + v_{x0}t + \tfrac{1}{2}a_x t^2 \qquad\qquad y = y_0 + v_{y0}t + \tfrac{1}{2}a_y t^2$$

However, $a_x = 0$, and $a_y = -32.2$ ft/sec^2. Therefore,

$$x = v_{x0}t \qquad\qquad y = v_{y0}t - \tfrac{1}{2}gt^2$$

$$\begin{aligned} \mathbf{v_{x0}} = v_0 \cos\theta \qquad\qquad & \mathbf{v_{y0}} = v_0 \sin\theta \\ = 180\cos 45^\circ \qquad\qquad & = 180\sin 45^\circ \\ = 127 \text{ ft/sec} \qquad\qquad & = 127 \text{ ft/sec} \end{aligned}$$

$$y = 127t - 16.1t^2$$

When the projectile is on the ground, $y = 0$. Thus,

$$\begin{aligned} 0 &= t(127 - 16.1t) \\ t &= 7.89 \text{ sec} \\ x &= 127t \\ &= (127)(7.89) \\ &= 1000 \text{ ft} \end{aligned}$$

Answer is (B)

DYNAMICS–21

A projectile is fired with a velocity, v, perpendicular to a surface that is inclined at an angle, θ, with the horizontal. Determine the expression for the distance R to the point of impact.

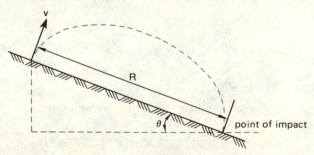

(A) $R = \dfrac{2v^2 \sin\theta}{g\cos^2\theta}$
   (B) $R = \dfrac{2v^2 \sin\theta}{g\cos\theta}$
   (C) $R = \dfrac{2v\cos\theta}{g\sin\theta}$

(D) $R = \dfrac{2v\sin\theta}{g\cos\theta}$
   (E) $R = \dfrac{2v\cos\theta}{g}$

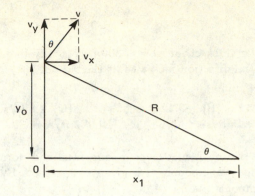

Using the notation in the figure, $x_0 = 0$, $a_x = 0$, and $y_0 = R \sin \theta$. Therefore,

$$x = x_0 + v_x t + \tfrac{1}{2} a_x t^2 \qquad\qquad y = y_0 + v_y t + \tfrac{1}{2} a_y t^2$$
$$\quad = v_x t \qquad\qquad\qquad\qquad = R \sin \theta + v_y t + \tfrac{1}{2}(-g)t^2$$
$$\quad = v \sin \theta t \qquad\qquad\qquad\quad = R \sin \theta + v \cos \theta t - \tfrac{1}{2} g t^2$$

At impact, let $t = t_1$, $x_1 = R \cos \theta$, and $y_1 = 0$. The two equations above give:

$$R \cos \theta = v \sin \theta t_1$$

$$t_1 = \frac{R \cos \theta}{v \sin \theta} \qquad\qquad\qquad\qquad 1$$

$$0 = R \sin \theta + v \cos \theta t_1 - \tfrac{1}{2} g t_1^2 \qquad\qquad 2$$

Equations 1 and 2 give:

$$0 = R \sin \theta + v \cos \theta \left(\frac{R \cos \theta}{v \sin \theta} \right) - \frac{1}{2} g \left(\frac{R \cos \theta}{v \sin \theta} \right)^2$$

$$0 = \sin^2 \theta + \cos^2 \theta - \frac{1}{2} g \left(\frac{R \cos^2 \theta}{v^2 \sin \theta} \right)$$

$$1 = \frac{g \cos^2 \theta R}{2 v^2 \sin \theta}$$

$$R = \frac{2 v^2 \sin \theta}{g \cos^2 \theta}$$

Answer is (A)

DYNAMICS-22

A cyclist on a circular track of radius $r = 800$ feet is traveling at 27 ft/sec. His speed in the tangential direction increases at the rate of 3 ft/sec². What is the cyclist's total acceleration?

(A) -5.1 ft/sec²   (B) -3.12 ft/sec²   (C) 2.8 ft/sec²
      (D) 3.13 ft/sec²   (E) 4.2 ft/sec²

The total acceleration is made up of tangential and normal components. The tangential component is given as 3 ft/sec². By definition, the normal component is:

$$a_n = \frac{v^2}{r} = \frac{(27)^2}{800} = 0.91 \text{ ft/sec}^2$$

$$a = \sqrt{a_n^2 + a_t^2}$$
$$= \sqrt{(0.91)^2 + (3)^2}$$
$$= 3.13 \text{ ft/sec}^2$$

Answer is (D)

DYNAMICS-23

A motorcycle moves at a constant speed of $v = 40$ ft/sec around a curved road of radius $r = 300$ feet. What is the magnitude and general direction of the motorcycle's acceleration?

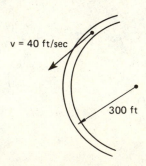

v = 40 ft/sec

300 ft

(A) 0
(B) 2.63 ft/sec² away from the center of curvature
(C) 2.63 ft/sec² towards the center of curvature
(D) 5.33 ft/sec² away from the center of curvature
(E) 5.33 ft/sec² towards the center of curvature

The normal acceleration, a_n, is:

$$a_n = \frac{v^2}{r} = \frac{(40)^2}{300} = 5.33 \text{ ft/sec}^2$$

Since the velocity in the tangential direction is constant, $a_t = 0$. Thus, only the normal component of acceleration contributes to total acceleration, $a = 5.33 \text{ ft/sec}^2$. The normal component is always directed toward the center of curvature.

Answer is (E)

DYNAMICS–24

A pendulum of mass m and length L rotates about the vertical axis. If the angular velocity is ω, determine the expression for the height h.

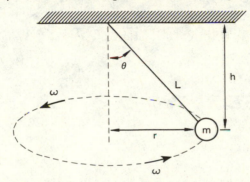

(A) $h = \dfrac{g \cos \theta}{\omega^2}$ ✓(B) $h = \dfrac{g}{\omega^2}$ (C) $h = \dfrac{g}{\omega^2 \cos \theta}$

(D) $h = \dfrac{gL \cos \theta}{\omega^2}$ (E) $h = \dfrac{r}{2}$

A free-body diagram of the pendulum is:

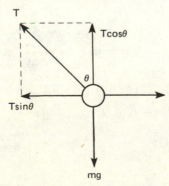

Since the pendulum undergoes uniform circular motion,

$$T \sin \theta = ma_n = mr\omega^2 \qquad\qquad 1$$

Assuming the pendulum is not accelerating in the vertical direction, a force balance gives:

$$T \cos \theta = mg \qquad\qquad 2$$

Combining equations 1 and 2,

$$\tan \theta = \frac{r\omega^2}{g}$$

However, $\tan \theta = \frac{r}{h}$. Therefore,

$$h = \frac{r}{\tan \theta} = \frac{g}{\omega^2}$$

Answer is (B)

DYNAMICS–25

A 3 kg block is moving at a speed of 5 m/s. What is the force required to bring the block to a stop in 8×10^{-4} seconds?

(A) 9.2 N   (B) 13.2 N   (C) 15.4 N
         (D) 18.8 N   (E) 23.2 N

Newton's second law gives:

$$F = ma = m\frac{dv}{dt} = m\frac{\Delta v}{\Delta t}$$
$$= (3 \text{ kg}) \left(\frac{-5 \text{ m/s}}{8 \times 10^{-4}\text{s}} \right)$$
$$= 18.8 \text{ N}$$

Answer is (D)

DYNAMICS-26

A rope is used to tow an 800 kg car on a level surface. The rope will break if the tension exceeds 2000 newtons. What is the greatest acceleration that the car can reach without breaking the rope?

(A) 1.2 m/s^2 (B) 2.5 m/s^2 (C) 3.8 m/s^2

(D) 4.5 m/s^2 (E) 6.8 m/s^2

$$F_{\max} = m a_{\max}$$
$$2000 = 800 a_{\max}$$
$$a_{\max} = \frac{2000\,\text{N}}{800\,\text{kg}} = 2.5\ \text{m/s}^2$$

Answer is (B)

DYNAMICS-27

A force of 5 pounds acts on a 32.2 lbm body for 2 seconds. If the body is initially at rest, how far is it displaced by the force?

(A) 2 ft (B) 5 ft (C) 10 ft (D) 12 ft (E) 15 ft

A 32.2 lbm body has a mass of 1 slug. The acceleration is found using Newton's second law.

$$a = \frac{F}{m} = \frac{5\,\text{lbf}}{1\,\text{slug}} = 5\ \text{ft/sec}^2$$

For a body undergoing constant acceleration, with $v_0 = 0$ and $t = 2$ seconds,

$$\Delta x = \tfrac{1}{2}at^2 = \tfrac{1}{2}(5)(2)^2$$
$$= 10\ \text{ft}$$

Answer is (C)

DYNAMICS-28

A car of mass $m = 1000$ slugs accelerates in 10 seconds from rest at a constant rate to a speed of $v = 20$ ft/sec. What is the resultant force on the car due to this acceleration?

(A) 1000 lbf   (B) 2000 lbf   (C) 3000 lbf

         (D) 4000 lbf   (E) 5000 lbf

For constant acceleration,

$$v = v_0 + at$$

$$a = \frac{v - v_0}{t} = \frac{20 - 0}{10}$$

$$= 2 \text{ ft/sec}^2$$

$$F = ma = (1000)(2) = 2000 \text{ lbf}$$

Answer is (B)

DYNAMICS-29

A man weighs himself in an elevator. When the elevator is at rest, he weighs 185 pounds; when the elevator starts moving upwards, he weighs 210 pounds. How fast is the elevator accelerating, assuming constant acceleration?

(A) 1 ft/sec^2 (B) 2.17 ft/sec^2 (C) 4.35 ft/sec^2

       (D) 9.81 ft/sec^2 (E) 32.2 ft/sec^2

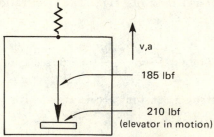

v,a

185 lbf

210 lbf
(elevator in motion)

The mass of the man can be determined from his weight at rest:

$$W = mg$$

$$m = \frac{W}{g} = \frac{185}{32.2}$$

$$= 5.75 \text{ slugs}$$

At constant acceleration,

$$F = ma$$

$$a = \frac{F}{m} = \frac{210 - 185}{5.75}$$

$$= 4.35 \text{ ft/sec}^2$$

Answer is (C)

DYNAMICS-30

A truck weighing 3 kips moves up a slope of 15°. What is the force generated by the engine if the truck is accelerating at a rate of 9 ft/sec^2? Assume the coefficient of friction is $\mu = 0.1$.

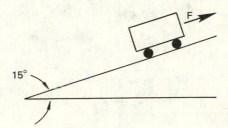

(A) 1700 lbf   (B) 1900 lbf   (C) 1920 lbf
        (D) 4140 lbf   (E) 28100 lbf

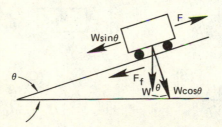

In the direction parallel to the slope, a force balance gives:

$$\sum F_x = ma = F - (W \sin \theta + F_f)$$

F_f is the friction force which is equal to $\mu N = \mu W \cos \theta$.

$$F = W(\sin \theta + \mu \cos \theta) + ma$$

$$= 3000 \left[\sin 15° + (0.1)\cos 15°\right] + \left(\frac{3000}{32.2}\right)(9)$$

$$= 1900 \text{ lbf}$$

Answer is (B)

DYNAMICS-31

In the figure shown, the two pulleys are well lubricated, and the horizontal surface is frictionless. The cord connecting the masses M_A and M_B is weightless. What is the ratio of the acceleration of mass A to the acceleration of mass B? Assume the system is released from rest.

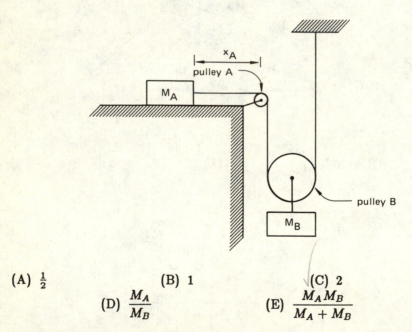

(A) $\frac{1}{2}$ (B) 1 (C) 2

(D) $\dfrac{M_A}{M_B}$ (E) $\dfrac{M_A M_B}{M_A + M_B}$

Assuming that the accelerations of both masses are constant, their respective displacement equations are:

$$x_A = \tfrac{1}{2}a_A t^2 \qquad\qquad x_B = \tfrac{1}{2}a_B t^2$$

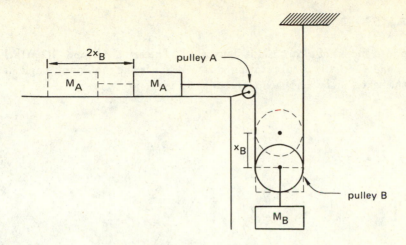

Taking the ratio of x_A to x_B,

$$\frac{x_A}{x_B} = \frac{a_A}{a_B}$$

Since mass B is supported by a two-segment rope section, and mass A is pulled by only one rope, the displacement of mass B is half the displacement of mass A. Therefore,

$$x_A = 2x_B$$
$$\frac{a_A}{a_B} = \frac{2x_B}{x_B}$$
$$= 2$$

Answer is (C)

DYNAMICS–32

A simplified model of a carousel is illustrated. The arms AB and AC attach the seats B and C to a vertical rotating shaft. What is the maximum angle of tilt, θ, for the seats, if the carousel operates at 12 rpm?

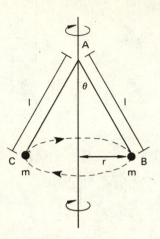

(A) 36°   (B) 39°   (C) 45°   (D) 51°   (E) 54°

The free-body diagram is:

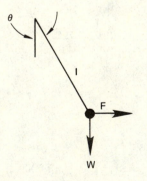

The angular velocity, ω, of the carousel is:

$$\omega = \frac{(12\,\text{rpm})\left(2\pi\,\frac{\text{rad}}{\text{rev}}\right)}{60\,\frac{\text{sec}}{\text{min}}}$$

$$= 1.257 \text{ rad/s}$$

The rotational force, F, expressed in terms of θ is:

$$F = ma = mr\omega^2$$
$$= ml \sin \theta \omega^2$$
$$= (200)(25)(\sin \theta)(1.257)^2$$
$$= 7900 \sin \theta$$

From the free-body diagram,

$$\tan \theta = \frac{F}{W} = \frac{7900 \sin \theta}{(200)(32.2)}$$
$$= 1.23 \sin \theta$$
$$\cos \theta = \frac{1}{1.23}$$
$$\theta = \cos^{-1}(0.813)$$
$$= 36°$$

Answer is (A)

DYNAMICS-33

In the ballistic balance shown, a bullet is fired into a block of mass M, which can swing freely. Which of the following is true for the system after impact?

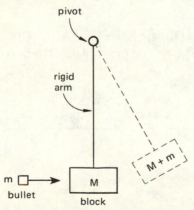

(A) Both mechanical energy and momentum are conserved.
(B) Mechanical energy is conserved; momentum is not conserved.
(C) Momentum is conserved; mechanical energy is not conserved.
(D) Neither mechanical energy nor momentum is conserved.
(E) Mechanical energy is zero, and the momentum is constant.

Momentum is not conserved since an external force, gravity, acts on the bullet-block mass. Only mechanical energy is conserved.

Answer is (B)

DYNAMICS-34

Three masses are attached by a weightless cord. If mass m_2 is exactly halfway between the other masses and is located at the center of the flat surface when the masses are released, what is its acceleration? Assume there is no friction in the system.

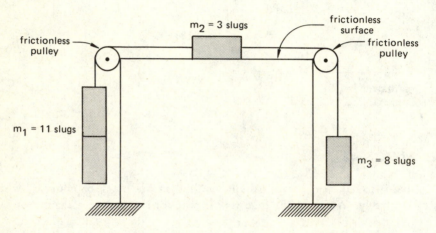

(A) 0   (B) 4.4 ft/sec²   ✓(C) 32.2 ft/sec²
        (D) 204 ft/sec²   (E) 236 ft/sec²

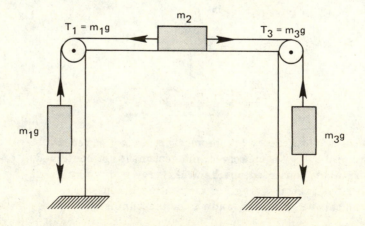

Since $m_1 > m_3$, m_1 will move downward, and m_2 will be displaced to the left. A force balance on m_2 gives:

$$T_1 - T_3 = m_2 a_2$$

T_1 and T_3 are the tensions in the cord due to the masses m_1 and m_3.

$$a_2 = g\left(\frac{m_1 - m_3}{m_2}\right)$$
$$= 32.2 \left(\frac{11 - 8}{3}\right)$$
$$= 32.2 \text{ ft/sec}^2$$

Answer is (C)

DYNAMICS–35

The maximum capacity of an elevator is 2000 pounds. The elevator starts from rest, and its velocity varies with time as shown in the graph. What is the maximum tension, T, in the elevator cable at full capacity? Neglect the mass of the elevator.

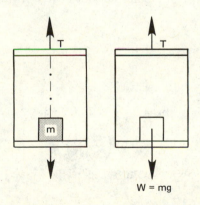

(A) 373 lbf   (B) 1630 lbf   (C) 2370 lbf

(D) 14,000 lbf   (E) 76,000 lbf

PROFESSIONAL PUBLICATIONS, INC. • Belmont, CA

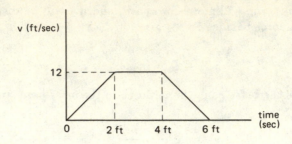

The maximum tension occurs during the period of maximum acceleration. This occurs for $0 < t < 2$ seconds, with acceleration, a, equal to $\frac{12}{2} = 6$ ft/sec^2. The mass of the occupants is $m = 2000/322$ slugs. During this time,

$$\sum F = ma = T - W$$

$$T = W + ma = 2000 + \left(\frac{2000}{32.2}\right)(6)$$

$$= 2370 \text{ lbf}$$

Answer is (C)

DYNAMICS-36

What is the kinetic energy of a 900 lbm motorcycle traveling at 60 mph?

(A) 22,500 ft-lbf
(B) 52,300 ft-lbf
(C) 80,700 ft-lbf
(D) 108,000 ft-lbf
(E) 348,000 ft-lbf

$$m = \frac{900}{32.2} = 28.0 \text{ slugs}$$

$$v = (60)\left(\frac{5280}{3600}\right)$$

$$= 88 \text{ ft/sec}$$

The kinetic energy is:

$$E_k = \tfrac{1}{2}mv^2 = \tfrac{1}{2}(28)(88)^2$$
$$= 108,000 \text{ ft-lbf}$$

Answer is (D)

DYNAMICS-37

A lead hammer weighs 10 lbf. In one swing of the hammer, the nail is driven .6 inches into a wood block. The velocity of the hammer's head at impact is 8 ft/sec. What is the average resistance of the wood block?

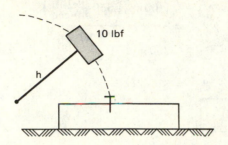

(A) 1010 lbf   (B) 1800 lbf   (C) 2010 lbf
          (D) 3600 lbf   (E) 32,400 lbf

Because energy is conserved, the kinetic energy of the hammer before impact is equal to the work done by the resistance force of the wood block. $m = \frac{10}{32.2} = 0.311$ slugs, $v = 18$ ft/sec, and $x_{nail} = \frac{0.6}{12} = 0.05$ ft.

$$\tfrac{1}{2}mv^2 = Fx$$
$$F = \frac{\tfrac{1}{2}mv^2}{x} = \frac{\tfrac{1}{2}(0.311)(18)^2}{0.05}$$
$$= 1010 \text{ lbf}$$

Answer is (A)

DYNAMICS-38

An automobile uses 100 hp to move at a uniform speed of 60 mph. What is the thrust force provided by the engine?

(A) 530 lbf   (B) 625 lbf   (C) 1100 lbf
          (D) 1800 lbf   (E) 2200 lbf

Power is defined as work done per unit time, which, for a linear system, is equivalent to force × velocity. Therefore,

$$P = F\mathbf{v}$$

$$F = \frac{P}{\mathbf{v}}$$

$$= \frac{(100\,\text{hp})\left(550\,\frac{\text{ft-lbf}}{\text{hp-sec}}\right)}{(60\,\text{mph})\left(\dfrac{5280\,\frac{\text{ft}}{\text{mi}}}{3600\,\frac{\text{sec}}{\text{hr}}}\right)}$$

$$= 625\,\text{lbf} \checkmark$$

Answer is (B)

DYNAMICS-39

A 2000 lbf weight is initially suspended on a 500 foot long cable. The weight is then raised 400 feet. If the cable weighs 5 lbf/ft, how much work is done?

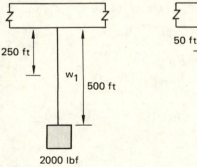

250 ft w_1 500 ft   2000 lbf

50 ft w_2 100 ft   2000 lbf

(A) 400,000 ft-lbf
(B) 1,400,000 ft-lbf
(C) 2,000,000 ft-lbf
(D) 2,750,000 ft-lbf
(E) 3,850,000 ft-lbf

The weight of the extended cable for the two situations is:

$$W_1 = (500)(5) = 2500\,\text{lbf}$$
$$W_2 = (100)(5) = 500\,\text{lbf}$$

These weights may be considered to be concentrated at the midpoints of the extended cables. Choosing the datum to be at the top of the cable, the work done is equal to the difference in potential energies of the two situations.

$$E_{p1} = (2000)(-500) + (2500)(-250) = -1,625,000 \text{ ft-lbf}$$
$$E_{p2} = (2000)(-100) + (500)(-50) = -225,000 \text{ ft-lbf}$$
$$W = E_{p2} - E_{p1} = -225,000 - (-1,625,000)$$
$$= 1,400,000 \text{ ft-lbf}$$

Answer is (B)

DYNAMICS-40

A 130 pound man is standing on the top of a building 120 feet above the ground. What is his potential energy relative to the ground?

(A) 2380 ft-lbf   (B) 7200 ft-lbf   (C) 9480 ft-lbf
       (D) 14,600 ft-lbf   (E) 15,600 ft-lbf

$$E_p = Wh = (130)(120) = 15,600 \text{ ft-lbf}$$

Answer is (E)

DYNAMICS-41

A 0.05 kg mass is attached to a spring $(k = 0.5 \text{ N/m})$. What is the total energy, E, for the body in the following diagram? Neglect the spring mass.

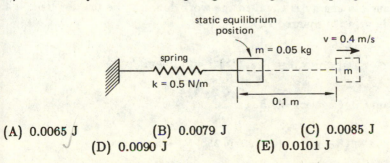

(A) 0.0065 J   (B) 0.0079 J   (C) 0.0085 J
       (D) 0.0090 J   (E) 0.0101 J

The total energy is the sum of the kinetic and potential energies.

$$E = E_k + E_p$$
$$= \tfrac{1}{2}mv^2 + \tfrac{1}{2}kx^2$$
$$= \tfrac{1}{2}(0.05)(0.4)^2 + \tfrac{1}{2}(0.5)(0.1)^2$$
$$= 0.0065 \text{ J}$$

Answer is (A)

DYNAMICS–42

A 1000 kg car is traveling down a 25° slope. At the instant that the speed is 13 m/s, the driver applies the brakes. What constant force, F, parallel to the road, must be provided by the brakes if the car is to stop in 90 meters?

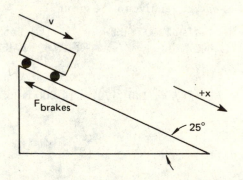

(A) 1290 N   (B) 2900 N   (C) 5080 N

(D) 8630 N   (E) 9040 N

The change in energy is equal to the work done by the brakes. The change in velocity squared is:

$$(v^2 - v_0^2) = \left[0 - (13)^2\right] = -169 \text{ m}^2/\text{s}^2$$

The change in elevation of the car is:

$$(h - h_0) = (0 - 90\sin 25°) = -38 \text{ meters}$$

Therefore,

$$\Delta E_k + \Delta E_p = Fx$$
$$\tfrac{1}{2}m(v^2 - v_0^2) + mg(h - h_0) = Fx$$
$$\tfrac{1}{2}(1000)(-169) + (1000)(9.81)(-38) = F(90)$$
$$F = -5080 \text{ N}$$

Answer is (C)

DYNAMICS-43

A bullet of mass 0.1 slug is fired at a wooden block resting on a horizontal surface. A spring with stiffness $k = 300$ kips/inch resists the motion of the block. If the maximum displacement of the block produced by the impact of the bullet is 1.33 inches, what is the velocity of the bullet at impact? Assume there are no losses at impact, and the spring has no mass.

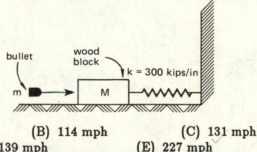

(A) 66 mph (B) 114 mph (C) 131 mph
(D) 139 mph (E) 227 mph

Due to the conservation of energy, the kinetic energy of the bullet before impact is equal to the potential energy of the spring-mass-bullet system at maximum compression.

$$E_{k,\text{bullet}} = E_{p,\text{system}}$$
$$\tfrac{1}{2}m_{\text{bullet}}v^2 = \tfrac{1}{2}kx^2$$

$$v = \sqrt{\frac{kx^2}{m_{\text{bullet}}}}$$

$$= \sqrt{\frac{(300,000)(1.33)^2}{0.1}} = 2300 \text{ in/sec}$$
$$= 131 \text{ mph}$$

Answer is (C)

DYNAMICS–44

A simple pendulum consists of a 100 slug mass attached to a weightless cord. If the mass is moved laterally such that $h = 2$ inches, and then is released, what is the maximum tension in the cord, T?

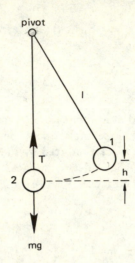

(A) 1.97 kips   (B) 3.58 kips   (C) 3.94 kips

(D) 5.90 kips   (E) 11.8 kips

The maximum tension will occur when the pendulum is at its lowest point, position 2 in the figure. The force balance in the vertical y direction gives:

$$ma_y = T - mg$$
$$T = ma_y + mg$$
$$= \frac{mv^2}{l} + mg \qquad\qquad 1$$

From the conservation of energy, with $h = 0$,

$$E_{p1} = E_{k2}$$
$$mgh = \tfrac{1}{2}mv^2$$
$$v = \sqrt{2gh} \qquad\qquad 2$$

Equations 1 and 2 give:

$$T_{\max} = \frac{m\left(\sqrt{2gh}\right)^2}{l} + mg$$

$$= mg\left(\frac{2h + l}{l}\right)$$

$$= (1000)(32.2)\left[\frac{(2)(\frac{2}{12}) + 1.5}{1.5}\right]$$

$$= 3940 \text{ lbf}$$

$$= 3.94 \text{ kips}$$

Answer is (C) ✓

DYNAMICS–45

A stationary passenger car of a train is set into motion by the impact of a moving locomotive. What is the impulse delivered to the car if it has a velocity of 38 ft/sec immediately after the collision? The weight of the car is 13 kips.

(A) 0.34 kip-sec   (B) 7.7 kip-sec   (C) 15.3 kip-sec

(D) 342 kip-sec   (E) 492 kip-sec

For the impulse-momentum principle,

$$Imp = \Delta(mv)$$
$$mv_1 + Imp = mv_2$$
$$Imp = m(v_2 - v_1)$$
$$= \left(\frac{13,000}{32.2}\right)\left(38 - 0\right)$$
$$= 15.3 \text{ kip-sec}$$

Answer is (C) ✓

DYNAMICS-46

Which of the following statements is false?

(A) The time rate of change of the angular momentum about a fixed point is equal to the total moment of the external forces acting on the system about the point.

✓(B) The coefficient of restitution can be less than zero.

(C) The frictional force always acts to resist motion.

(D) Momentum is conserved during elastic collisions.

(E) The period of a simple pendulum is independent of the mass of the pendulum.

The coefficient of restitution is defined as the ratio of the impulses corresponding to the period of restitution and to the period of deformation of a body, respectively. Its value is always between 0 and 1.

Answer is (B)

DYNAMICS-47

Two identical balls hit head-on in a perfectly elastic collision. Given that the initial velocity of one ball is 0.85 m/s and the initial velocity of the other is −0.53 m/s, what is the relative velocity of each ball after the collision?

(A) 0.85 m/s and −0.53 m/s
(B) 1.2 m/s and −0.72 m/s
(C) 1.2 m/s and −5.1 m/s
(D) 1.8 m/s and −0.98 m/s
(E) 2.3 m/s and −6.1 m/s

The conservation of momentum gives:

$$m v_{01} + m v_{02} = m v_1 + m v_2$$
$$0.85 + (-0.53) = v_1 + v_2$$
$$v_1 + v_2 = 0.32 \text{ m/s}$$

1

Since kinetic energy is conserved,

$$\tfrac{1}{2}mv_{01}^2 + \tfrac{1}{2}mv_{02}^2 = \tfrac{1}{2}mv_1^2 + \tfrac{1}{2}mv_2^2$$
$$(0.85)^2 + (-0.53)^2 = v_1^2 + v_2^2$$
$$v_1^2 + v_2^2 = 1 \qquad\qquad 2$$

Combining equations 1 and 2,

$$v_2^2 - 0.32v_2 - 0.4488 = 0$$

$$v_2 = \frac{0.32 \pm \sqrt{(-0.32)^2 - (4)(1)(-0.4488)}}{2}$$
$$= 0.85 \text{ m/s}$$

$$v_1 = 0.32 - v_2 = 0.32 - 0.85$$
$$= -0.53 \text{ m/s}$$

Answer is (A)

DYNAMICS–48

A steel ball weighing 200 pounds strikes a stationary wooden ball weighing 200 pounds. If the steel ball has a velocity of 16.7 ft/sec at impact, what is its velocity immediately after impact? Assume the collision is perfectly elastic.

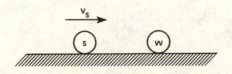

(A) −16.7 ft/sec   (B) −8.35 ft/sec   (C) 0

(D) 8.35 ft/sec   (E) 16.7 ft/sec

Since the balls have the same weight, they have equal mass. Denoting the instances before and after the collision, respectively, by the subscripts 1 and 2, $v_{s1} = 16.7$ ft/sec and $v_{w1} = 0$. Conservation of momentum gives:

$$m_s v_{s1} + m_w v_{w1} = m_s v_{s2} + m_w v_{w2}$$
$$v_{s2} + v_{w2} = v_{s1} = 16.7 \text{ ft/sec} \qquad \qquad 1$$

Conservation of energy gives:

$$\tfrac{1}{2} m_s v_{s1}^2 + \tfrac{1}{2} m_w v_{w1}^2 = \tfrac{1}{2} m_s v_{s2}^2 + \tfrac{1}{2} m_w v_{w2}^2$$
$$v_{s2}^2 + v_{w2}^2 = v_{s1}^2 = (16.7)^2$$
$$= 279 \qquad \qquad 2$$

Solving 1 and 2 simultaneously,

$$v_{s2}^2 + (279 - 33.4 v_{s2} + v_{s2}^2) = 279$$
$$2 v_{s2}^2 - 33.4 v_{s2} = 0$$
$$v_{s2}^2 - 16.7 v_{s2} = 0$$
$$v_{s2} = 0 \text{ or } v_{s2} = 16.7 \text{ ft/sec}$$

If $v_{s2} = 16.7$ ft/sec, then $v_{w2} = 0$, and no change has occurred during the collision. This is physically impossible, so $v_{s2} = 0$.

Answer is (C)

DYNAMICS-49 ✓

Two masses collide in a perfectly inelastic collision. Given the data in the figure, find the velocity and direction of the resulting combined mass.

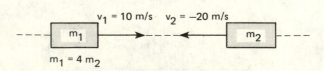

$v_1 = 10$ m/s $v_2 = -20$ m/s

m_1 m_2

$m_1 = 4\,m_2$

(A) The mass is stationary.
(B) 4 m/s to the right
(C) 5 m/s to the left
(D) 10 m/s to the right
(E) 10 m/s to the left

Let the positive direction of motion be defined to the right. Also, let m_3 be the resultant combined mass moving at velocity v_3 after the collision. Since momentum is conserved,

$$m_1v_1 + m_2v_2 = m_3v_3$$

However, $m_3 = m_1 + m_2 = 5m_2$. Therefore,

$$4m_2(10) + m_2(-20) = 5m_2v_3$$
$$40m_2 - 20m_2 = 5m_2v_3$$
$$v_3 = 4 \text{ m/s to the right} \checkmark$$

Answer is (B) $\checkmark$

DYNAMICS–50

A ball is dropped onto a solid floor from an initial height, h_0. If the coefficient of restitution, e, is 0.90, how high will the ball rebound?

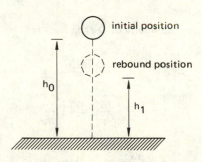

initial position

rebound position

h_0

h_1

(A) $0.45h_0$ (B) $0.81h_0$ (C) $0.85h_0$
          (D) $0.9h_0$ (E) h_0

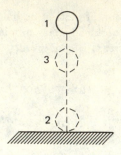

The subscripts 1, 2, and 3 denote the positions shown above. Conservation of energy gives, before impact:

$$E_{1,\text{total}} = E_{2,\text{total}}$$

Since the kinetic energy at position 1 and the potential energy at position 2 is zero,

$$mgh_0 = \tfrac{1}{2}mv_0^2$$
$$v_0 = \sqrt{2gh_0}$$

v_0 is the velocity of the ball before impact. After impact, the kinetic energy at position 3 is zero. Denoting the velocity of the ball after impact by v,

$$\tfrac{1}{2}mv^2 = mgh$$
$$v = \sqrt{2gh}$$

By definition, the coefficient of restitution is:

$$e = \frac{v_{\text{ball}} - v_{\text{floor}}}{v_{0,\text{floor}} - v_{0,\text{ball}}} = -\frac{v}{v_0}$$

$$= -\frac{\sqrt{2gh}}{\sqrt{2gh_0}} = -\sqrt{\frac{h}{h_0}}$$

$$h = e^2 h_0 = (0.9)^2 h_0$$
$$= 0.81 h_0$$

Answer is (B)

DYNAMICS–51

A ball suspended in space explodes into three pieces whose masses, initial velocities, and directions are given in the figure. Find the velocity of m_3.

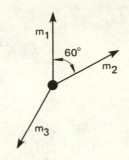

(A) 20 m/s   (B) 22.9 m/s   (C) 34.6 m/s

(D) 40 m/s   (E) 45.8 m/s

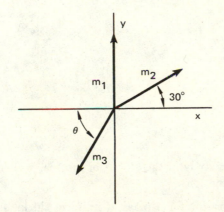

Defining the x and y axes as shown, conservation of momentum for the x direction gives:

$$m_2 v_2 \cos 30° + m_3 v_3 \cos \theta = 0$$
$$2m_1 v_2 \cos 30° + 4m_1 v_3 \cos \theta = 0$$
$$2m_1 (40) \cos 30° = -4m_1 v_3 \cos \theta$$
$$20 \cos 30° = -v_3 \cos \theta$$

$$v_3 = -\frac{17.32}{\cos \theta}$$

1

For the y direction,

$$m_1 v_1 + m_2 v_2 \sin 30° + m_3 v_3 \sin \theta = 0$$
$$m_1(20) + 2m_1(40) \sin 30° = -4m_1 v_3 \sin \theta$$
$$-4v_3 \sin \theta = 60$$
$$v_3 = -\frac{60}{4 \sin \theta} \qquad\qquad 2$$

Equations 1 and 2 give:

$$\tan \theta = \frac{60}{(17.32)(4)}$$
$$\theta = 40.9°$$
$$v_3 = \frac{-17.32}{-\cos 40.9°}$$
$$= 22.9 \text{ m/s}$$

$$\boxed{\text{Answer is (B)}}$$

DYNAMICS-52

Which of the following is false?

(A) Kinematics is the study of the effects of motion, while kinetics is the study of the causes of motion.

(B) The radius of gyration for a mass of uniform thickness is identical to that for a planar area of the same shape.

(C) Angular momentum for rigid bodies may be regarded as the product of angular velocity and inertia.

(D) The acceleration of a body rotating with a constant angular velocity is zero.

(E) The number of generalized coordinates cannot exceed the number of degrees of freedom.

A body rotating at a constant angular velocity has no angular acceleration. It does have a linear acceleration.

$$\boxed{\text{Answer is (D)}}$$

DYNAMICS-53

A uniform beam of weight W is supported by a pin joint and a wire. What will be the angular acceleration, α, at the instant that the wire is cut?

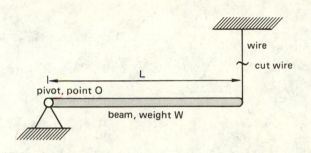

(A) $\dfrac{g}{L}$

(B) $\dfrac{3g}{2L}$

(C) $\dfrac{2g}{L}$

(D) $\dfrac{Wg}{L}$

(E) $\dfrac{2Wg}{L}$

The only force on the beam is its weight acting at the center, a distance of $\frac{L}{2}$ from point O. Taking the moment about O,

$$W\left(\frac{L}{2}\right) = I_O\alpha$$

For a slender beam rotating about its end,

$$I_O = \tfrac{1}{3}mL^2$$

$$\alpha = \frac{WL}{2I_O} = \frac{WL}{2(\frac{1}{3}mL^2)}$$

$$= \frac{3g}{2L}$$

Answer is (B)

DYNAMICS-54

A thin circular disk of mass 25 kg and radius 1.5 meters is spinning about its axis with an angular velocity of $\omega = 1800$ rpm. It takes 2.5 minutes to stop the motion by applying a constant force, F, to the edge of the disk. What force is required?

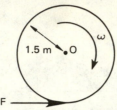

(A) 7.24 N   (B) 16.2 N   (C) 23.6 N   (D) 31.7 N   (E) 49.8 N

The effective force caused by the spinning of the disk is a couple $I_O \alpha$ acting at the axis, where α is the angular acceleration. Therefore,

$$Fr = -I_O \alpha$$

Designating the positive rotational direction as counter-clockwise, $\omega = -1800$ rpm. Therefore,

$$F = -\frac{I_O \alpha}{r} = -\frac{\left(\frac{1}{2}mr^2\right)(\alpha)}{r}$$

$$= -\frac{1}{2}mr\left(\frac{\Delta\omega}{\Delta t}\right)$$

$$= -\frac{1}{2}(25)(1.5)\left[\left(\frac{-1800}{2.5}\right)(2\pi)\left(\frac{1}{3600}\right)\right]$$

$$= 23.6 \text{ N}$$

Answer is (C)

DYNAMICS-55

What is the period of an oscillating body whose mass, m, is 0.025 kg if the spring constant, k, is 0.44 N/m?

(A) 0.5 s   (B) 1 s   (C) 1.5 s

(D) 2 s   (E) 2.5 s

By definition, the period T is given by:

$$T = 2\pi\sqrt{\frac{m}{k}} = 2\pi\sqrt{\frac{0.025}{0.44}}$$
$$= 1.5 \text{ s}$$

Answer is (C)

DYNAMICS-56

What is the frequency of oscillation of a body if its mass, m, is 0.015 kg, and k is 0.5 N/m?

(A) 0.51 Hz   (B) 0.66 Hz   (C) 0.78 Hz
          (D) 0.92 Hz   (E) 1.1 Hz

By definition, the frequency, f, is:

$$f = \frac{1}{2\pi}\sqrt{\frac{k}{m}} = \frac{1}{2\pi}\sqrt{\frac{0.5}{0.015}}$$
$$= 0.92 \text{ Hz}$$

Answer is (D)

DYNAMICS-57

What is the natural frequency, ω, of an oscillating body whose period of oscillation is 1.8 s?

(A) 1.8 rad/s   (B) 2.65 rad/s   (C) 3.49 rad/s
          (D) 4.24 rad/s   (E) 5 rad/s

$$\omega = \frac{2\pi}{T} = \frac{2\pi}{1.8}$$
$$= 3.49 \text{ rad/s}$$

Answer is (C)

DYNAMICS–58

A one-story frame is subject to a sinusoidal forcing function $q(t) = Q \sin \omega t$ at the transom. Find the frequency of $q(t)$, (in hertz) if the frame is in resonance with the force.

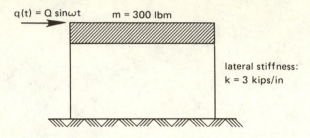

(A) 2.57 Hz   (B) 2.92 Hz   (C) 3.58 Hz
          (D) 7.62 Hz   (E) 9.90 Hz

Resonance occurs when the forced frequency, ω, equals the natural frequency, ω_n.

$$m = \frac{300}{32.2} = 9.32 \text{ slugs}$$

$$k = (3000)(12) = 36{,}000 \text{ lbf/ft}$$

$$f = \frac{1}{2\pi}\sqrt{\frac{k}{m}} = \frac{1}{2\pi}\sqrt{\frac{36{,}000}{9.32}}$$

$$= 9.90 \text{ Hz}$$

Answer is (E)

DYNAMICS–59

In the mass-spring system shown, the mass, m, is displaced 3 inches to the right and then released. Find the maximum velocity of m.

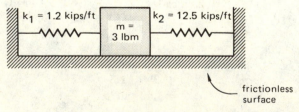

(A) 34.3 ft/sec   (B) 59.5 ft/sec   (C) 86.3 ft/sec
          (D) 87.1 ft/sec   (E) 95.9 ft/sec

The kinetic energy before the mass is released is zero. The maximum velocity will occur when the mass returns to the point of static equilibrium, where the deflection is zero and hence the potential energy equals zero. Therefore, since the total energy of the system is constant,

$$E_{p,1} = E_{k,2}$$
$$\tfrac{1}{2}k_1 x_1^2 + \tfrac{1}{2}k_2 x_2^2 = \tfrac{1}{2}mv^2$$

The displacement is:

$$x = \frac{3 \text{ in}}{12 \frac{\text{in}}{\text{ft}}}$$
$$= 0.25 \text{ ft}$$

$$v = \sqrt{\frac{k_1 x_1^2 + k_2 x_2^2}{m}}$$

$$= \sqrt{\frac{(1.2)(0.25)^2 + (12.5)(0.25)^2}{\frac{3}{32.2}}}$$

$$= 95.9 \text{ ft/sec}$$

Answer is (E)

DYNAMICS-60

A cantilever beam with an end mass, $m = 3000$ slugs, deflects 2 inches when a force of 3 kips is applied at the end. The beam is subsequently mounted on a spring of stiffness, $k_s = 2$ kips/in. What is the natural frequency of the beam-spring system?

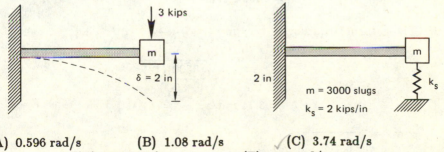

(A) 0.596 rad/s   (B) 1.08 rad/s   (C) 3.74 rad/s
          (D) 4.03 rad/s   (E) 6.20 rad/s

A cantilever with an end mass m can be modeled:

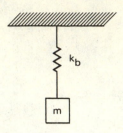

where $k_b = 3000/2 = 1500$ lbf/in. Therefore, for the beam-spring system, the overall model is:

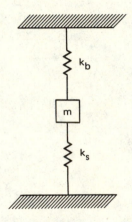

For this model, both springs undergo the same deflection. Hence,

$$k = k_b + k_s = 1500 + 2000$$
$$= 3500 \text{ lbf/in}$$

The natural frequency is, therefore:

$$\omega = \sqrt{\frac{k}{m}} = \sqrt{\frac{(3500)(12)}{3000}}$$
$$= 3.74 \text{ rad/s}$$

Answer is (C)

12 DC ELECTRICITY

DC ELECTRICITY-1

Which statement about a charge placed on a dielectric material is true?

(A) The charge is instantly carried to the material's surface.
(B) The charge is immediately lost to the atmosphere.
(C) The charge is confined to the region in which the charge was placed.
(D) The charge increases the conductivity of the material.
(E) The charge reduces the conductivity of the material.

In a dielectric, all charges are attached to specific atoms or molecules.

Answer is (C)

DC ELECTRICITY-2

The coulomb force, F, acts on two charges a distance, r, apart. What is F proportional to?

(A) r
(B) r^2
(C) $1/r^2$
(D) $1/r^3$
(E) F is independent of r.

The coulomb force is:

$$F = \frac{q_1 q_2}{4\pi \epsilon r^2}$$

q_1 and q_2 are the charges, and ϵ is the permittivity of the surrounding medium. Hence, F is proportional to the inverse of r^2.

Answer is (C)

DC ELECTRICITY-3

The force between two electrons in a vacuum is 1×10^{-15} N. How far apart are the electrons?

(A) 1.42×10^{-12} m   (B) 5.05×10^{-12} m   (C) 4.8×10^{-7} m
        (D) 1.7×10^{-6} m   (E) 0.48 m

Coulomb's law is:

$$F = \frac{q_1 q_2}{4\pi\epsilon_o r^2}$$

$\epsilon_o = 8.85 \times 10^{-12}$ $C^2/N \cdot m^2$. Also, for an electron, $q = 1.6 \times 10^{-19}$ C. Solving for r,

$$r = q\sqrt{\frac{1}{4\pi\epsilon F}}$$

$$= (1.6 \times 10^{-19})\sqrt{\frac{1}{(4\pi)(8.85 \times 10^{-12})(1 \times 10^{-15})}}$$

$$= 4.8 \times 10^{-7} \text{ m}$$

Answer is (C)

DC ELECTRICITY-4

Two solid spheres have charges of 1 coulomb and −8 coulombs, respectively. The permittivity, ϵ_o, is 8.85×10^{-12} $C^2/N \cdot m^2$, and the distance between the sphere centers, r, is 0.3 meters. Determine the force on the spheres.

(A) -1.01×10^{13} N   (B) -8×10^{11} N   (C) 0 N
        (D) 8×10^{11} N   (E) 1.01×10^{13} N

Because of their symmetry, charged spheres may be treated as point charges. Coulomb's law states:

$$F = \frac{q_1 q_2}{4\pi\epsilon_o r^2} = \frac{(1)(-8)}{(4\pi)(8.85 \times 10^{-12})(0.3)^2}$$

$$= -8 \times 10^{11} \text{ N}$$

Answer is (B)

DC ELECTRICITY-5

Two spheres have charges of 1 and 2 coulombs, respectively. If the spheres are each 1 meter in diameter, and the distance between the centers of the spheres is 10 meters, what is the force on the spheres? (The permittivity is 8.85×10^{-12} $C^2/N{\cdot}m^2$.)

(A) -2.2×10^8 N   (B) -1.8×10^8 N   (C) -1.49×10^8 N
       (D) 1.8×10^8 N   (E) 2.2×10^8 N

The spheres act as point charges located at their centers. Their diameters do not affect the force between them.

$$F = \frac{q_1 q_2}{4\pi\epsilon_o r^2}$$

$$= \frac{(1)(2)}{(4\pi)(8.85 \times 10^{-12})(10)^2}$$

$$= -1.8 \times 10^8 \text{ N}$$

Answer is (B)

DC ELECTRICITY-6

A 0.001 coulomb charge is separated from a 0.003 coulomb charge by 10 meters. If P denotes the point of zero electric field between the charges, determine the distance, x, from the 0.001 coulomb charge.

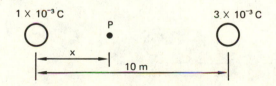

(A) -13.7 m   (B) 2.24 m   (C) 3.66 m
              (D) 6.34 m   (E) 7.78 m

The force that is experienced by a point object with a charge, q, in an electric field of strength, E, is:

$$E = \frac{F}{q_2} = \frac{q_1}{4\pi\epsilon r^2}$$

At the point where E is zero,

$$\frac{0.001}{4\pi\epsilon_o x^2} = \frac{0.003}{4\pi\epsilon_o(10-x)^2}$$
$$(10-x)^2 = 3x^2$$
$$x^2 + 10x - 50 = 0$$

Solving for the positive x value,

$$x = \frac{-10 + \sqrt{(10)^2 - (4)(1)(-50)}}{2}$$
$$= 3.66 \text{ m}$$

Answer is (C)

DC ELECTRICITY–7

A 3 coulomb charge and a 5 coulomb charge are 10 meters apart. A 7 coulomb charge is placed on a line adjoining the two charges, x meters away from the 3 coulomb charge. If the 7 coulomb charge is in equilibrium, find the value of x

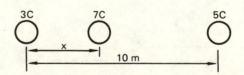

(A) 3.87 m   (B) 4.36 m   (C) 5 m
          (D) 5.64 m   (E) 6.13 m

At equilibrium, $F_{37} = F_{75}$. Using Coulomb's law,

$$\frac{(3)(7)}{4\pi\epsilon_o x^2} = \frac{(7)(5)}{4\pi\epsilon_o (10 - x)^2}$$

$$21(10 - x)^2 = 35x^2$$

$$x^2 + 30x - 150 = 0$$

Solving for a positive value of x,

$$x = \frac{-30 + \sqrt{(30)^2 - (4)(1)(-150)}}{2}$$

$$= 4.36 \text{ m}$$

Answer is (B)

DC ELECTRICITY-8

Two charges, A and B, of equal and opposite value are separated by a distance, d. r is the distance from a charge to any point, P, lying on the normal plane that bisects the length, d. What is the electric field at point P if K is a constant equal to $1/4\pi\epsilon$?

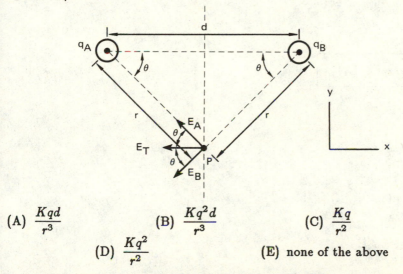

(A) $\dfrac{Kqd}{r^3}$

(B) $\dfrac{Kq^2 d}{r^3}$

(C) $\dfrac{Kq}{r^2}$

(D) $\dfrac{Kq^2}{r^2}$

(E) none of the above

The total electric field will be in the x direction only, since the y components of the charges cancel each other out. By definition, with $\mathbf{a_r}$ denoting the unit radial vector,

$$\mathbf{E} = \left(\frac{Kq}{r^2}\right)\mathbf{a_r}$$

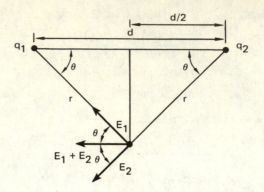

Therefore,

$$\mathbf{E}_T = \mathbf{E}_A + \mathbf{E}_B = \left(\frac{Kq}{r^2}\right)\cos\theta + \left(\frac{Kq}{r^2}\right)\cos\theta$$

$$= \left(\frac{2Kq}{r^2}\right)\left(\frac{\frac{1}{2}d}{r}\right)$$

$$= \frac{Kqd}{r^3}$$

Answer is (A)

DC ELECTRICITY–9

A metallic spherical shell has a charge of 0.001 coulombs. The shell is 2 meters
in diameter. Which of the following correctly shows the variation of electric field
with respect to the distance from the center of the sphere, r?

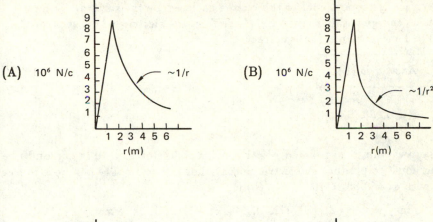

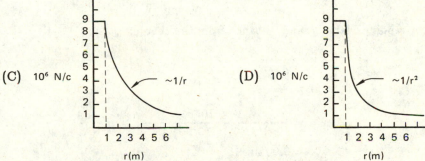

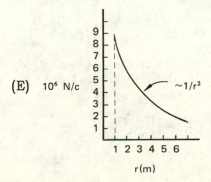

Outside the sphere, Coulomb's law can be used to find the electric field. Thus, the electric field varies as $1/r^2$ for $r > 1$ m. On the surface of the sphere, $r = 1$ m.

$$E = \frac{1}{4\pi\epsilon_o} \frac{q}{(1)^2} = 9 \times 10^6 \text{ N/C}$$

Gauss' law states that the electric flux passing through a given closed surface is proportional to the charge enclosed by the surface. There is no charge within the sphere. Therefore, the electric field is zero for $r < 1$ meter. Only (E) is correct.

Answer is (E)

DC ELECTRICITY–10

How far away must an isolated positive point charge of 1×10^{-8} coulombs be in order for it to produce an electric potential of 100 volts? The charge is in free space with $\epsilon_o = 8.85 \times 10^{-12}$ C^2/N·m^2.

(A) 0.90 m   (B) 1.23 m   (C) 5.31 m

(D) 8.56 m   (E) 22.8 m

For a point charge q, and distance from the charge, r,

$$V = \frac{q}{4\pi\epsilon r}$$

$$r = \frac{q}{4\pi\epsilon V}$$

$$= \frac{1 \times 10^{-8}}{(4\pi)(8.85 \times 10^{-12})(100)}$$

$$= 0.90 \text{ m}$$

Answer is (A)

DC ELECTRICITY–11

A point charge, q, in a vacuum creates a potential, V, at a distance, r. A reference voltage of zero is arbitrarily selected when $r = a$. If $K = 1/4\pi\epsilon_o$, which of the following is the correct expression for V?

(A) $Kq\left(\dfrac{1}{r^2} - \dfrac{1}{a^2}\right)$ (B) $Kq\left(\dfrac{a}{r^2}\right)$ (C) $Kq\left(\dfrac{1}{r} - \dfrac{1}{a}\right)$

(D) $Kq\left(\dfrac{r}{a^2}\right)$ (E) none of the above

From Coulomb's law for a point charge,

$$E = \frac{Kq}{r^2}$$

The total voltage is measured between the reference voltage, a, and r.

$$V = -\int E \, dr = -\int_a^r \frac{Kq}{r^2} \, dr$$
$$= Kq \left(\frac{1}{r} - \frac{1}{a} \right)$$

Answer is (C)

DC ELECTRICITY–12

What accelerating voltage is required to accelerate an electron to a kinetic energy of 5×10^{-15} joules? The charge of an electron is 1.6×10^{-19} coulombs.

(A) 8000 V   (B) 13,200 V   (C) 19,150 V

          (D) 24,900 V   (E) 31,250 V

For an electron, kinetic energy is:

$$E_k = qV$$
$$V = \frac{E_k}{q} = \frac{5 \times 10^{-15}}{1.6 \times 10^{-19}}$$
$$= 31,250 \text{ V}$$

Answer is (E)

DC ELECTRICITY–13

A certain potential variation in the xy plane is given by the expression:

$$\nabla V = \frac{1}{\sqrt{x^2 + 4y^2}}(\mathbf{i} + \mathbf{j})$$

Which of the following gives the magnitude and direction (angle made with the x-axis) of the electric field intensity at the point $(2,1)$?

(A) $-\dfrac{\sqrt{2}}{4}, \pi$ (B) $\dfrac{1}{2}, -\dfrac{\pi}{4}$ (C) $\dfrac{\sqrt{2}}{2}, \dfrac{\pi}{4}$

(D) $\dfrac{1}{2}, \dfrac{\pi}{4}$ (E) $\dfrac{3}{4}, \dfrac{\pi}{2}$

$$V = \int E \, dr$$

$$\mathbf{E} = \nabla V = \frac{\partial V}{\partial x}\mathbf{i} + \frac{\partial V}{\partial y}\mathbf{j}$$

Since this is equivalent to the expression given,

$$E = \sqrt{E_x^2 + E_y^2} = \sqrt{\left(\frac{\partial V}{\partial x}\right)^2 + \left(\frac{\partial V}{\partial y}\right)^2}$$

$$= \sqrt{\left(\frac{1}{\sqrt{x^2 + 4y^2}}\right)^2 + \left(\frac{1}{\sqrt{x^2 + 4y^2}}\right)^2}$$

$$= \sqrt{\frac{2}{x^2 + 4y^2}}$$

Evaluating at the point $(2,1)$,

$$E = \sqrt{\frac{2}{(2)^2 + 4(1)^2}}$$

$$= \frac{1}{2}$$

The angle from horizontal that the **E** field is directed is:

$$\tan \theta = \frac{E_x}{E_y}$$

$$= \frac{1/\sqrt{x^2 + 4y^2}}{1/\sqrt{x^2 + 4y^2}} = 1$$

$$\theta = \frac{\pi}{4}$$

Answer is (D)

DC ELECTRICITY–14

A conducting spherical shell 0.2 meters in diameter carries a charge of 0.001 coulombs. Which of the following is a correct plot of electric potential, V, versus distance from the sphere center, r?

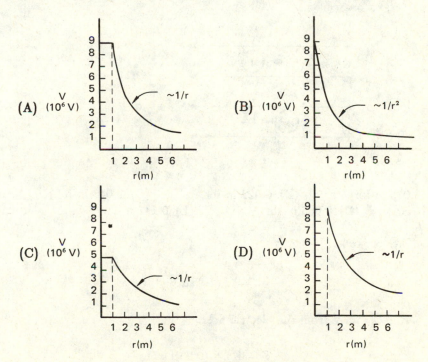

(A) $(10^6\,V)$ ~1/r

(B) $(10^6\,V)$ ~1/r²

(C) $(10^6\,V)$ ~1/r

(D) $(10^6\,V)$ ~1/r

(E) none of the above

PROFESSIONAL PUBLICATIONS, INC. • Belmont, CA

Outside the shell, $V = \dfrac{q}{4\pi\epsilon_o r}$

There are no charges inside the shell. Therefore, the electric potential within the sphere is constant and equal to the electric potential at the surface of the sphere. At the surface of the sphere, at $r = 1$ meter,

$$V = \frac{q}{4\pi\epsilon_o r}$$
$$= \frac{0.001}{(4\pi)(8.85 \times 10^{-12})(1)}$$
$$= 9 \times 10^6 \text{ V}$$

Answer is (A)

DC ELECTRICITY–15

An electric dipole is placed in a uniform electric field of intensity, E. Given the information in the figure, what is the torque acting on the dipole?

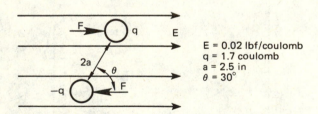

E = 0.02 lbf/coulomb
q = 1.7 coulomb
a = 2.5 in
θ = 30°

(A) 0.015 in-lbf   (B) 0.029 in-lbf   (C) 0.043 in-lbf
          (D) 0.085 in-lbf   (E) 0.147 in-lbf

The torque is:

$$T = F(2a)\sin\theta$$

F is the force from the electric field.

$$F = Eq = (0.02)(1.7)$$
$$= 0.034 \text{ lbf}$$

Solving for T,

$$T = (0.034)(2)(2.5)(\sin 30°)$$
$$= 0.085 \text{ in-lbf}$$

Answer is (D)

DC ELECTRICITY–16

The north poles of two bar magnets have strengths of 500 poles and 250 poles, respectively. If the poles are 25.4 inches apart, what is the magnitude and direction of the force between them?

(A) 0.125 dynes; the magnets attract each other
(B) 3.00 dynes; the magnets repel each other
(C) 30 dynes; the magnets repel each other
(D) 194 dynes; the magnets repel each other
(E) 1250 dynes; the magnets attract each other

The magnitude of the force between two bar magnets of strengths M_1 and M_2 is:

$$F = \frac{M_1 M_2}{\mu r^2}$$

In the cgs system, μ has a value of 1. The force is:

$$F = \frac{(500)(250)}{(1)(25.4)^2(2.54)^2}$$
$$= 30 \text{ dynes}$$

Since the magnitude of the force is positive, the magnets repel each other.

Answer is (C)

DC ELECTRICITY–17

Which of the following is not a property of magnetic field lines?

(A) The field is stronger where the lines are closer together.
(B) The lines intersect surfaces of equal intensity at right angles.
(C) Magnetic field lines have no beginnings and no ends.
(D) The lines cross themselves only at right angles.
(E) The lines exist near a straight wire carrying a direct current.

Magnetic field lines do not cross. Their direction at any given point is unique.

Answer is (D)

DC ELECTRICITY–18

What is the tesla a unit of?

(A) permittivity
(B) capacitance
(C) inductance
(D) magnetic induction
(E) magnetic flux

The tesla is a unit of magnetic induction.

Answer is (D)

DC ELECTRICITY–19

The south poles of two bar magnets are 7.5 centimeters apart. The magnets are of equal strength, and repel each other with a force of 49 dynes. What is the strength of each magnet?

(A) 0.056 maxwells
(B) 0.86 maxwells
(C) 11.3 maxwells
(D) 52.5 maxwells
(E) 680 maxwells

The force between two magnets of strength M_1 and M_2 is:

$$F = \frac{M_1 M_2}{\mu r^2}$$

$M_1 = M_2$, and $\mu = 1$ in the cgs system. Therefore,

$$m = \sqrt{F \mu r^2}$$
$$= \sqrt{(49)(1)(7.5)^2}$$
$$= 52.5 \text{ maxwells}$$

Answer is (D)

DC ELECTRICITY-20

A charge of 3×10^{-5} coulombs moves along the axis of a cylinder that has a diameter of 0.02 meters. What is the maximum magnetic intensity on the surface of the cylinder if the charge has speed of 1×10^6 m/s?

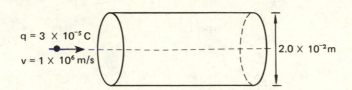

(A) 0 (B) 7500 μT (C) 15,000 μT

(D) 22,500 μT (E) 30,000 μT

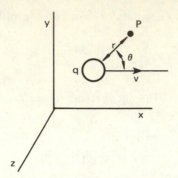

The magnitude of the magnetic field, B, at a point, P, for an electric charge, q, moving at a speed, v, is:

$$B = \frac{qv \sin \theta}{10r^2} \quad \text{(microteslas)}$$

The maximum value of B occurs when $\theta = 90°$ on the surface of the cylinder. Thus, $r = 0.01$ meters. Evaluating for B,

$$B = \frac{(3 \times 10^{-5})(1 \times 10^6)(\sin 90°)}{10(0.01)^2}$$
$$= 30,000 \ \mu\text{T}$$

Answer is (E)

DC ELECTRICITY–21

For a field given by $B = \mu H$ Wb/m^2, what is the energy storage per unit volume?

(A) $U = \dfrac{B^2}{2\mu}$ 　　　(B) $U = \dfrac{H^2}{2}$ 　　　(C) $U = \dfrac{H^2}{2\mu}$

　　(D) $U = \dfrac{H^2}{2\mu^2}$ 　　　　(E) $U = \dfrac{HB^2}{2}$

The energy stored in a coil is $U = \frac{1}{2}BH$. Since $B = \mu H$, $H = B/\mu$. Therefore,

$$U = \tfrac{1}{2}B\left(\frac{B}{\mu}\right) = \frac{B^2}{2\mu}$$

Answer is (A)

DC ELECTRICITY–22

The magnetic flux density, B, and the magnetic field intensity, H, have the following relationship:

$$B = \mu_o(H + M) \quad \text{Wb/m}$$

μ_o is the permeability of free space (in henrys/meter), and μ is the magnetic polarization of the material (in amperes/meter). If B is increasing, which of the following may be true about the state of metal X at a value of $H = 100$ amps/m?

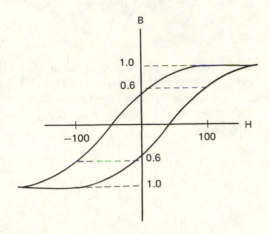

(A) $B = 0.6$ Wb/m; metal X is nonferrous
(B) $B = 0.6$ Wb/m; metal X is ferrous
(C) $B = 1.0$ Wb/m; metal X is nonferrous
(D) $B = 1.0$ Wb/m; metal X is ferrous
(E) none of the above

Nonferrous metals do not exhibit hysteresis; hence, metal X is ferrous. The hysteresis curve follows a counterclockwise path. Therefore, for B to be increasing at an H value of 100 amps/m, $B = 0.6$ Wb/m.

Answer is (B)

DC ELECTRICITY–23

Two identical coils of radius r are placed at a distance r apart, as shown. Which of the following describes the magnetic field created by passing a uniform current through the assembly?

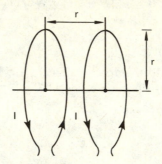

(A) The magnetic field is negligible regardless of the magnitude of I.
(B) The magnetic field is zero midway between the two coils.
(C) The magnetic field is uniform between the two coils.
(D) The magnetic field is zero at the centers of the coils.
(E) There is no magnetic field.

The magnetic field is:

$$B = \frac{\mu I}{2\pi r}$$

Since the two coils are circular with their centers aligned, the field between them will be fairly uniform.

Answer is (C)

DC ELECTRICITY–24

Which statement is true?

(A) Magnetic flux lines have sources only.
(B) Magnetic flux lines have sinks only.
(C) Magnetic flux lines have both sources and sinks.
(D) Magnetic flux lines do not have sources or sinks.
(E) Magnetic flux lines are produced by stationary charges.

Magnetic flux lines are closed loops with no sources or sinks. There is no known particle that gives lines of magnetism.

Answer is (D)

DC ELECTRICITY–25

A charge of 0.75 coulombs passes through a wire every 15 seconds. What is the current in the wire?

(A) 5 mA   (B) 9.38 mA   (C) 20 mA   (D) 50 mA   (E) 75 mA

Current equals the charge per unit time passing through the wire.

$$I = \frac{q}{t} = \frac{0.75\,\text{C}}{15\,\text{s}}$$
$$= 0.05\ \text{A}$$
$$= 50\ \text{mA}$$

Answer is (D)

DC ELECTRICITY–26

What is the total resistance between points A and B?

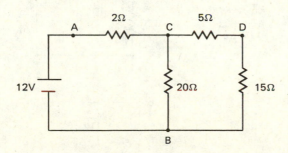

(A) 12 Ω   (B) 15.6 Ω   (C) 18.7 Ω   (D) 22 Ω   (E) 27 Ω

The total resistance is the sum of the resistance between points A and C, plus the equivalent resistance of the resistors in parallel between points C and B.

$$R_{total} = 2 + \cfrac{1}{\cfrac{1}{20} + \cfrac{1}{5 + 15}}$$
$$= 2 + 10$$
$$= 12 \ \Omega$$

Answer is (A)

DC ELECTRICITY–27

What is the total resistance (as seen by the battery) of the following network?

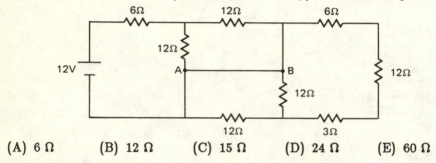

(A) 6 Ω   (B) 12 Ω   (C) 15 Ω   (D) 24 Ω   (E) 60 Ω

AB is a short circuit. Therefore, the rest of the circuit does not contribute to the resistance. The effective circuit is, then,

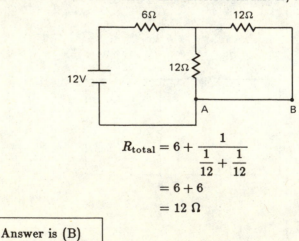

$$R_{total} = 6 + \cfrac{1}{\cfrac{1}{12} + \cfrac{1}{12}}$$
$$= 6 + 6$$
$$= 12 \ \Omega$$

Answer is (B)

DC ELECTRICITY–28

In the circuit shown, $R = 10$ ohms, and the electrostatic potential, V, is 2 volts. What is the current, I?

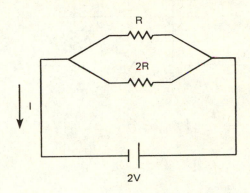

(A) 0.10 A   (B) 0.30 A   (C) 0.67 A   (D) 3.33 A   (E) 13.3 A

Current is equal to voltage divided by total resistance.

$$I = \frac{V}{R_{\text{total}}}$$

$$R_{\text{total}} = \frac{1}{\dfrac{1}{R} + \dfrac{1}{2R}}$$

$$= \frac{2R}{3}$$

$$= 6.67 \ \Omega$$

$$I = \frac{2}{6.67}$$

$$= 0.30 \text{ A}$$

Answer is (B)

DC ELECTRICITY–29

Find the current passing through the 3 ohm resistor.

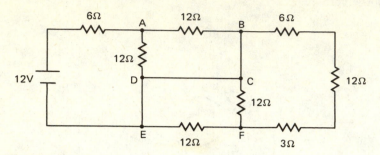

(A) 0 A
(B) 0.333 A
(C) 1 A
(D) 4 A
(E) none of the above

Current from a battery will always follow a path of zero resistance in a circuit. Instead of flowing through the 3 ohm resistor and its neighboring resistors, the current will follow the path BCDE, a short circuit. There will be no current in the resistor.

Answer is (A)

DC ELECTRICITY–30

What is the current passing through the 30 ohm resistor?

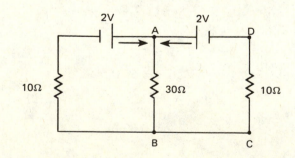

(A) 0 A   (B) 29 mA   (C) 50 mA
        (D) 57 mA   (E) 80 mA

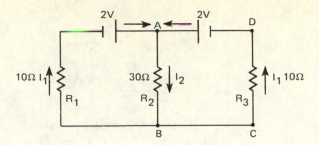

The circuit is symmetrical. Therefore, a current, I_1, flows through the resistors, R_1, and R_3. Another current, I_2, flows through resistor R_2. From Kirchoff's current laws at point A,

$$\sum I = 0$$
$$2I_1 = I_2$$

Using Kirchoff's voltage law around the loop ABCDA,

$$V = R_2 I_2 + R_3 I_1$$
$$2 = 30 I_2 + 10 I_1$$
$$2 = 30(2I_1) + 10 I_1$$
$$2 = 70 I_1$$
$$I_1 = 0.0286 \text{ A}$$
$$I_2 = 2I_1 = 2(0.0286)$$
$$= 0.057 \text{ A}$$
$$= 57 \text{ mA}$$

Answer is (D)

DC ELECTRICITY–31

What is the current through AB?

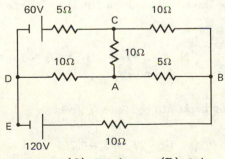

(A) 0.5 A   (B) 1.2 A   (C) 3.4 A   (D) 4 A   (E) 6.2 A

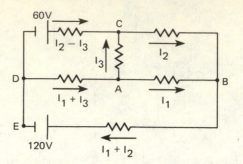

By redrawing the circuit and designating the currents as shown in loop ACBA, the currents through the remaining loops can be expressed in terms of I_1, I_2, and I_3. Since voltage equals resistance multiplied by current, for loop CDAC,

$$60 = -5(I_2 - I_3) + 10(I_1 + I_3) + 10I_3$$
$$0 = 10I_1 - 5I_2 + 25I_3 - 60$$
$$I_2 = 2I_1 + 5I_3 - 12$$

For loop BEDAB,

$$120 = 10(I_1 + I_2) + 10(I_1 + I_3) + 5I_1$$
$$0 = 25I_1 + 10I_2 + 10I_3 - 120$$
$$I_3 = -2.5I_1 - I_2 + 12$$

Around loop ACBA,

$$0 = -5I_1 + 10I_2 + 10I_3$$

Solving the three equations for the three unknowns, equations 1 and 2 produces:

$$I_3 = -2.5I_1 - (2I_1 + 5I_3 - 12) + 12$$
$$= -0.75I_1 + 4$$

Substituting back into equation 1 gives:

$$I_2 = 2I_1 + 5(-0.75I_1 + 4) - 12$$
$$= -1.75I_1 + 8$$

PROFESSIONAL PUBLICATIONS, INC. • Belmont, CA

Equation 3 is:

$$0 = -5I_1 + 10(-1.75I_1 + 8) + 10(-0.75I_1 + 4)$$
$$30I_1 = 120$$
$$I_1 = 4 \text{ A}$$

Answer is (D)

DC ELECTRICITY-32

In the circuit shown, what is the current through CD?

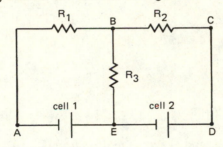

(A) 0.2 A   (B) 0.6 A   (C) 1 A   (D) 1.9 A   (E) 2.8 A

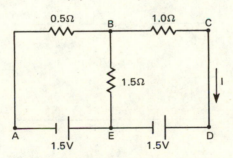

The method of superposition is used to find the current, I. Let I_1 be the current from cell 1, and let I_2 be the current from cell 2. Then, $I = I_1 + I_2$. Short-circuiting cell 2 to find I_1, the equivalent resistance through CD is:

$$R_{CD,1} = \left[0.5 + \frac{(1.5)(1)}{2.5} \right]$$
$$= 1.1 \ \Omega$$
$$I_1 = \left(\frac{1.5}{2.5} \right) \left(\frac{1.5}{1.1} \right)$$
$$= 0.820 \text{ A}$$

Short-circuiting cell 1 to find I_2,

$$R_{CD,2} = 1 + \frac{(0.5)(1.5)}{2}$$
$$= 1.375 \; \Omega$$
$$I_2 = \frac{1.5}{1.375}$$
$$= 1.09 \text{ A}$$

The total current is, therefore:

$$I = I_1 + I_2 = 0.82 + 1.09$$
$$= 1.9 \text{ A}$$

Answer is (D)

DC ELECTRICITY–33

For the network shown, find the voltage drop from C to D.

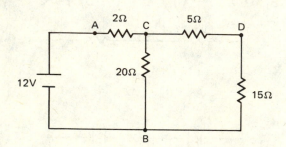

(A) 2 V   (B) 3 V   (C) 8 V
          (D) 10 V   (E) none of the above

The total resistance is:

$$R_{\text{total}} = 2 + \frac{1}{\dfrac{1}{20} + \dfrac{1}{15 + 5}}$$
$$= 12 \; \Omega$$
$$I_{\text{total}} = \frac{V}{R_{\text{total}}} = \frac{12}{12} = 1 \text{ A}$$

Use a current divider to find the current in section CDB.

$$I_{CDB} = (1)\left(\frac{20}{40}\right)$$

$$= 0.5 \text{ A}$$

$$V_{CD} = (0.5)(5)$$

$$= 2.5 \text{ V}$$

None of the numerical answer choices are correct.

Answer is (E)

DC ELECTRICITY–34

Find the voltage drop across the 4 ohm resistor in the network shown.

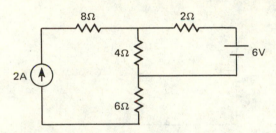

(A) 4.3 V   (B) 6.67 V   (C) 12.1 V
          (D) 24.4 V   (E) none of the above

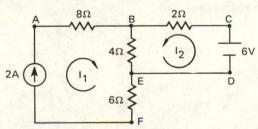

The network is redrawn with the currents and circuit points labeled as shown. The current through BE is equal to the sum of currents from AB and CB:

$$I_{BE} = I_1 + I_2 = 2 + I_2$$

PROFESSIONAL PUBLICATIONS, INC. • Belmont, CA

Kirchoff's voltage law around loop DCBE gives the equation:

$$6 = 2I_2 + 4I_{BE} = 2I_2 + 4(2 + I_2)$$
$$I_2 = -0.333 \text{ A (opposite to the direction that it was defined)}$$
$$I_{BE} = 2 - 0.333 = 1.67 \text{ A}$$
$$V_{BE} = (1.67)(4)$$
$$= 6.67 \text{ V}$$

Answer is (B)

DC ELECTRICITY–35

What is the voltage at point A in the network shown?

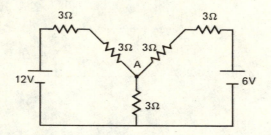

(A) 1 V   (B) 2.25 V   (C) 3 V   (D) 4.5 V   (E) 6 V

The circuit is redrawn:

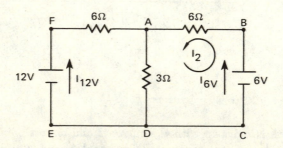

Superposition is used to find I_2:

$$I_2 = I_{6V} - I_{12V}$$

I_{6V} is the current through BA from the 6 volt source, and I_{12V} is the current through BA from the 12 volt source. The resistances are calculated:

$$R_{6V} = 6 + \frac{1}{\frac{1}{6} + \frac{1}{3}} = 8\,\Omega$$

$$I_{6V} = \frac{6}{8} = 0.75\text{ A}$$

$$R_{12V} = 6 + \frac{1}{\frac{1}{6} + \frac{1}{3}} = 8\,\Omega$$

$$I_{2V} = \left(\frac{3}{9}\right)\left(\frac{12}{8}\right) = 0.5\text{ A}$$

Therefore,
$$\begin{aligned}
I_2 &= I_{6V} - I_{12V} \\
&= 0.75 - 0.5 \\
&= 0.25\text{ A} \\
V_A &= 6 - 2(3)(0.25) \\
&= 4.5\text{ V}
\end{aligned}$$

Answer is (D)

DC ELECTRICITY–36

What is the voltage drop across the 8 ohm resistor in the following circuit?

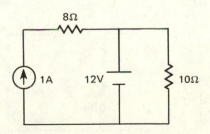

(A) 8 V   (B) 12 V   (C) 20 V   (D) 22 V   (E) 36 V

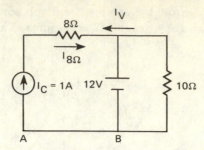

Redrawing the circuit as shown, with I_C equal to the component of the current through the 8 ohm resistor due to the current source, and I_V equal to the component of the current through the resistor due to the voltage source,

$$I_{8\Omega} = I_C - I_V$$

But, $I_C = 1$ amp, and $I_V = 0$ amps. Therefore,

$$I_{8\Omega} = 1 \text{ A}$$
$$V_{8\Omega} = IR = (1)(8)$$
$$= 8 \text{ V}$$

Answer is (A)

DC ELECTRICITY–37

Determine the voltage drop across the 6 ohm resistor.

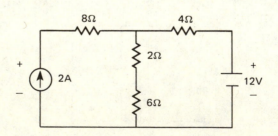

(A) 6 V   (B) 9 V   (C) 10 V   (D) 18 V   (E) 20 V

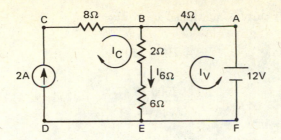

From superposition, with I_C designating the current through the resistor from the current source, and I_V designating the current through the resistor from the voltage source,

$$I_{6\Omega} = I_C + I_V$$

I_C and I_V are:

$$I_C = 2v\left(\frac{4\Omega}{(4+2+6)\Omega}\right) = 0.667 \text{ A}$$

$$I_V = \frac{12v}{(4+2+6)\Omega} = 1 \text{ A}$$

$$I_{6\Omega} = 0.667 + 1 = 1.67 \text{ A}$$

Therefore,

$$V_{6\Omega} = IR = (1.67)(6) = 10 \text{ V}$$

Answer is (C)

DC ELECTRICITY–38

The voltage drop across a device is 50 volts, and the current drawn is 30 amps. What is the power rating of this device?

(A) 0.659 hp   (B) 1 hp   (C) 1.50 hp

(D) 2.01 hp   (E) 3.02 hp

Power equals current multiplied by voltage.

$$P = IV = (30)(50)$$
$$= 1500 \text{ W}$$
$$= \frac{1500 \text{ W}}{746 \dfrac{\text{W}}{\text{hp}}}$$
$$= 2.01 \text{ hp}$$

Answer is (D)

DC ELECTRICITY-39

In the circuit shown, $V = 6$ volts, and the internal resistance of the source, r, is 1 ohm. For a specific value of R, the power output is the maximum possible. What is this maximum power?

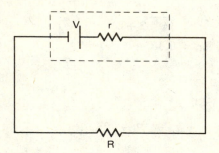

(A) 4.5 W   (B) 6 W   (C) 9 W

(D) 18 W   (E) 36 W

Power is given by the following expression:

$$P = I^2 R$$
$$I = \frac{V}{R + r}$$
$$P = \frac{RV^2}{(R + r)^2}$$

Maximizing this equation,

$$\frac{dP}{dR} = \frac{V^2(R+r)^2 - 2(R+r)RV^2}{(R+r)^4} = 0$$
$$V^2(R+r)^2 = 2(R+r)RV^2$$
$$R + r = 2R$$
$$R = r = 1\ \Omega$$

Therefore, the maximum power is:

$$P_{max} = \frac{RV^2}{(R+r)^2}$$
$$= \frac{(1)(6)^2}{(1+1)^2}$$
$$= \frac{36}{4} = 9\ W$$

Answer is (C)

DC ELECTRICITY–40

What is the time constant of the network?

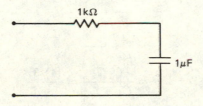

(A) 1 sec   (B) 10 sec   (C) 100 sec
        (D) 1000 sec   (E) 3600 sec

The time constant, τ, is:

$$\tau = \frac{1}{RC} = \frac{1}{(1000)(1 \times 10^{-6})}$$
$$= 1000\ sec$$

Answer is (D)

DC ELECTRICITY-41

For the two capacitors shown, $C_1 = 1$ microfarad and $C_2 = 3$ microfarads. What is the equivalent capacitance between A and B?

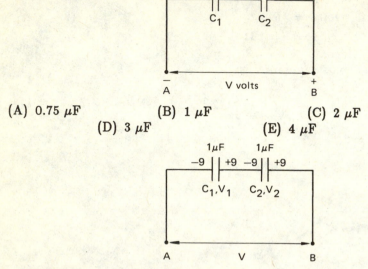

(A) 0.75 μF   (B) 1 μF   (C) 2 μF

          (D) 3 μF   (E) 4 μF

By definition, $q = CV$. For capacitors in series, the charge, q, is the same on each capacitor. Therefore,

$$V = V_1 + V_2$$

$$\frac{q}{C} = \frac{q}{C_1} + \frac{q}{C_2}$$

$$\frac{1}{C} = \frac{1}{C_1} + \frac{1}{C_2}$$

$$C = \frac{C_1 C_2}{C_1 + C_2} = \frac{(1)(3)}{1 + 3}$$

$$= 0.75 \ \mu\text{F}$$

Answer is (A)

DC ELECTRICITY-42

The equivalent capacitance of capacitors C_1 and C_2 connected in series is 7.3 microfarads. If the capacitance of $C_1 = 9.6$ microfarads, what is the capacitance of C_2?

(A) 2.3 μF   (B) 30.5 μF   (C) 35 μF   (D) 49.3 μF   (E) 84.5 μF

For capacitors in series, the equivalent capacitance, C, is:

$$\frac{1}{C} = \frac{1}{C_1} + \frac{1}{C_2}$$

$$C_2 = \frac{C_1 C}{C_1 - C} = \frac{(9.6)(7.3)}{9.6 - 7.3}$$

$$= 30.5 \ \mu F$$

Answer is (B)

DC ELECTRICITY-43

Find the voltage at point A at the instant the switch is closed.

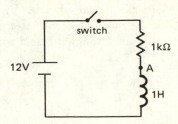

(A) 0 V   (B) 1 V   (C) 3 V   (D) 6 V   (E) 12 V

The current in the circuit is:

$$I(t) = \frac{E}{R}\left(1 - e^{-Rt/L}\right)$$

$$= \frac{12}{1000}\left(1 - e^{-1000t/1}\right)$$

Therefore, at time $t = 0$ seconds, $I(0) = 0.012$ A.

$$V_A = 12 - IR = 12 - (0.012)(1000)$$

$$= 0 \ V$$

Answer is (A)

DC ELECTRICITY–44

Which of the following statements regarding the motion of a conductor through a changing magnetic field is false?

(A) The lines of magnetic flux pass from the north pole to the south pole of the magnet.

(B) When a conductor is "open circuited," no current flows despite its motion through the field.

(C) The conductor must move at constant velocity in order to generate a current.

(D) A current forced opposite the direction of the conductor's motion will create a torque.

(E) If more lines of flux are cut in one area than another, more current will be generated in that area.

A varying amount of flux gives rise to a current at any speed. Flux through a conducting loop can be varied either by changing the magnetic field or by changing the speed of a conductor in the field. Thus, a conductor accelerating through a magnetic field will generate current.

Answer is (C)

DC ELECTRICITY–45

Diesel trains are driven by DC motors powered by DC generators. These are, in turn, driven by diesel engines. Which of the following is the reason for using such a configuration instead of AC generator-motor sets?

(A) The DC configuration provides high torque and good incremental power at low speeds, and performs equally well at high speeds.

(B) The DC equipment is significantly less expensive.

(C) Historically, the DC engine configuration has been used. There is no reason to change this.

(D) By using the DC equipment, the power factor problems associated with AC equipment are avoided.

(E) Although the AC units provide high torque at low speeds, they do not perform well enough at high speeds.

At low speeds the DC system is best because it gives high torque and excellent control. DC motors or generators generally cost more than AC units because they have windings on the armature, which the AC units lack. The power factor problem has nothing to do with the failure to use AC systems. At low speeds, the torque delivered by AC units is poor.

Answer is (A)

DC ELECTRICITY–46

In a DC motor, what is the definition of "field resistance"?

(A) It is the load resistance seen by a generator without considering inductance.
(B) It is the resistance of the excitation circuit.
(C) It is the resistance of the armature windings plus the load resistance.
(D) It is the static resistance of the motor.
(E) It is the resistance of the stator windings.

The field circuit is the circuit that excites the pole pieces, thereby producing the flux cut by the armature windings. The field resistance is the resistance of this circuit.

Answer is (B)

DC ELECTRICITY–47

In a DC motor, which of the following does not cause sparking at a commutator?

(A) no load on the output leads
(B) frozen armature
(C) high brush contact resistance
(D) the use of graphite brushes with good contact pressure
(E) the nonlinear relationship between current and resistance at the brushes

Sparking at the commutators results from having very low resistance at that point. It is normally associated with the use of copper or mainly copper brushes because the resistance of the brushes is non-linear and drops as current increases.

Answer is (C)

DC ELECTRICITY–48

The armature in a DC generator has one or more pairs of conductors or coils in which current is produced. In general, which of the following is true about the amount of power produced?

(A) No gain in power is achieved beyond four pairs of coils.
(B) More coils give more power.
(C) Power is related only to the number of poles.
(D) Power is a function only of the output voltage and current.
(E) None of the above are true.

Every coil that cuts the flux lines gives more power. It is desirable to place as many coils on the armature as possible.

Answer is (B)

DC ELECTRICITY–49

Which of the following limits the number of coils that may be placed on the armature of a DC motor or generator?

(A) the type of winding used
(B) the number of poles
(C) the size of the load on the motor
(D) coil to coil arcing due to the breakdown of insulation
(E) nothing

Coils must be well insulated from each other when spaced close together, otherwise failure will occur due to arcing. There is also a physical limit on the number of coils that will fit on an armature. However, this is due to the volume of insulation material needed to prevent shorts.

Answer is (D)

DC ELECTRICITY-50

The overall torque of a DC motor is:

$$T = K_T \phi I_a z$$

K_T is a constant for the particular machine, ϕ is the total magnetic flux per pole, I_a is the armature current, and z is the number of conductors on the surface of the armature. Which of the following is false regarding the above equation?

(A) It applies to both motors and generators.
(B) It applies only to a machine having an even number of pairs of poles.
(C) It applies since torque is indirectly dependent upon the pole winding current.
(D) It applies whether or not there is a load on the system.
(E) It applies to rotational and nonrotational DC systems.

The equation is valid for generators or motors having any number of poles. (B) is false.

Answer is (B)

DC ELECTRICITY-51

A DC system is protected from lightning by putting a thyrite tube in the circuit that connects the high voltage line with the ground. Which of the following is not true regarding a thyrite tube?

(A) It maintains a very high resistance at or below the system operating voltage.
(B) Its resistance becomes low at very high voltages.
(C) It has a very fast recovery time with regard to voltage change.
(D) It vaporizes at high voltages.
(E) Its exceptional stability and long life are due to its ceramic composition.

The resistance of thyrite tubes drops significantly near the operating voltage, allowing large currents to be discharged. They do not vaporize at high voltages, but instead recover immediately after passing large currents to the ground.

Answer is (D)

DC ELECTRICITY-52

The saturation curve limits the voltage at which a generator or motor can operate. Which of the following statements is incorrect regarding saturation curves?

(A) As field current increases, the hysteresis effect limits the increase in the flux produced.

(B) Poles that allow the production of more flux permit higher operating voltages.

(C) More flux at a constant field voltage can be produced by increasing the number of poles.

(D) Saturation does not depend upon the type of steel used in the poles.

(E) The saturation curve for a machine at one speed is proportional to the saturation curve at another speed.

Saturation occurs at a particular driving voltage for any steel.

Answer is (D)

DC ELECTRICITY-53

Series and shunt motors are connected like series and shunt generators, respectively. The terms refer to the manner in which the self-excitation of the poles is connected to the unit. Which of the following statements is false?

(A) The torque curve of a shunt motor is linear.

(B) Field coils of the shunt motor or generator are in parallel with the armature windings.

(C) Field coils of the series motor or generator are in series with the armature windings.

(D) The torque curve for a series motor is parabolic.

(E) The torque curves of both shunt and series motors are not affected by the value of the armature current.

The torque for these motors is directly related to the armature current.

Answer is (E)

DC ELECTRICITY–54

In terms of efficiency, shunt and series motors or generators have similar characteristics. Which of the following statements is false?

(A) Series motors have low torque at low speeds.

(B) Shunt and series motors have approximately 80% efficiency above one-third of the rated load.

(C) Efficiency decreases with lower speeds for both types of motors.

(D) The two types of motors have identical or very similar efficiency curves.

(E) Core losses make up a substantial portion of the losses that contribute to the inefficiency of both motor types.

Although the efficiency curves for the two types of motors both drop at lower operating speeds, they are still quite different.

Answer is (D)

DC ELECTRICITY–55

What is the pole pitch?

(A) the mica used to insulate the poles from each other
(B) the space on the stator allocated to one pole
(C) the space on the stator allocated to two poles
(D) the angle at which the pole windings are wound
(E) none of the above

Pole pitch is defined as the periphery of the armature divided by the number of poles. Thus, it is the space on the stator allocated to one pole.

Answer is (B)

DC ELECTRICITY–56

Which of the following statements regarding a compound motor is false?

(A) It has a shunt winding.
(B) It has a series winding.
(C) It has commutators, armature windings, and field windings.
(D) It operates poorly when subjected to sudden loads.
(E) It has both shunt and series exciter windings.

The advantage of a compound motor is its ability to respond to sudden heavy loads. The motor's speed is reduced quickly when the load is applied, transferring the kinetic energy of the system to the work area.

Answer is (D)

DC ELECTRICITY–57

Which of the following does not contribute to core losses in DC motors?

(A) eddy currents in the armature
(B) hysteresis losses in the armature
(C) commutator losses
(D) eddy currents and hysteresis losses in the armature
(E) none of the above

The induced eddy currents together with the hysteresis losses constitute the core losses. Commutator losses do not contribute to the core losses.

Answer is (C)

DC ELECTRICITY–58

Which of the following are power losses in a DC motor?

(A) I^2R losses
(B) gear and frictional losses
(C) core losses
(D) hysteresis losses
(E) all of the above

All of the choices are types of power losses in a DC motor. Note that core losses include hysteresis losses.

Answer is (E)

DC ELECTRICITY–59

Which of the following statements is false regarding large DC motors?

(A) To avoid flashover, the voltage difference between adjacent commutators should not exceed 15 volts DC.

(B) Two, four, six, or eight coils may be laid in a slot in the armature to make maximum use of the flux and to generate as much power as possible.

(C) The pitch of the winding of a given coil is the number of slots spanned by the coil.

(D) Lap winding utilizes several full pathways directly under two poles.

(E) Wave winding does not pass under all poles at one time.

The lap winding loops several times under the same poles, while the wave winding lies under all poles at the same time.

Answer is (E)

DC ELECTRICITY–60

Which of the following statements is not true about the operation of parallel shunt generators?

(A) The drooping load or decreasing terminal voltage characteristic of shunt generators makes two or more units operating in parallel more stable.

(B) The use of several units makes maintenance and repair easier.

(C) The parallel configuration makes it possible to add units as needed, and to shut off unnecessary units at low load demands.

(D) One large unit would be more expensive than the use of several smaller ones, even if it ran at full load at all times.

(E) It is unlikely that all the generators would produce the same amount of power.

Although it is rare to be able to load a unit fully at all times due to maintenance considerations, one large unit operating as such would be more economical.

Answer is (D)

DC ELECTRICITY–61

How many commutators does a DC machine require to produce the wave form shown?

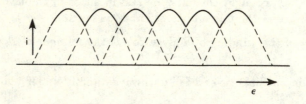

(A) two   (B) three   (C) four   (D) five   (E) six

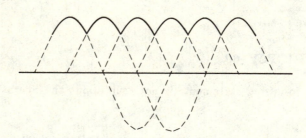

Two commutators are needed for each coil. Each coil produces a single sine wave. If the alternate peaks in the figure are reversed, two full sine waves are described. Therefore, since there are two coils, four commutators are needed.

Answer is (C)

DC ELECTRICITY–62

A generator supplies a 160 kW load over 2500 feet of two-conductor copper feeder having a resistance of 0.078 ohms per 1000 feet. The bus voltage is constant at 800 volts. What is the voltage delivered at the specified load?

(A) 612 V   (B) 652 V   (C) 702 V

     (D) 712 V   (E) 752 V

Since it is a two-conductor feeder, the total resistance of the feeder is:

$$R = \frac{(2)(2500)(0.078)}{1000} = 0.39\ \Omega$$

By definition, $V = RI$.

$$V_{\text{load}} = 800 - 0.39I$$
$$I = \frac{800 - V_{\text{load}}}{0.39}$$

Using this equation for I, the expression $P = VI$ gives:

$$160,000\ \text{W} = V_{\text{load}}I$$
$$= V_{\text{load}}\left(\frac{800 - V_{\text{load}}}{0.39}\right)$$

$$62,400 = 800V_{\text{load}} - V_{\text{load}}^2$$
$$0 = V_{\text{load}}^2 - 800V_{\text{load}} + 62,400$$

Solving for V_{load},

$$V_{\text{load}} = \frac{800 \pm \sqrt{(800)^2 - (4)(1)(62,400)}}{2}$$

$$= 87.5\ \text{V or } 712\ \text{V}$$

Of the choices, the only reasonable voltage is 712 V.

Answer is (D)

DC ELECTRICITY-63

The horsepower developed by a pony brake is given by the following equation:

$$P = \frac{2\pi F L n}{33{,}000}$$

F is the force in pounds produced at a distance, L, (in feet) from the center of the shaft, and n is the shaft speed in rpm. For $L = 3$ feet, what brake force will be produced at 200 rpm by a 40 hp motor?

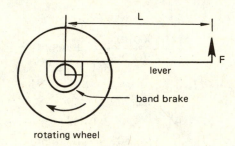

(A) 250 lbf   (B) 350 lbf   (C) 360 lbf

        (D) 370 lbf   (E) none of the above

Solving for F,

$$
\begin{aligned}
F &= \frac{33{,}000 P}{2\pi L n} \\
&= \frac{(33{,}000)(40)}{2\pi(3)(200)} \\
&= 350 \text{ lbf}
\end{aligned}
$$

Answer is (B)

13 AC ELECTRICITY

AC ELECTRICITY–1

An alternating current with a frequency of 60 Hz is passed through a moving coil galvanometer which measures DC current. What will the galvanometer reading be equal to?

(A) the peak value of the AC current
(B) the average value of the AC current
(C) a value between the peak and average values of the AC current
(D) the rms value
(E) a negligible amount

If the galvanometer is designed to measure DC current, it will not be able to respond quickly enough to measure an alternating current of 60 Hz. The reading will be negligible.

Answer is (E)

AC ELECTRICITY–2

Which of the following effects are generally less for an alternating current than for a direct current?

(A) heating effects
(B) chemical effects
(C) magnetic effects
(D) impedance
(E) both chemical and magnetic effects

Chemical effects are generally less for an AC current than for a DC current. Heating and magnetic effects are generally greater for an AC current than for a DC current. Impedance for an AC current is either larger than or the same as a DC current.

Answer is (B)

AC ELECTRICITY-3

The following two sine waves, K and M, are plotted as phasors. Determine which of the numbered vectors, $\mathbf{v}$, corresponds to K.

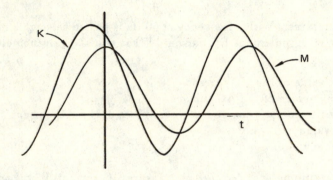

$$K = K_0 \cos (\omega t + \theta)$$
$$M = M_0 \cos (\omega t)$$

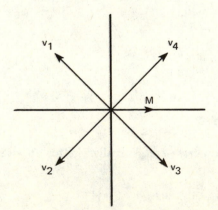

(A) $K = \mathbf{v}_1$ (B) $K = \mathbf{v}_2$ (C) $K = \mathbf{v}_3$

(D) $K = \mathbf{v}_4$ (E) $K = \mathbf{v}_2$ or $\mathbf{v}_3$

The magnitude of the vector corresponds to the amplitude of the wave. Thus, the vector, K, is longer than M. All angles are measured from the positive x-axis, with a leading angle measured counterclockwise by convention. Therefore, since the peak of K leads that of M by less than 90°, the K vector lies in the first quadrant. The only choice satisfying these conditions is (D).

Answer is (D)

AC ELECTRICITY–4

A wire carries an AC current of $3\cos 100\pi t$ amps. What is the average current over 6 seconds?

(A) 0 A   (B) $\frac{\pi}{6}$ A   (C) 1.5 A   (D) $\frac{6}{\pi}$ A   (E) 3 A

If T is the total period of time, and $I(t)$ is the current as a function of time, the average current is:

$$I_{\text{average}} = \int_0^T \frac{I(t)}{T}\,dt$$

Therefore, for the particular AC current,

$$
\begin{aligned}
I_{\text{average}} &= \int_0^6 \frac{3\cos 100\pi t}{6}\,dt \\
&= \frac{1}{2}\int_0^6 \cos 100\pi t\,dt \\
&= \frac{1}{200\pi}\left[\sin 100\pi t\right]_0^6 \\
&= 0 \text{ A}
\end{aligned}
$$

Answer is (A)

AC ELECTRICITY-5

What is the I_{rms} value for the wave form shown?

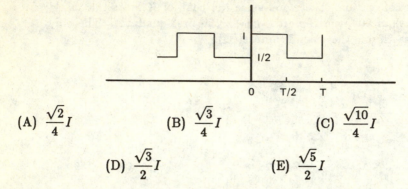

(A) $\dfrac{\sqrt{2}}{4}I$

(B) $\dfrac{\sqrt{3}}{4}I$

(C) $\dfrac{\sqrt{10}}{4}I$

(D) $\dfrac{\sqrt{3}}{2}I$

(E) $\dfrac{\sqrt{5}}{2}I$

If T is the period, and I is the current, the rms (effective) value is:

$$I_{rms} = \sqrt{\frac{1}{T}\int_0^T [I(t)]^2\, dt}$$

For the square wave shown,

$$I(t) = I \qquad 0 \le t \le \frac{T}{2}$$
$$I(t) = \frac{I}{2} \qquad \frac{T}{2} \le t \le T$$

Therefore,

$$I_{rms} = \sqrt{\frac{1}{T}\int_0^{T/2} I^2\, dt + \frac{1}{T}\int_{T/2}^T \left(\frac{I}{2}\right)^2 dt}$$
$$= \sqrt{\frac{1}{T}\left(\frac{I^2 T}{2}\right) + \frac{1}{T}\left(\frac{I^2 T}{8}\right)}$$
$$= \sqrt{\frac{5}{8}I^2}$$
$$= \frac{\sqrt{10}}{4}I$$

Answer is (C)

AC ELECTRICITY–6

A sinusoidal AC voltage with an rms value of $E_{rms} = 60$ volts is applied to a purely resistive circuit as shown. What steady voltage generates the same power as the alternating voltage?

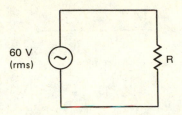

(A) 38.2 V   (B) 42.4 V   (C) 60 V   (D) 84.9 V   (E) 120 V

By definition of average power,

$$P_{average} = \frac{E_{rms}^2}{R} = \frac{E^2}{R}$$
$$E = E_{rms}$$
$$= 60 \text{ V}$$

Answer is (C)

AC ELECTRICITY–7

What is the average current through the resistor, R, in the rectifier shown?

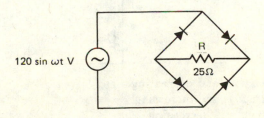

(A) 0 A   (B) 0.76 A   (C) 3.06 A   (D) 4.80 A   (E) 6.12 A

The type of rectifier shown is a "full wave" rectifier, with an average current of:

$$I_{average} = \frac{E_{average}}{R}$$

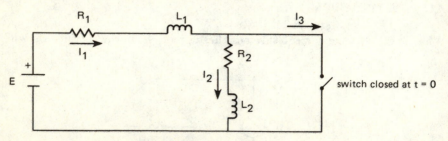

$E_{average}$ for a full wave rectifier is:

$$E_{average} = \frac{1}{\pi} \int_0^\pi 120 \sin \omega t \, dt$$

$$= (2) \left(\frac{120}{\pi} \right) = \frac{240}{\pi}$$

Therefore,

$$I_{average} = \left(\frac{240}{\pi} \right) \left(\frac{1}{25} \right)$$

$$= 3.06 \text{ A}$$

Answer is (C)

AC ELECTRICITY–8

For the circuit shown, $I_1 = I_2$ before the switch is closed. If the switch is closed at time t_0, what is the behavior of I_1 at t_0?

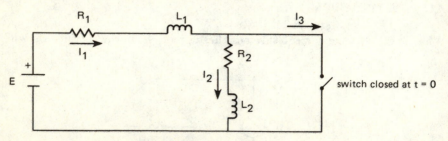

(A) I_1 is discontinuous and decreasing.
(B) I_1 is discontinuous and increasing.
(C) I_1 is continuous and decreasing.
(D) I_1 is continuous and increasing.
(E) The system is stable and not transient.

For $t < t_0$, the current travels through L_1 and L_2. After the switch is closed, the current will slowly stop flowing through L_2 and, as t goes to infinity, it will travel around the outer loop with $I_1 = I_3 = E/R_1$ and $I_2 = 0$. At $t = 0$, $I_2 = E/(R_1 + R_2)$, but for $t > 0$,

$$I_1(t) = \frac{E}{R_1}\left(1 - Ke^{-R_1 t/L}\right)$$

$$= \frac{E}{R_1}\left(1 - \frac{R_2}{R_1(R_1 + R_2)}e^{-R_1 t/L}\right)$$

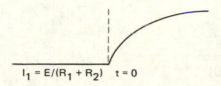

$$I_1 = E/(R_1 + R_2) \quad t = 0$$

Therefore, I_1 is continuous and increasing at $t = 0$.

Answer is (D)

AC ELECTRICITY-9

A $2\mu F$ capacitor in the circuit shown has a reactance of $X_C = 1500$ ohms. What is the frequency of the AC source?

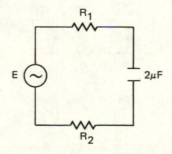

(A) 3 Hz   (B) 53 Hz   (C) 60 Hz

        (D) 119 Hz   (E) 150 Hz

The reactance is:

$$X_C = \frac{1}{\omega C}$$

PROFESSIONAL PUBLICATIONS, INC. ● Belmont, CA

Therefore,

$$\omega = \frac{1}{CX_C}$$

Thus, the frequency is:

$$f = \frac{\omega}{2\pi} = \frac{1}{2\pi C X_C}$$
$$= \frac{1}{2\pi(2 \times 10^{-6})(1500)}$$
$$= 53 \text{ Hz}$$

Answer is (B)

AC ELECTRICITY–10

If the capacitor and the inductor in the circuit shown have the same reactance, what is the frequency of the AC source?

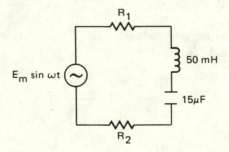

(A) 27 Hz   (B) 184 Hz   (C) 212 Hz
          (D) 1150 Hz   (E) 1160 Hz

If the inductor and capacitor have the same reactance, then,

$$\frac{1}{\omega C} = \omega L$$
$$\omega^2 = \frac{1}{CL}$$
$$\omega = \frac{1}{\sqrt{CL}}$$

The frequency, f, is:

$$f = \frac{\omega}{2\pi} = \frac{1}{2\pi\sqrt{CL}}$$

$$= \frac{1}{2\pi\sqrt{(15 \times 10^{-6})(50 \times 10^{-3})}}$$

$$= 184 \text{ Hz}$$

Answer is (B)

AC ELECTRICITY–11

An alternating voltage of $E = 10\sin\omega t$ volts is applied to the RCL circuit shown. What is the effective current, I_{rms}, if the circuit is in resonance with the driving voltage? The magnitude of the current is $I = \left[R^2 + \left(\omega L - \frac{1}{\omega C}\right)^2\right]^{-1/2}$.

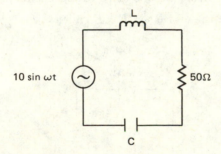

(A) 0.141 A
(B) 0.200 A
(C) 7.07 A
(D) 7.14 A
(E) The current cannot be computed with the given data.

At resonance, the impedance of the circuit is equal to the impedance of the resistor. Therefore,

$$\omega L = \frac{1}{\omega C}$$

and

$$I = E(R^2 + 0)^{-\frac{1}{2}} = \frac{E}{R}$$

$$I_{\text{rms}} = \frac{E_{\text{rms}}}{R} = \frac{\dfrac{E}{\sqrt{2}}}{R}$$

$$= \frac{\dfrac{10}{\sqrt{2}}}{50}$$

$$= 0.141 \text{ A}$$

Answer is (A)

AC ELECTRICITY–12

In the RCL circuit shown, $R = 10\ \Omega$, $C = 30\ \mu\text{F}$, and $L = 0.5$ H. At what frequency will the rms current be one-third of the maximum possible rms current? The magnitude of the current is $I = E\left[R^2 + \left(\omega L - \frac{1}{\omega C}\right)^2\right]^{-1/2}$

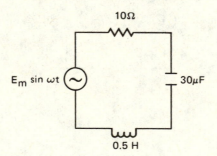

(A) 36.8 Hz   (B) 41.1 Hz   (C) 45.7 Hz

(D) 160 Hz   (E) 258 Hz

The maximum I_{rms} occurs at resonance. That is,

$$I_{\text{rms,max}} = \frac{E_{\text{rms}}}{R}$$

For the rms current to be one-third of the maximum,

$$\frac{1}{3}\frac{E_{rms}}{R} = \frac{E_{rms}}{\left[R^2 + \left(\omega L - \frac{1}{\omega C}\right)^2\right]^{\frac{1}{2}}}$$

$$\frac{1}{3R} = \frac{1}{\left[R^2 + \left(\omega L - \frac{1}{\omega C}\right)^2\right]^{\frac{1}{2}}}$$

$$R^2 + \left(\omega L - \frac{1}{\omega C}\right)^2 = 9R^2$$

$$\left(\omega L - \frac{1}{\omega C}\right)^2 = 8R^2$$

$$\omega L - \frac{1}{\omega C} = 2\sqrt{2}R$$

$$\omega^2 - \frac{2\sqrt{2}R\omega}{L} - \frac{1}{LC} = 0$$

Solving for the positive ω value,

$$\omega = \frac{\sqrt{2}R}{L} + \sqrt{\frac{2R}{L} + \frac{1}{LC}}$$

$$= 20\sqrt{2} + \sqrt{40 + \frac{1}{(0.5)(30 \times 10^{-6})}}$$

$$= 287 \text{ sec}^{-1}$$

$$f = \frac{\omega}{2\pi}$$

$$= \frac{287}{2\pi}$$

$$= 45.7 \text{ Hz}$$

Answer is (C)

AC ELECTRICITY-13

Which of the following statements regarding transformers is false?

(A) The copper losses $(I^2 R)$ in the primary and secondary coils are equal.

(B) Transformer power losses are generally low, approximately 1 to 3 percent.

(C) Power losses in transformers are converted to heat, which is then dissipated.

(D) One three-phase transformer weighs more than three equivalent single-phase transformers.

(E) An auto transformer taps into the primary coil to obtain a reduced voltage. Therefore, it does not have a secondary coil.

Power conversion using a three-phase transformer is more efficient than conversion using three separate single-phase units. Reduced weight and space requirements are obtained for the three-phase transformer.

Answer is (D)

AC ELECTRICITY-14

An ideal step-up transformer with a power factor of 1.0 is used in the circuit shown. The turns ratio is 70, and the primary rms voltage is 120 volts. What is the average power dissipated due to the resistance, R?

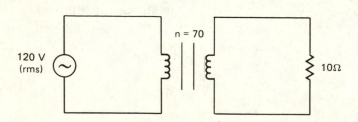

(A) 17.1 W   (B) 29.4 W   (C) 8.4×10^4 W

(D) 1.44×10^5 W   (E) 7.06×10^6 W

For an ideal transformer, the turns ratio is:

$$\frac{V_{rms,2}}{V_{rms,1}} = \frac{N_2}{N_1} = 70$$

$$V_{rms,2} = 70V_{rms,1}$$
$$= (70)(120)$$
$$= 8400 \text{ V}$$

Since power is given by $P = I^2R = \dfrac{V^2}{R}$,

$$P_{average,2} = V_{rms,2}^2 / R$$
$$= (8400)^2/10$$
$$= 7.06 \times 10^6 \text{ W}$$

There is no power loss in the primary circuit.

Answer is (E)

AC ELECTRICITY–15

In a transformer, the total voltage induced in each winding is proportional to the number of turns in that winding:

$$\frac{E_1}{E_2} = \frac{N_1}{N_2}$$

Disregarding all losses, determine E_3.

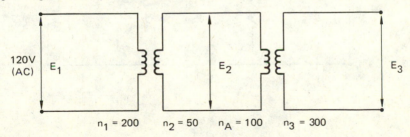

(A) 45 V
(B) 65 V
(C) 75 V
(D) 95 V
(E) none of the above

From the ratio given,

$$E_2 = \frac{N_2}{N_1} E_1$$

$$E_3 = \frac{N_3}{N_A} E_2$$

$$= \left(\frac{N_3}{N_A}\right)\left(\frac{N_2}{N_1}\right) E_1$$

$$= \left(\frac{300}{100}\right)\left(\frac{50}{200}\right) 120$$

$$= 90 \text{ V}$$

Answer is (E)

AC ELECTRICITY–16

Determine the resonant frequency, ω, of the circuit shown.

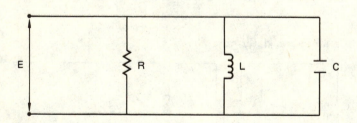

(A) $\dfrac{1}{\sqrt{LC}}$

(B) $\dfrac{2}{\sqrt{LC}}$

(C) $\sqrt{\dfrac{LC}{3}}$

(D) $\sqrt{\dfrac{LC}{2}}$

(E) $\sqrt{LC}$

Using vector notation as specified in the figure,

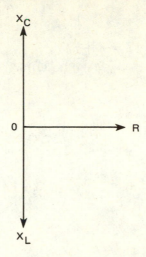

Resonance occurs when $X_C = X_L$, and $Z = R$. Since $X_C = 1/j\omega C$, and $X_L = j\omega L$,

$$\frac{1}{j\omega C} = j\omega L$$

$$\omega^2 = \frac{1}{LC}$$

$$\omega = \frac{1}{\sqrt{LC}}$$

Answer is (A)

AC ELECTRICITY–17

Determine the power angle, ϕ, in the circuit if $R = 25$ ohms, $L = 0.2$ henrys, $V = 200$ volts, and $f = 30$ Hz.

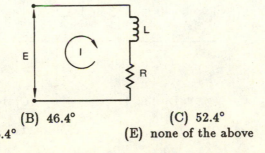

(A) 36.4°   (B) 46.4°   (C) 52.4°

        (D) 56.4°   (E) none of the above

PROFESSIONAL PUBLICATIONS, INC. • Belmont, CA

The power angle is the impedance angle. The impedance for the circuit is:

$$\mathbf{Z} = R + jX_L$$
$$= 25 + j2\pi(30)(0.2)$$
$$= 25 + j37.7$$
$$\tan\phi = \frac{X_L}{R}$$
$$\phi = \tan^{-1}\left(\frac{37.7}{25}\right)$$
$$= 56.4°$$

Answer is (D)

AC ELECTRICITY–18

Determine the impedance of the circuit. The line current is **I**, and the line voltage lies along the real axis.

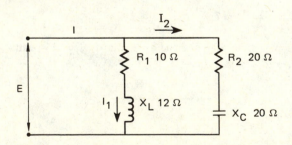

(A) 12.2 Ω   (B) 13.2 Ω   (C) 14.3 Ω

(D) 15.2 Ω   (E) 16.2 Ω

$$I_1 = \frac{E}{R_1 + jX_L} = \frac{E}{R_1 + jX_L}\left(\frac{R_1 - jX_L}{R_1 - jX_L}\right)$$

$$= E\left(\frac{R_1 - jX_L}{R_1^2 + X_L^2}\right)$$

$$= 120\left[\frac{10 - j12}{(10)^2 + (12)^2}\right]$$

$$= 4.91 - j5.9$$

$$I_2 = \frac{E}{R_2 - jX_C} = \frac{E}{R_2 - jX_C}\left(\frac{R_2 + jX_C}{R_2 + jX_C}\right)$$

$$= E\left(\frac{R_2 + jX_C}{R_2^2 + X_C^2}\right)$$

$$= 120\left[\frac{20 + j20}{(20)^2 + (20)^2}\right]$$

$$= 3 + j3.0$$

The total current is:

$$I = I_1 + I_2$$
$$= (4.91 - j5.90) + (3 + j3.0)$$
$$= 7.91 - j2.90$$

The impedance is:

$$Z = \frac{E}{I} = \frac{120 + j0}{7.91 - j2.9}$$
$$= 13.4 + j4.90$$
$$Z = \sqrt{(13.4)^2 + (4.90)^2}$$
$$= 14.3\ \Omega$$

Answer is (C)

AC ELECTRICITY–19

What is the power factor for problem 18?

(A) 73.7%   (B) 78.7%   (C) 83.7%

       (D) 93.7%   (E) 98.7%

The power factor is:

$$\cos\phi = \frac{P_{\text{real}}}{P_{\text{apparent}}} = \frac{R}{Z}$$

Since $R = 13.4$ and $Z = 14.3$,

$$\cos\phi = \frac{13.4}{14.3} = 93.7\%$$

Answer is (D)

AC ELECTRICITY–20

What is the input impedance of the amplifier shown, assuming it is an ideal op-amp?

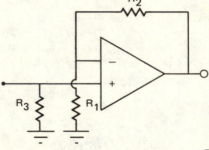

(A) R_1 (B) R_3 (C) $\dfrac{R_2}{R_1} + R_3$

(D) $\dfrac{R_1 R_3}{R_1 + R_3}$ (E) $\dfrac{1}{R_3}$

To find the input impedance, a test voltage is applied to the input. The resistance seen by the test voltage will be equal to the impedance: $R_{\text{input}} = V_{\text{test}}/I_{\text{test}}$. The circuit can be replaced with its equivalent:

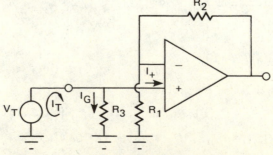

By Kirchoff's law, $I_{test} = I_G + I_{(+)}$. In an ideal op-amp, there is no current drawn by the positive and negative terminals. Therefore, $I_{(+)} = 0$ and $I_{test} = I_G$. Around that loop,

$$I_{test} = \frac{V_{test}}{R_3}$$

$$\frac{V_{test}}{I_{test}} = R_3$$

$$R_{input} = \frac{V_{test}}{I_{test}}$$

$$= R_3$$

Answer is (B)

AC ELECTRICITY–21

What is the input impedance of the following amplifier, assuming it is an ideal op-amp?

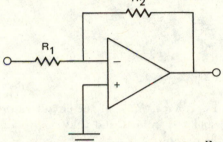

(A) R_1

(B) R_2

(C) $\dfrac{R_2}{R_1}$

(D) $\dfrac{R_1}{R_1 + R_2}$

(E) $\dfrac{R_2}{R_1 + R_2}$

To find the input impedance or resistance, the circuit is examined using a source voltage, V_s, and a source current, I_s. The circuit diagram becomes:

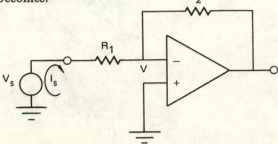

The input resistance will be:

$$R_{in} = \frac{V_s}{I_s}$$

For an op-amp to be in the range of linear operation, the voltage at the (+) terminal must equal the voltage of the (−) terminal. Therefore, $V = 0$, and,

$$I_s = \frac{V_s - V}{R_1} = \frac{V_s}{R_1}$$

$$R_{in} = (V_s)\left(\frac{R_1}{V_s}\right)$$

$$= R_1$$

Answer is (A)

AC ELECTRICITY–22

What is the output impedance, R_{out}, of the circuit shown? Assume no current is drawn at the (+) and (−) inputs, and that the op-amp has a small internal resistance, R_{int}, at the output.

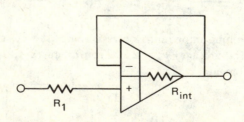

(A) R_{int}

(B) R_1

(C) $\dfrac{R_1 R_{int}}{R_1 + R_{int}}$

(D) $R_{out} \approx 0$

(E) R_{out} is infinite.

In terms of its operation, the op-amp diagram is like the solid part in the following figure:

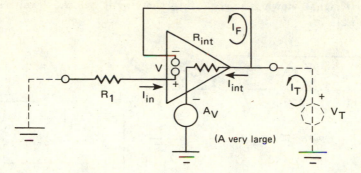

To find the output resistance, a test voltage, V_t, is attached to the output and the input is grounded. Then,

$$R_{\text{out}} = \frac{V_t}{I_t}$$

The test current, I_t, is equal to the internal current, I_{int}, plus the forced current, I_f. Since the inputs draw no current, $I_f = 0$. Therefore,

$$I_t = I_{\text{int}} = \frac{V_{\text{int}}}{R_{\text{int}}}$$
$$= \frac{V_t - (-AV)}{R_{\text{int}}}$$

Since no current is drawn at the inputs, the (+) input must be at V_t volts, and the (−) input must be at 0 volts, so that $V = V_t$. Thus,

$$I_{\text{int}} = \frac{V_t(1 + AV)}{R_{\text{int}}}$$
$$\frac{V_t}{I_{\text{int}}} = \frac{R_{\text{int}}}{1 + A}$$
$$R_{\text{out}} = \frac{V_{\text{int}}}{I_{\text{int}}} = \frac{R_{\text{int}}}{1 + A}$$

Since A is very large, and R_{int} is very small,

$$R_{\text{out}} \approx 0$$

Answer is (D)

AC ELECTRICITY–23

The 700 Hz signal shown is injected into the circuit below. What will be the output signal?

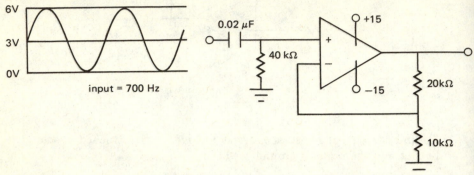

input = 700 Hz

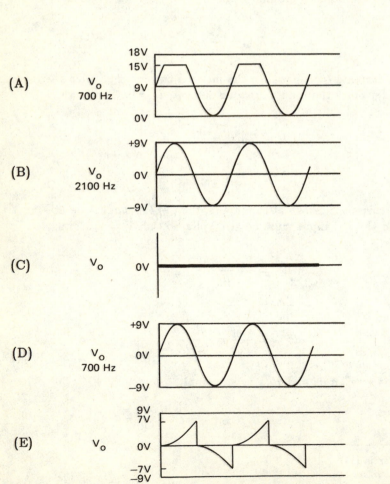

(A) V_o
       700 Hz

(B) V_o
       2100 Hz

(C) V_o

(D) V_o
       700 Hz

(E) V_o

The op-amp part of the circuit is a simple non-inverting amplifier with a gain of:

$$\frac{V_{\text{out}}}{V_{\text{in}}} = \frac{10 + 20}{10} = 3$$

The input into the amplifier is a high-pass filter with a cutoff frequency of:

$$f_c = \frac{1}{2\pi RC} = \frac{1}{2\pi (40 \times 10^3)(0.02 \times 10^{-6})}$$
$$= 200 \text{ Hz}$$

Thus, the AC component of the signal will pass through and be amplified three times, while the DC component will be cut out, resulting in a 9 volt amplitude sinusoid centered about 0 volts. This is known as an active high-pass filter. Thus, the correct output is shown in choice (D).

Answer is (D)

AC ELECTRICITY-24

The signal shown is the source for the amplifier. Which of the choices is the output signal?

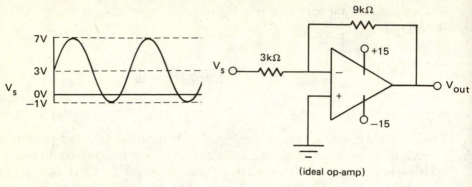

(ideal op-amp)

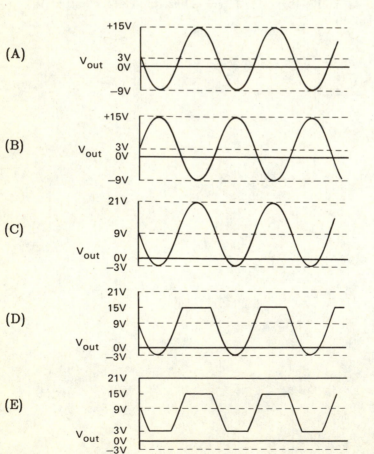

The amplifier is an inverting amplifier, so,

$$V_{out} = -\frac{9}{3}V_{in} = -3V_{in}$$

Both the DC and the AC components will be amplified. The DC component is $(3)(3) = 9$ volts, so the new waveform is centered at $V = 9$ volts. The AC component is $(3)(8) = 24$ volts peak-to-peak. Since the amplifier has only a 15 volt source, though, the voltage will be clipped at ± 15 volts. Since it never goes to -15 volts, the lower half of the output signal will be intact, and the upper half will be clipped at $+15$ volts.

Answer is (D)

AC ELECTRICITY–25

Two AC signals, V_1 and V_2, are to be combined such that:

$$V_{out} = \frac{3}{2}V_2 - \frac{5}{2}V_1$$

The subtracting amplifier circuit shown is used. What must be the values of R_1, R_2, R_3, and R_4?

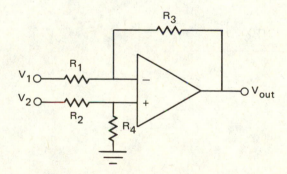

(A) $R_1 = 2\,k\Omega$, $R_2 = 2\,k\Omega$, $R_3 = 5\,k\Omega$, $R_4 = 3\,k\Omega$
(B) $R_1 = 2\,k\Omega$, $R_2 = 4\,k\Omega$, $R_3 = 5\,k\Omega$, $R_4 = 3\,k\Omega$
(C) $R_1 = 4\,k\Omega$, $R_2 = 8\,k\Omega$, $R_3 = 10\,k\Omega$, $R_4 = 2\,k\Omega$
(D) $R_1 = 5\,k\Omega$, $R_2 = 3\,k\Omega$, $R_3 = 4\,k\Omega$, $R_4 = 2\,k\Omega$
(E) $R_1 = 10\,k\Omega$, $R_2 = 7\,k\Omega$, $R_3 = 8\,k\Omega$, $R_4 = 6\,k\Omega$

The output for this op-amp configuration is:

$$V_{out} = \left(\frac{R_1 + R_3}{R_1}\right)\left(\frac{R_4}{R_2 + R_4}\right)V_2 - \frac{R_3}{R_1}V_1$$

The ratio $R_3{:}R_1$ must be 5:2. Therefore, the initial values $R_1 = 2$ kΩ and $R_3 = 5$ kΩ are chosen. Thus, the coefficient of V_2 is:

$$\left(\frac{R_1 + R_3}{R_1}\right)\left(\frac{R_4}{R_2 + R_4}\right) = \left(\frac{2 + 5}{2}\right)\left(\frac{R_4}{R_2 + R_4}\right) = \frac{3}{2}$$

$$\left(\frac{7}{2}\right)\left(\frac{R_4}{R_2 + R_4}\right) = \frac{3}{2}$$

$$\frac{R_4}{R_2 + R_4} = \frac{3}{7}$$

$$7R_4 = 3R_2 + 3R_4$$

$$4R_4 = 3R_2$$

$R_2 = 4$ kΩ and $R_4 = 3$ kΩ are chosen. Checking the results,

$$V_{out} = \left(\frac{2 + 5}{2}\right)\left(\frac{3}{4 + 3}\right)V_2 - \frac{5}{2}V - 1$$

$$= \frac{3}{2}V_2 - \frac{5}{2}V_1$$

Answer is (B)

AC ELECTRICITY-26

A sinusoidal signal with maximum voltage, $V_0 = 30$ mV, is to be amplified to at least 0.3 V, but not inverted. Which of the following operational amplifier configurations will best achieve this? Assume ideal op-amps.

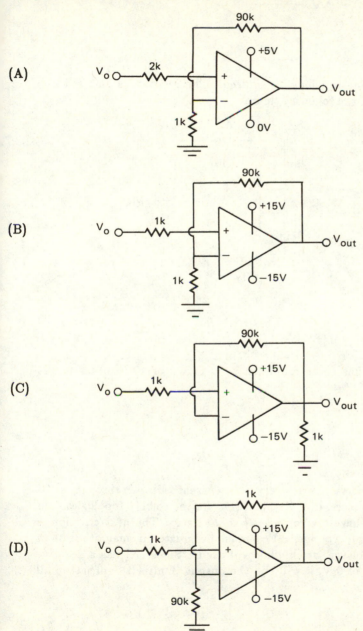

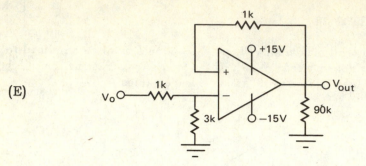

(E)

A non-inverting topology is required. The resistances and voltages are labeled in the following figure:

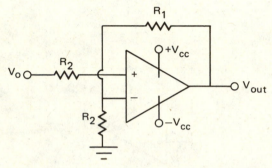

For this topology,

$$V_{\text{out}} = \frac{R_1 + R_2}{R_2} V_0$$

The voltage has to be amplified by a factor of 10 in order to get 30 mV up to 0.3 V. So,

$$\frac{R_1 + R_2}{R_2} = 10$$
$$R_1 = 9R_2$$

R_0 is not important, since very little current is drawn through it. Of the answer choices, (A), (B), and (D) are the correct topologies, and (A) and (B) have the correct R_1 to R_2 ratio. The next criterion is that the supply voltage, $\pm V_{cc}$, must be greater in magnitude than the output voltage, or clipping will occur. Since (A) has a negative input supply of 0 V, it will clip the output. Only (B) will satisfy all requirements.

Answer is (B)

AC ELECTRICITY-27

In the ideal op-amp configuration shown, $V_{out} = 12$ V sinusoidal, $R_f = 60$ kΩ, $R_1 = 30$ kΩ, and $R_2 = 10$ kΩ. Nothing is known about the inputs except that $V_1 = 5V_2$, and that they are 180° out of phase with the output. From this information, what are the maximum voltages of the inputs?

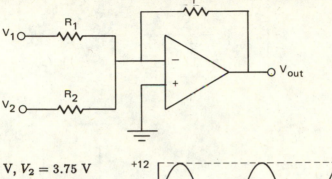

A) $V_1 = 0.75$ V, $V_2 = 3.75$ V
B) $V_1 = 1.00$ V, $V_2 = 5.00$ V
C) $V_1 = 2.50$ V, $V_2 = 12.5$ V
D) $V_1 = 3.75$ V, $V_2 = 0.75$ V
E) $V_1 = 5.00$ V, $V_2 = 1.00$ V

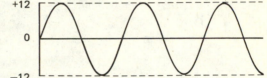

This is an adding amplifier with an output of:

$$V_{out} = -\left(\frac{R_f}{R_1}V_1 + \frac{R_f}{R_2}V_2\right)$$

It is known that $V_1 = 5V_2$. Substituting this into the equation for V_{out}, and evaluating with the given R values and V_{out},

$$12 = -\left[\left(\frac{60}{30}\right)5V_2 + \frac{60}{10}V_2\right]$$
$$V_2 = -0.75 \text{ volts}$$
$$V_1 = 5(-0.75)$$
$$= -3.75 \text{ volts}$$

Since the output is a sinusoid, the inputs must also be sinusoids. It is known that they are 180° out of phase, which is confirmed by the negative sign of the voltage. Thus, the maximum voltages are $V_1 = 3.75$ V and $V_2 = 0.75$ V.

Answer is (D)

AC ELECTRICITY-28

A zero crossing detector is needed for the noisy circuit shown below. The phase of the detector output is not important. (That is, the detector can show a time lag.) Which of the following op-amp configurations would be best?

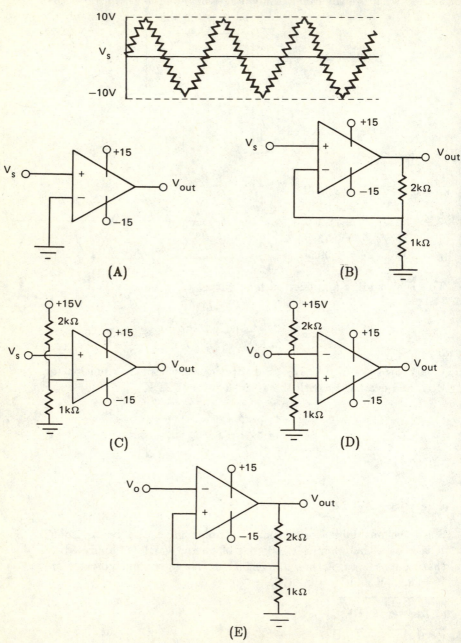

Because of the noise, a simple comparator such as in (A) will not work. There will be false zero crossings where the noise crosses zero at the signal crossing.

The configurations shown in (C) and (D) will not work for the same reason as the comparator. The device shown in (B) is a non-inverting amplifier, which will amplify the entire signal. A device with hysteresis, such as the Schmitt trigger, is needed. Such a device will not change until a threshold is reached, and will not change again until the negative threshold is reached. The configuration is:

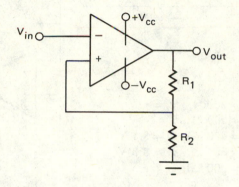

$$V_{\text{threshold}} = \frac{\pm R_2}{R_1 + R_2} V_{cc}$$

The only Schmitt trigger circuit is given in (E). With the R_1 and R_2 resistances shown, it will trigger at $V = \left(\frac{1}{2+1}\right) 15 = 5$ volts, which will work. There will be some delay, and the signal will be inverted.

Answer is (E)

AC ELECTRICITY–29

A 30 mV sinusoidal signal must be inverted, amplified to 6 volts, and chopped at 4 volts. If the following circuit is used, what are the values of R_1, R_2, and the avalanche voltage of the zener diodes, Z? There is a forward voltage drop of -0.7 V for the diodes.

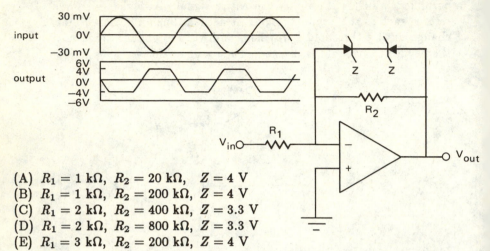

(A) $R_1 = 1$ kΩ, $R_2 = 20$ kΩ, $Z = 4$ V
(B) $R_1 = 1$ kΩ, $R_2 = 200$ kΩ, $Z = 4$ V
(C) $R_1 = 2$ kΩ, $R_2 = 400$ kΩ, $Z = 3.3$ V
(D) $R_1 = 2$ kΩ, $R_2 = 800$ kΩ, $Z = 3.3$ V
(E) $R_1 = 3$ kΩ, $R_2 = 200$ kΩ, $Z = 4$ V

The amplification is similar to that found in the normal inverting amplifier.

$$V_{\text{out}} = -\frac{R_2}{R_1}V_{\text{in}}$$

$$\frac{R_2}{R_1} = \frac{V_{\text{out}}}{V_{\text{in}}} = \frac{6}{30 \times 10^{-3}}$$

$$= 200$$

When $\pm V_{\text{out}} < Z$, one of the diodes will be reverse biased, but not avalanched. There is essentially an open circuit across the diodes, and the amplifier is a simple inverting amplifier. When $\pm V_{\text{out}} \geq Z$, one of the diodes is forward biased with a voltage drop of 0.7 V, while the other diode is avalanched at the Zener voltage. This keeps V_{out} constant. Thus, $V_{\text{out}} = 4$ V is equal to the Zener voltage plus the 0.7 V voltage drop of the other diode.

$$V_{\text{out}} = Z + 0.7$$
$$Z = 4 - 0.7 = 3.3 \text{ V}$$

So, $R_2 = 200R_1$, and $Z = 3.3$ V.

Answer is (C)

AC ELECTRICITY–30

An AC alternator operated as a motor is called a synchronous motor. Which of the following statements regarding synchronous motors is false?

(A) The average speed, regardless of load, does not decrease, since the motor must operate at a constant speed.

(B) When a load is increased, the increased torque is a result of the shift in the relative positions of the fields on the rotor and stator.

(C) The relationship between speed, frequency, and number of poles is the same for the rotating field of the induction motor and for the alternator.

(D) The poles of a synchronous motor must be salient.

(E) If a synchronous motor has salient poles, it will usually operate as an induction motor if there is no field current.

Salient poles are laminated pole pieces. Although salient poles are generally used, either salient or nonsalient poles can be used in a synchronous motor.

Answer is (D)

AC ELECTRICITY–31

Which of the following statements about induction motors is false?

(A) They are used to increase the line power factor.
(B) They have no slip rings, no poles, and no excited field current.
(C) They have no commutators and no windings on the armature.
(D) Squirrel cage induction motors operate at essentially constant speeds.
(E) The rotor frequency is equal to the slip multiplied by the stator frequency.

An induction motor degrades the power factor. All the other answer choices are true.

Answer is (A)

AC ELECTRICITY-32

A single-phase induction motor is not self-starting. Instead, auxiliary methods must be used, such as varying inductance, resistance, and capacitance. Which of the following is not true regarding this situation?

(A) A capacitor motor uses capacitance to split the phase, resulting in two phases almost 90° apart.

(B) Capacitor motors have lower starting torque than comparably sized single-phase induction motors.

(C) To obtain a higher reactance, a capacitor can be used when starting, and then be switched out of the circuit by mechanical means.

(D) If the capacitor remains in the circuit, the power factor will have a value close to unity.

(E) When properly split, the currents in the two windings of a single-phase motor are nearly 90° out of phase, which is an optimum phase relationship for a two-phase motor.

Due to the favorable phase relationship, the torque is higher for a capacitor motor than for other types of single-phase motors. For example: a capacitive phase split motor gives better torque than a resistively split motor. Therefore, choice (B) is false.

Answer is (B)

AC ELECTRICITY-33

A squirrel cage motor has such low resistance that it draws excessive currents when starting. Which of the following will not reduce this problem?

(A) connecting the windings as in a three-phase Y transformer, taking 58% of the normal line voltage. When up to sufficient motor speed, switching to a delta connection.

(B) use of an in-line rheostat

(C) use of an autotransformer to reduce line voltage

(D) use of a resistor in series with or without a reactor

(E) use of a class A motor

A class A motor draws a heavy starting current, usually 200 to 300 percent of the normal load. The other alternatives reduce the effective voltage across the windings, thus reducing the problem of excessive currents.

Answer is (E)

AC ELECTRICITY–34

Which of the following statements about AC generators is not true?

(A) The poles of an AC generator are located on the rotor.

(B) The three main types of AC generator are: direct-connect engine driven, water driven, and turbine driven.

(C) An AC generator uses commutators.

(D) Large turbine driven generators usually have two pole rotors to accommodate the high speed of the turbine.

(E) In an AC generator, the armature is usually stationary, and the poles rotate on the rotor.

Commutators are not used in AC machines. It is the relative motion between the rotor and the stationary armature located on the stator that generates the power.

Answer is (C)

AC ELECTRICITY–35

Which of the following is false regarding rotating machinery?

(A) The avoidance of harmonics in the production of a sine wave can be achieved by using a coil having multiple loops passing through adjacent slots rather than using only one pair of slots.

(B) Uniformity in the production of flux on a pole can be obtained by using distributed field windings over a portion of the rotor surface.

(C) In order to have a ground or neutral, three-phase alternator windings are commonly connected in wye, rather than in delta to avoid short-circuiting the third and higher harmonics.

(D) AC generator ratings are usually given in units of kVA (kilovolt·amps).

(E) At zero power factor, the generator delivers real power to a load.

A power factor of 1 delivers only real power and no reactive power to a resistive load. A zero power factor is associated with a non-resistive load or with no load conditions.

> **Answer is (E)**

AC ELECTRICITY–36

"Hunting" is an oscillation effect resulting from the improper generation of power. Which of the following is not a cause of hunting?

(A) inherent irregularities found in the production of power from a gas engine

(B) uneven input from a flywheel

(C) system having irregular or systematically irregular loads

(D) failure to dampen systematic variations in the power source driving an alternator

(E) use of several engine-driven units not run in parallel

No matter how irregular the load, it does not cause hunting unless the load dominates the line voltage, which is very unlikely.

> **Answer is (C)**

AC ELECTRICITY–37

A single-phase transmission line is 50 miles long, and consists of two solid conductors spaced 5 feet apart center to center. The diameter of the conductors is 0.46 inches. The reactance for each conductor is:

$$X = 2\pi f \ell \left(80 + 741 \log_{10} \frac{d}{r} \right) (1 \times 10^{-6}) \text{ ohms}$$

f is the frequency in hertz, d is the distance between the conductors in inches, r is the radius of the conductors in inches, and ℓ is the length of the conductors in miles. At a frequency of 60 Hz, determine the total reactance for the conductors.

(A) 23.5 Ω   (B) 35.3 Ω   (C) 50.5 Ω
        (D) 70.5 Ω   (E) none of the above

Evaluating the equation for X,

$$X_{total} = 2\pi(60)(50)\left[80 + 741\log_{10}\left(\frac{60}{0.23}\right)\right](1 \times 10^{-6})(2 \text{ conductors})$$

$$= 70.5 \ \Omega$$

Answer is (D)

AC ELECTRICITY–38

A 10-pole synchronous motor operates on a 60 cycle voltage. What is the speed of the motor?

(A) 520 rpm   (B) 620 rpm   (C) 660 rpm   (D) 720 rpm   (E) 740 rpm

A p-pole machine produces $\frac{1}{2}p$ cycles per revolution. Since the armature turns at a constant speed, n, the generated frequency is:

$$f = \left(\frac{n}{60}\right)\left(\frac{p}{2}\right)$$

$$n = \frac{120f}{p} = \frac{(120)(60)}{10}$$

$$= 720 \text{ rpm}$$

Answer is (D)

AC ELECTRICITY–39

The core of a 400 cycle aircraft transformer has a net cross section of 2 in². The maximum flux density is 60,000 maxwells/in², and there are 70 turns in the secondary coil. What is the voltage induced in the secondary coil?

(A) 129 V   (B) 149 V   (C) 169 V

(D) 1490 V   (E) 1690 V

The formula for induced voltage is:

$$E = 4.44fNB_mA(1 \times 10^{-8}) \text{ volts}$$

f is the frequency in Hz, N is the number of turns, B_m is the maximum flux density in maxwells/in^2, and A is the cross-sectional area of the core in in^2.

$$E = 4.44(400)(70)(60,000)(2)(1 \times 10^{-8})$$
$$= 149 \text{ V}$$

Answer is (B)

AC ELECTRICITY–40

A 150 kVA, 1000 volt single-phase alternator has an open circuit emf of 750 volts. When the alternator is short-circuited, the armature current is 460 amps. What is the synchronous impedance?

(A) 1.63 Ω   (B) 2.23 Ω   (C) 2.63 Ω
              (D) 3.23 Ω   (E) 3.63 Ω

Synchronous impedance, Z, is:

$$Z = \frac{V_{oc}}{I_a}$$

V_{oc} is the open circuit voltage, and I_a is the armature current when the alternator is short-circuited. Therefore,

$$Z = \frac{750}{460} = 1.63 \text{ Ω}$$

Answer is (A)

AC ELECTRICITY–41

In a balanced three-phase system with a power factor of unity, the line voltage, E_l, and the line current, I_l, deliver normal AC power. What is the expression for the power, P?

(A) $P = E_l I_l$ (B) $P = \frac{1}{2} E_l I_l$ (C) $P = \frac{1}{\sqrt{2}} E_l I_l$

          (D) $P = \frac{\sqrt{3}}{2} E_l I_l$ (E) $P = \sqrt{3} E_l I_l$

The power developed by a three-phase generator is three times the coil voltage, E_c, multiplied by the coil current, I_c.

$$P = 3E_cI_c$$

The line voltage has the following relationship with the coil voltage:

$$E_l = \sqrt{3}E_c$$

Therefore, since $I_c = I_l$ for a power factor of 1,

$$P = \frac{3}{\sqrt{3}E_lI_l}$$
$$= \sqrt{3}E_lI_l$$

Answer is (E)

AC ELECTRICITY–42

A three-phase alternator has three armature coils each rated at 1200 volts and 120 amps. What is the kVA rating of this unit?

(A) 432 kVA   (B) 442 kVA   (C) 522 kVA

(D) 540 kVA   (E) 580 kVA

The kVA rating is equal to the power output.

$$\begin{aligned}
\text{kVA} &= 3E_cI_c \\
&= 3(1200)(120) \\
&= 432{,}000 \text{ VA} \\
&= 432 \text{ kVA}
\end{aligned}$$

Answer is (A)

AC ELECTRICITY–43

What is the relationship between the line current, I_l, and the coil current, I_c, in a balanced delta system?

(A) $\dfrac{I_c}{\sqrt{3}}$ (B) $\dfrac{I_c}{\sqrt{2}}$ (C) $I_l = I_c$

       (D) $I_l = \sqrt{2}\,I_c$ (E) $I_l = \sqrt{3}\,I_c$

In a balanced system, the current is 30° out of phase with the line voltage when the power factor is 1. The vector representation is:

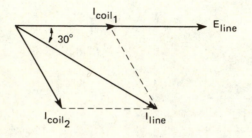

Therefore,

$$I_l = \sqrt{3}\,I_c$$

Answer is (E)

14

PHYSICS

PHYSICS-1

The figure shown is used to indicate combinations of color primaries for subtractive mixing of colors. Which one of the following is true?

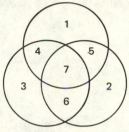

(A) 1 = green, 6 = magenta, 4 = yellow
(B) 1 = yellow, 4 = blue, 7 = white
(C) 1 = cyan, 5 = green, 7 = white
(D) 1 = red, 6 = cyan, 3 = yellow
(E) 1 = magenta, 5 = red, 7 = black

This figure could be rotated so that 1 = magenta, yellow, or cyan. However, the only choice that has all three colors in the proper places is (E).

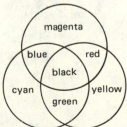

Answer is (E)

PHYSICS–2

Which of the following statements is false?

(A) wavelength of visible light > wavelength of microwaves
(B) frequency of radio waves < frequency of infrared waves
(C) wavelength of x-rays > wavelength of gamma rays
(D) frequency of ultraviolet > frequency of infrared
(E) wavelength of radiowaves > wavelength of visible light

The electromagnetic spectrum is:

radio waves	micro-waves	infra-red	visible	ultra-violet	x-rays	gamma rays

increasing frequency $\longrightarrow$

$\longleftarrow$ increasing wavelength

The wavelength of microwaves is greater than the wavelength of visible light. Therefore, (A) is false.

Answer is (A)

PHYSICS–3

A light source emits a total luminous flux of 1000 lumens distributed uniformly over a quarter of a sphere. What is the luminous intensity at 2.5 meters?

(A) 270 candles   (B) 318 candles   (C) 357 candles
            (D) 400 candles   (E) 442 candles

$$I = \frac{F}{\Omega}$$

where I = luminous intensity

Ω = solid angle

= π steradians

F = luminous flux

= 1000 lum

$$I = \frac{1000}{\pi}$$

= 318 candles

Answer is (B)

PHYSICS-4

A 100 watt light bulb emits a total luminous flux of 1500 lumens, distributed uniformly over a hemisphere. What is the illuminance at a distance of 2 meters?

(A) 10.6 lm/m²   (B) 21.2 lm/m²   (C) 33.5 lm/m²
                    (D) 46.8 lm/m²   (E) 59.7 lm/m²

$$E = \frac{F}{A}$$

E = illuminance

F = luminous flux

A = area

$$= \frac{4\pi r^2}{2}$$
$$= 2\pi(2)^2$$
$$= 25.13 \text{ m}^2$$
$$E = \frac{1500}{25.13}$$
$$= 59.7 \text{ lm/m}^2$$

Answer is (E)

PHYSICS-5

A light bulb is used to light a stage 8 feet below. A chair sits on the stage 3 feet from a spot directly below the bulb. If the bulb has a luminous intensity of $I = 150$ candles, what is the illumination on the floor around the chair?

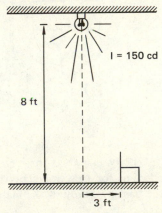

I = 150 cd

8 ft

3 ft

(A) 0.73 footcandles   (B) 1.9 footcandles   (C) 2.7 footcandles
                    (D) 5.5 footcandles   (E) 7.3 footcandles

The illumination, E, is given by the following formula:

$$E = \frac{I \cos \theta}{r^2}$$

$I =$ luminous intensity of the source

$\theta =$ angle from the normal to the surface the light strikes

$r =$ distance from the light source

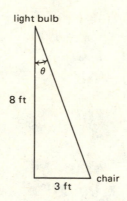

$$\cos \theta = \frac{8}{8.5}$$

$$r = \sqrt{8^2 + 3^2}$$

$$= \sqrt{73}$$

$$E = \frac{(150)\left(\frac{8}{8.5}\right)}{73}$$

$$= 1.9 \text{ footcandles}$$

Answer is (B)

PHYSICS–6

Light of wavelength λ and intensity I_0 passes through a 0.05 meter thick slab of glass whose absorption coefficient for that wavelength is 15 m^{-1}. What is the intensity, I, of the light after passing through the slab?

(A) 0.34 I_0 (B) 0.47 I_0 (C) 0.62 I_0

          (D) 0.75 I_0 (E) 0.91 I_0

$$I = I_0 e^{-\alpha x}$$

where α = absorption coefficient

$$= 15 \text{ m}^{-1}$$

$$I = I_0 e^{-(15)(0.05)}$$

$$= I_0 e^{-0.75}$$

$$= 0.47 \, I_0$$

Answer is (B)

PHYSICS–7

A light ray passing through air $(n = 1)$ strikes a glass surface $(n_{glass} = 1.52)$ at an angle of 60° from the normal to the surface. What is the angle between the reflected light and the surface?

(A) 7.5°   (B) 15°   (C) 30°

            (D) 45°   (E) 60°

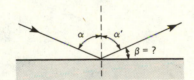

The reflection law states that the angle of incidence is equal to the angle of reflection $(\alpha = \alpha')$. Therefore, $\alpha' = 60°$ and $\beta = 30°$.

Answer is (C)

PHYSICS–8

What type of materials usually has a higher index of refraction?

(A) lighter materials   (B) heavier materials   (C) denser materials

            (D) less dense materials   (E) gaseous materials

In general, denser materials have higher indices of refraction.

Answer is (C)

PHYSICS–9

What is the path of the refracted ray in the following figure?

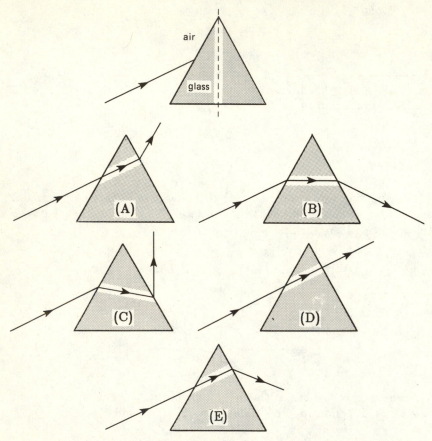

The light ray is refracted at interface 1. It is not normal to the surface of the glass, so its path changes direction. This eliminates choices (A), (D), and (E). Since $n_{glass} > n_{air}$, the ray is bent towards the normal to the surface of the glass at interface 1, and away from the normal to the surface at interface 2. Thus, (B) is the correct path of the ray.

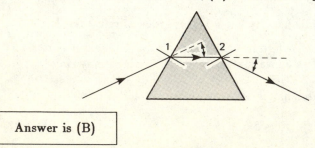

Answer is (B)

PHYSICS–10

How can the index of refraction of material x be defined?

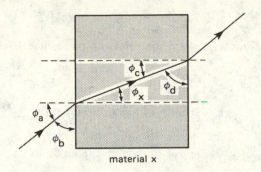

material x

(A) $n_x = \dfrac{\sin \phi_a}{\sin \phi_b}$

(B) $n_x = \dfrac{\sin \phi_a}{\sin \phi_c}$

(C) $n_x = \dfrac{\sin \phi_c}{\sin \phi_d}$

(D) $n_x = \dfrac{\sin \phi_b}{\sin \phi_c}$

(E) $n_x = \dfrac{\sin \phi_b}{\sin \phi_d}$

Indices of refraction are defined such that $n_a \sin \phi_a = n_x \sin \phi_x$, where n_a and n_x are the indices of refraction of materials a and x, and ϕ_a and ϕ_b are the angles between the light ray and the normal to the interface between the two materials.

$$n_a = \text{reference index for air}$$
$$= 1$$

Therefore, $\quad n_x = \dfrac{\sin \phi_a}{\sin \phi_x}$

From the figure, $\quad \phi_x = \phi_c$

$$n_x = \dfrac{\sin \phi_a}{\sin \phi_c}$$

Answer is (B)

PHYSICS–11

What is the index of refraction of a material if the speed of light through the material is 2.37×10^8 m/s?

(A) 1.10   (B) 1.19   (C) 1.27   (D) 1.34   (E) 1.52

The index of refraction of a material can be defined:

$$n = \frac{c}{v}$$

c = speed of light in a vacuum

$\quad = 3.0 \times 10^8$ m/s

v = speed of light in the medium

$\quad = 2.37 \times 10^8$ m/s

$$n = \frac{3 \times 10^8}{2.37 \times 10^8}$$

$$= 1.27$$

Answer is (C)

PHYSICS–12

What is the speed of light through a glass whose index of refraction is 1.33?

(A) 1.8×10^8 m/s   (B) 2.26×10^8 m/s   (C) 2.51×10^8 m/s
(D) 2.79×10^8 m/s   (E) 3.11×10^8 m/s

$$n = \frac{c}{v}$$

$$v = \frac{c}{n}$$

$$= \frac{3.0 \times 10^8}{1.33}$$

$$= 2.26 \times 10^8 \text{ m/s}$$

Note: It is known that the speed of light through any medium is less than that through a vacuum. Therefore, choice (E) could be automatically eliminated.

Answer is (B)

PHYSICS-13

Light hits the surface of a trough of water at an angle of 30° from horizontal. The index of refraction of water is 1.333. What is the angle, α, indicated in the figure?

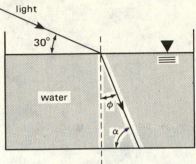

(A) 29.5°   (B) 34.0°   (C) 40.5°   (D) 45.0°   (E) 49.5°

First, use Snell's law to find ϕ. Then, use ϕ to find α. Since $n = 1$ for air,

$$1 \times \sin \phi_1 = n \sin \phi$$
$$n = 1.333$$
$$\phi_1 = 90 - 30$$
$$= 60°$$
$$\sin \phi = \frac{\sin 60°}{1.333}$$
$$= 0.650$$
$$\phi = 40.5°$$
$$\alpha = 90 - \phi$$
$$= 90 - 40.5$$
$$= 49.5°$$

Answer is (E)

PHYSICS-14

A light ray in air $(n_{air} = 1)$ is incident on a glass surface $(n_{glass} = 1.52)$ at an angle of 30° from the normal. What is the angle between the refracted light ray and the normal?

(A) 15.7°   (B) 19.2°   (C) 30°   (D) 45.3°   (E) 60°

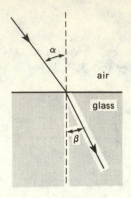

$$n_a \sin \alpha = n_g \sin \beta$$

$$\sin \beta = \frac{n_a \sin \alpha}{n_g}$$

$$= \frac{(1)(\sin 30°)}{1.52}$$

$$= 0.3289$$

$$\beta = 19.2°$$

Answer is (B)

PHYSICS–15

Given that $\alpha = 60°$, $n_{air} = 1$, and $n_{glass} = 1.52$, find the angle, γ, in the figure below.

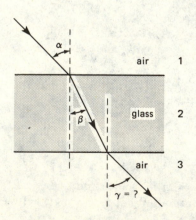

(A) 15°   (B) 30°   (C) 45°   (D) 60°   (E) 90°

Use Snell's law at each interface:

$$n_{\text{air}} \sin \alpha = n_{\text{glass}} \sin \beta$$
$$n_{\text{glass}} \sin \beta = n_{\text{air}} \sin \gamma$$
Therefore, $\quad n_{\text{air}} \sin \alpha = n_{\text{air}} \sin \gamma$
$$\sin \gamma = \sin \alpha$$
$$= \sin 60°$$
$$\gamma = 60°$$

Answer is (D)

PHYSICS–16

A student has a beaker of unknown liquid. When a beam of light of wavelength $\lambda = 0.59\,\text{Å}$ shines into it at an angle of 55° from the normal to the surface, the refracted beam exits at an angle of 33° from the normal to the surface. What is the liquid?

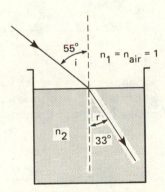

(A) acetic acid, $n = 1.30$
(B) water, $n = 1.33$
(C) nitric acid, $n = 1.40$
(D) sugar water at 68 °F, $n = 1.46$
(E) benzene, $n = 1.50$

$$n_1 \sin i = n_2 \sin r$$

$$i = \text{angle of incidence}$$

$$r = \text{angle of refraction}$$

$$n_2 = \left(n_1\right)\left(\frac{\sin i}{\sin r}\right)$$

$$= \left(1\right)\left(\frac{\sin 55°}{\sin 33°}\right)$$

$$= 1.50$$

The only liquid listed with an index of refraction of 1.50 is benzene. Therefore, (E) is the correct choice.

Answer is (E)

PHYSICS–17

A light ray in a medium ($n_{\text{medium}} = 1.7$) is totally reflected when it strikes the interface between the medium and air. What must the angle between the light ray and the normal to the surface be in order for this to occur?

(A) 0°   (B) 15°   (C) 18.2°   (D) 36°   (E) 45°

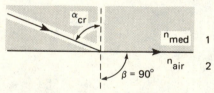

Since the index of refraction is greater for the medium than for air, total internal reflection may occur. For total internal reflection to occur, the angle of refraction (β) must be at least 90°.

$$n_{\text{medium}} \sin \alpha_{\text{cr}} = n_{\text{air}} \sin \beta$$

$$\sin \alpha_{\text{cr}} = \frac{n_{\text{air}} \sin 90°}{n_{\text{medium}}}$$

$$= \frac{(1)(1)}{1.7}$$

$$= 0.5882$$

$$\alpha_{\text{cr}} = 36°$$

Answer is (D)

PHYSICS–18

A diver underwater signals his partner in a boat using a very bright flashlight. Assuming the water is crystal clear and the surface is perfectly calm, at what angle from vertical, θ, can the diver hold his flashlight and still have its beam visible above the surface? (The indices of refraction are $n_{water} = 1.33$ and $n_{air} = 1.00$)

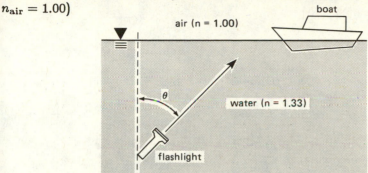

(A) $\theta < 32.5°$ (B) $\theta < 41.2°$ (C) $\theta < 45.0°$

(D) $\theta < 48.8°$ (E) $\theta < 51.4°$

Since $n_{air} < n_{water}$, total internal reflection can occur. This happens when the incident angle, θ, is large enough so that the refracted angle is at least 90°. The incident angle for which the refracted angle is exactly 90° is called the critical angle, θ_{cr}. The critical angle can be found from Snell's law:

$$\sin \theta_{cr} = \left(\frac{n_{air}}{n_{water}} \right) \sin 90°$$

$$= \frac{n_{air}}{n_{water}}$$

$$= \frac{1.00}{1.33}$$

$$\theta_{cr} = 48.8°$$

For the light to be seen above the surface, θ must be less than θ_{cr}. Therefore, $\theta < 48.8°$.

Answer is (D)

PHYSICS-19

The radius of curvature of a convex spherical mirror is 48 centimeters. What are the focal length and focal type?

(A) −12 cm, virtual focus
(B) −24 cm, virtual focus
(C) 0 cm, no focus
(D) 12 cm, real focus
(E) 24 cm, real focus

For a convex mirror, the radius of curvature, R, is negative. The equation for the focal length of a spherical mirror is:

$$\frac{1}{f} = \frac{2}{R}$$
$$f = \frac{R}{2}$$
$$= \frac{-48}{2}$$
$$= -24 \text{ cm}$$

The negative sign indicates that the mirror has a virtual focus. The focal length is −24 cm.

Answer is (B)

PHYSICS-20

An object 6 centimeters high is placed 12 centimeters away from a concave mirror whose focal length is 36 centimeters. What is the height of the image?

(A) 2 cm   (B) 4 cm   (C) 5 cm

        (D) 8 cm   (E) 9 cm

$$q = \text{image distance}$$
$$p = \text{object distance}$$
$$f = \text{focal length}$$
$$I = \text{image size}$$
$$O = \text{object size}$$

$$\frac{1}{f} = \frac{1}{p} + \frac{1}{q}$$

$$q = \frac{pf}{p - f}$$

$$= \frac{(12)(36)}{12 - 36}$$

$$= -18 \text{ cm}$$

$$\frac{I}{O} = \left| \frac{q}{p} \right|$$

$$= \frac{18}{12}$$

$$I = \left(\frac{3}{2} \right) O$$

$$= \left(\frac{3}{2} \right) \left(6 \right)$$

$$= 9 \text{ cm}$$

Answer is (E)

PHYSICS-21

An object is placed 10 centimeters away from a concave mirror with a focal length of 30 centimeters. What image is formed?

(A) a real image 10 cm in front of the mirror
(B) a virtual image 15 cm behind the mirror
(C) a real image 20 cm in front of the mirror
(D) a virtual image 25 cm behind the mirror
(E) a real image 30 cm behind the miror

$$\frac{1}{p} + \frac{1}{q} = \frac{1}{f}$$

p = object distance

q = image distance

f = focal length

$$q = \frac{pf}{p - f}$$
$$= \frac{(10)(30)}{10 - 30}$$
$$= -\frac{300}{20}$$
$$= -15 \text{ cm}$$

The negative sign indicates that this is a virtual image, 15 cm behind the mirror.

Answer is (B)

PHYSICS-22

A chess piece is placed 7 inches in front of a concave mirror that has a radius of curvature of 18 inches. Which of the following statements about the image are true?

I.   The image is larger than the object.
II.   The image is real.
III.   The image is upright.
IV.   The image is beyond the center of curvature.
V.   The object is at the focus.

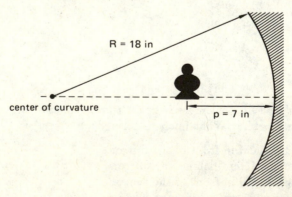

(A) V   (B) I and III   (C) II and IV

(D) I, III, and IV   (E) all are true

$$\frac{1}{p} + \frac{1}{q} = \frac{1}{f}$$

p = object distance

q = image distance

f = focal length

$$= \frac{R}{2}$$

$$= \frac{18}{2}$$

$$= 9 \text{ in}$$

$$q = \frac{pf}{p - f}$$

$$= \frac{(7)(9)}{7 - 9)}$$

$$= -31.5 \text{ in}$$

Thus, the image is virtual and behind the mirror. Statements II, IV, and V are all incorrect.

$$m = \text{magnification}$$

$$= -\frac{q}{p}$$

$$= -\frac{(-31.5)}{7}$$

$$= +4.5$$

The positive sign indicates that the image is upright. Therefore, III is true. Since $m > 1$, the image is larger than the object. Thus, I is also true.

Answer is (B)

PHYSICS–23

Consider the concave spherical mirror in the figure. What is the path of the reflected ray?

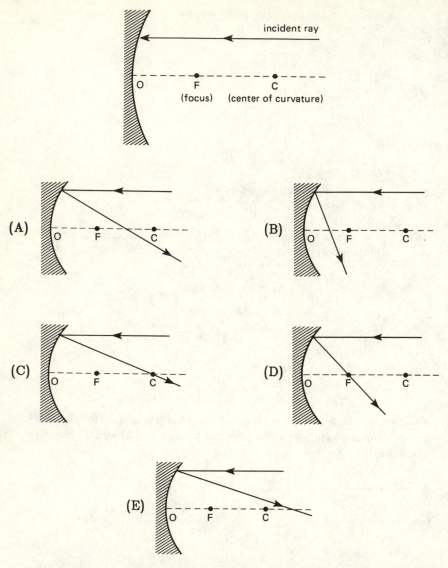

If an incident ray is parallel to the principal axis of a concave mirror, the reflected ray will pass through the focus point, F. Therefore, (D) is correct.

Answer is (D)

PHYSICS–24

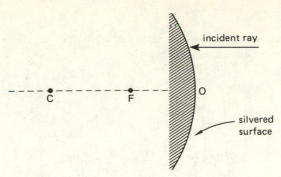

Consider the convex spherical mirror shown in the figure. What is the path of the reflected ray?

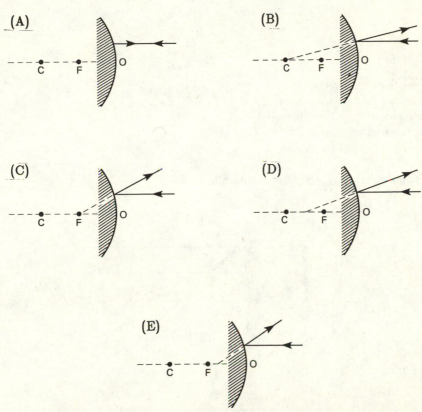

If the incident ray is parallel to the principal axis of a convex mirror, the reflected ray follows a path such that its extension passes through the focus. Therefore, (C) is correct.

Answer is (C)

PHYSICS–25

Which of the following describes the image of an object which is placed between the focus and the concave spherical mirror?

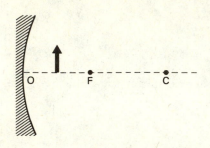

(A) real, larger, inverted
(B) virtual, larger, not inverted
(C) virtual, smaller, not inverted
(D) real, smaller, inverted
(E) real greater, not inverted

Construct a ray diagram to find the image.

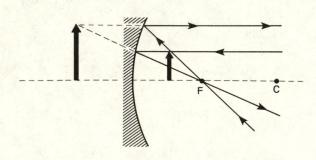

The image is virtual, behind the mirror, larger, and not inverted. Therefore, (B) is the correct answer.

Answer is (B)

PHYSICS–26

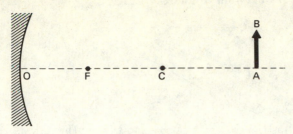

Which of the following is the image of the object as shown in the concave spherical mirror?

(A)

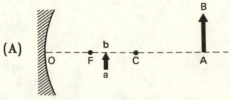

(B)

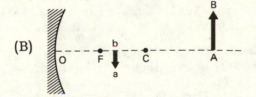

(C)

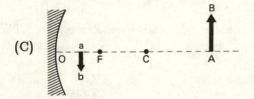

(D)

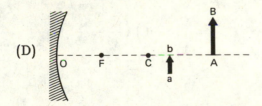

(E)

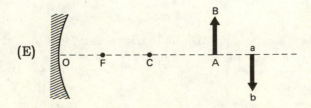

Construct a ray diagram to find the image.

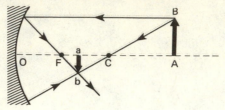

Thus, (B) gives the correct location and type of the image.

Answer is (B)

PHYSICS-27

A thin lens is made from glass with $n = 1.5$. It has a convex face with a 25 centimeter radius of curvature, and a concave face with a 35 centimeter radius of curvature. What is the focal length and type of the lens?

(A) diverging lens, virtual focus, focal length of 100 cm
(B) converging lens, real focus, focal length of 125 cm
(C) diverging lens, virtual focus, focal length of 150 cm
(D) converging lens, real focus, focal length of 175 cm
(E) diverging lens, virtual focus, focal length of 200 cm

It is important to have the sign conventions correct. The radius of curvature for the convex face, (R_1), is positive, but for the concave face, (R_2), it is negative.

$$\frac{1}{f} = \left(n - 1\right)\left(\frac{1}{R_1} + \frac{1}{R_2}\right)$$

$$= \left(1.5 - 1\right)\left(\frac{1}{25} - \frac{1}{35}\right)$$

$$= \left(0.5\right)\left(\frac{7 - 5}{175}\right)$$

$$= \frac{1}{175}$$

$$f = 175 \text{ cm}$$

The focal length is positive, indicating a real focus and, therefore, a converging lens with a focal length of 175 cm.

Answer is (D)

PHYSICS-28

An object 1 foot tall is placed 2 feet from a converging lens whose focal length is 1.5 feet. Where is the image formed?

(A) 2 feet from the lens
(B) 3 feet from the lens
(C) 4 feet from the lens
(D) 5 feet from the lens
(E) 6 feet from the lens

$$\frac{1}{f} = \frac{1}{p} + \frac{1}{q}$$

f = focal length
p = object distance
q = image distance

$$\frac{1}{q} = \frac{1}{f} - \frac{1}{p}$$

$$q = \frac{pf}{p-f}$$

$$= \frac{(2)(1.5)}{2-1.5}$$

$$= 6 \text{ ft}$$

Answer is (E)

PHYSICS-29

What is the image position of an object placed 15 centimeters away from a thin, spherical converging lens with a focal length of 10 centimeters?

(A) 15 cm beyond the lens
(B) 20 cm beyond the lens
(C) 25 cm beyond the lens
(D) 30 cm beyond the lens
(E) 35 cm beyond the lens

$$\frac{1}{p} + \frac{1}{q} = \frac{1}{f}$$

p = object distance

q = image distance

f = focal length

$$\frac{1}{q} = \frac{p - f}{pf}$$

$$q = \frac{pf}{p - f}$$

$$= \frac{(15)(10)}{15 - 10}$$

$$= \frac{150}{5}$$

$$= 30 \text{ cm}$$

The positive value of q means that there is a real image on the opposite side of the lens from the object (beyond the lens).

Answer is (D)

PHYSICS–30

Where is the image of an object placed 25 centimeters from a diverging lens with a focal length of 15 centimeters? What type of image is it?

(A) 9.375 cm behind the lens, virtual image
(B) 12.675 cm behind the lens, real image
(C) 15.225 cm behind the lens, virtual image
(D) 17.725 cm behind the lens, real image
(E) 20.275 cm behind the lens, virtual image

$$\frac{1}{p} + \frac{1}{q} = \frac{1}{f}$$

$p =$ object distance

$q =$ image distance

$f =$ focal length

$\quad = -15\,\text{cm} \quad (f$ is negative for a diverging lens)

$$\frac{1}{q} = \frac{p-f}{pf}$$

$$q = \frac{pf}{p-f}$$

$$= \frac{(25)(-15)}{25-(-15)}$$

$$= \frac{(25)(-15)}{40}$$

$$= -9.375\,\text{cm}$$

This is a virtual image, located 9.375 cm behind the lens.

Answer is (A)

PHYSICS–31

A magnifying glass has a plastic lens with an index of refraction of $n = 5.4$ and radii of curvature of 2.95 feet and 4.27 feet for the two faces. What is the magnification of the lens when it is held 2.36″ from an object being viewed?

(A) 1.21 (B) 1.60 (C) 1.98 (D) 2.16 (E) 2.78

A magnifying glass uses a biconvex lens. If f is the focal length of the lens, R_1 and R_2 are the radii of curvature, and n is the index of refraction, then,

$$\frac{1}{f} = \left(n - 1\right)\left(\frac{1}{R_1} + \frac{1}{R_2}\right)$$
$$= \left(5.4 - 1\right)\left(\frac{1}{2.95} + \frac{1}{4.27}\right)$$
$$= 2.52 \text{ ft}^{-1}$$
$$f = 0.397 \text{ ft}$$
$$= 4.76 \text{ in}$$
$$\frac{1}{p} + \frac{1}{q} = \frac{1}{f}$$
$$p = \text{object distance}$$
$$q = \text{image distance}$$
$$\frac{1}{q} = \frac{1}{f} - \frac{1}{p}$$
$$q = \frac{pf}{p - f}$$
$$= \frac{(2.36)(4.76)}{2.36 - 4.76}$$
$$= -4.68 \text{ ft}$$
$$m = \text{magnification}$$
$$= -\frac{q}{p}$$
$$= \frac{4.68}{2.36}$$
$$= 1.98$$

Answer is (C)

PHYSICS-32

An astronomer observing the night sky has a telescope with an 8″ objective lens. If two stars 700 light-years away are barely resolved by the telescope, how far apart are they? Assume the light from both stars has a wavelength of $\lambda = 5500$ angstroms.

(A) 123×10^6 miles   (B) 253×10^6 miles   (C) 1.36×10^9 miles
(D) 8.75×10^9 miles   (E) 13.6×10^9 miles

The minimum resolvable distance between two objects is:

$$d_0 = 1.22 \frac{\lambda L}{D}$$

d_0 = minimum resolvable distance

λ = wavelength of light

L = distance of the objects from the lens

D = diameter of the lens

In the above problem,

$$\lambda = 5500 \,\text{Å}$$
$$= 5.5 \times 10^{-6} \,\text{m}$$
$$L = 700 \text{ light-years}$$
$$D = 8 \text{ in}$$

$$d_0 = \left(1.22\right) \left[\frac{\left(5.5 \times 10^{-6} \,\text{m}\right) \left(\dfrac{39.4 \,\text{in}}{\text{m}}\right) \left(700 \text{ light-years}\right)}{8 \,\text{in}} \right]$$

$$= 0.00231 \text{ light-year}$$
$$1 \text{ light-year} = 5.87 \times 10^{12} \,\text{mi}$$
$$d_0 = 1.36 \times 10^9 \,\text{mi}$$

Answer is (E)

PHYSICS–33

A microscope has an eyepiece with a focal length, f_e, of 1 inch and a magnification of 5. If the objective lens is 0.25 inch from the object being viewed, and has a magnification of 10, what is the distance, d, between the two lenses?

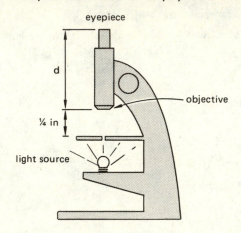

(A) $d = 1.3$ in

(B) $d = 2.2$ in

(C) $d = 3.3$ in

(D) $d = 3.7$ in

(E) $d = 3.9$ in

For a lens with p = object distance, q = image distance, and f = focal length, the lens equation states:

$$\frac{1}{p} + \frac{1}{q} = \frac{1}{f}$$

$$m = -\frac{q}{p}$$

$$= \text{magnification}$$

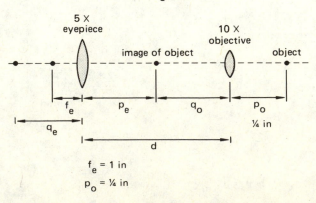

The sign on m indicates whether or not the image is inverted. From the figure, it can be seen that $d = |p_e| + |q_o|$. The subscripts denote the eyepiece, e, and the objective, o. First, find the distance between the image of the object and the objective lens.

$$m = -\frac{q_o}{p_o}$$
$$q_o = -m\,p_o$$
$$= -(10)(0.25)$$
$$= -2.5 \text{ in}$$
$$|q_o| = 2.5 \text{ in}$$

Next, find p_e for the eyepiece.

$$\frac{1}{p_e} + \frac{1}{q_e} = \frac{1}{f_e}$$
$$m_e = -\frac{q_e}{p_e}$$
$$|q_e| = m_e\,p_e$$
$$\frac{1}{p_e} - \frac{1}{m_e\,p_e} = \frac{1}{f_e}$$
$$p_e = f_e - \frac{f_e}{m_e}$$
$$= 1 - \frac{1}{5}$$
$$= 0.8 \text{ in}$$
$$d = |q_o| + |p_e|$$
$$= 2.5 + 0.8$$
$$= 3.3 \text{ in}$$

Answer is (C)

PHYSICS-34

A diffraction grating set up as shown is used to find the wavelength of the light emitted by the source. If the grating has 4000 lines per inch, the lens used to focus the light from the grating is 10″ from the screen where the light is focused, and the second-order spectral line is $1\frac{7}{8}″$ from the center position, what is the wavelength of the light?

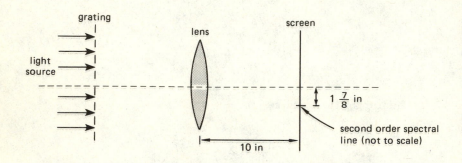

(A) 3260 Å   (B) 5840 Å   (C) 5950 Å

(D) 7320 Å   (E) 11,700 Å

The relationship between the wavelength, the position of the spectral line, and the diffraction grating is:

$$\sin \theta = \frac{n\lambda}{d}$$

where θ = angle from the center of lens straight to the screen and the first-order spectral line

   n = order of spectral line

   d = spacing of the grating

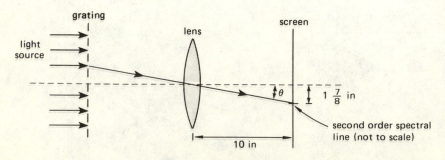

The following information can be determined from the diagram:

$$\sin \theta = \frac{1\frac{7}{8}}{\sqrt{(10)^2 + (1\frac{7}{8})^2}}$$

$$= 0.1843$$

$$d = \frac{1}{4000}$$

$$= 2.5 \times 10^{-4} \text{ in}$$

$n = 2$ for a second-order spectral line

$$\lambda = \frac{d \sin \theta}{n}$$

$$= \frac{(2.5 \times 10^{-4})(0.1843)}{2}$$

$$= \left(2.3 \times 10^{-5} \text{ in}\right)\left(\frac{1 \text{ m}}{39.4 \text{ in}}\right)$$

$$= 5.84 \times 10^{-7} \text{ m}$$

$$= 5840 \text{ Å}$$

Answer is (B)

PHYSICS–35

What is the photon energy associated with green-blue light of wavelength 500 nanometers? ($h = 6.626 \times 10^{-34}$ J·s, and $c = 3.0 \times 10^8$ m/s.)

(A) 1×10^{-19} J $\qquad$ (B) 2×10^{-19} J $\qquad$ (C) 3×10^{-19} J
$\qquad$ (D) 4×10^{-19} J $\qquad$ (E) 5×10^{-19} J

$$E = hf$$

$$= \frac{hc}{\lambda}$$

$$\lambda = 500 \text{ nm}$$

$$= 5.00 \times 10^{-7} \text{ m}$$

$$E = \frac{(6.626 \times 10^{-34})(3 \times 10^8)}{5 \times 10^{-7}}$$

$$= 3.97 \times 10^{-19} \text{ J}$$

$$E = 4 \times 10^{-19} \text{ J}$$

Answer is (D)

PHYSICS–36

What is the energy of a photon of wavelength 0.1 nanometer?

(A) 2.2×10^{-42} J   (B) 9.8×10^{-26} J   (C) 6.62×10^{-24} J

(D) 1.99×10^{-15} J   (E) 2.0×10^{-7} J

$$E = hf$$

$$= \frac{hc}{\lambda}$$

$$= \frac{\left(6.626 \times 10^{-34} \text{ J} \cdot \text{s}\right)\left(3.0 \times 10^{8} \frac{\text{m}}{\text{s}}\right)}{1 \times 10^{-10} \text{ m}}$$

$$= 1.99 \times 10^{-15} \text{ J}$$

Answer is (D)

PHYSICS–37

Light at wavelength 6493 Å is visible red light. It is, however, very close to the limits of the human eye. What is the energy of this light?

(A) 1.86×10^{-19} J   (B) 2.5×10^{-19} J   (C) 2.86×10^{-19} J

(D) 2.95×10^{-19} J   (E) 3.06×10^{-19} J

$$E = hf$$

$$= \frac{hc}{\lambda}$$

$$h = \text{Planck's constant} = 6.626 \times 10^{-34} \text{ J} \cdot \text{s}$$

$$\lambda = 6.493 \times 10^{-10} \text{ m}$$

$$E = \frac{(6.626 \times 10^{-34})(3.0 \times 10^{8})}{6493 \times 10^{-10}}$$

$$= 3.06 \times 10^{-19} \text{ J}$$

Answer is (E)

PHYSICS-38

To be effective, an alarm must be heard at a minimum level of 70 decibels. If it is to be effective for a man whose nearest neighbors live 200 feet down the street, what is the minimum power required?

(A) 1.4×10^{-4} W   (B) 0.0023 W   (C) 0.074 W

(D) 0.47 W   (E) 4.5 W

The intensity level, I, is related to distance and power according to:

$$I = \frac{P}{4\pi r^2}$$

P = power

r = distance from the source

In decibels, $I(\text{dB}) = 10 \log \dfrac{I}{I_0}$

$$I_0 = 1 \times 10^{-12} \text{ W/m}^2$$
$$= 9.29 \times 10^{-14} \text{ W/ft}^2$$
$$I = I_0 (10)^{I(\text{dB})/10}$$
$$P = (I_0)(10)^{I(\text{dB})/10}(4\pi r^2)$$
$$= (9.2905 \times 10^{-14})(10)^{70/10}(4\pi)(200)^2$$
$$= 0.47 \text{ W}$$

Answer is (D)

PHYSICS-39

A stationary observer hears a siren approaching. The siren has a sound frequency of 700 hertz, and is approaching the observer at 50 miles per hour. What is the frequency heard by the observer? Assume the velocity of sound in air is 1087 ft/sec.

(A) 600 Hz   (B) 650 Hz   (C) 700 Hz

(D) 750 Hz   (E) 800 Hz

$$f_o = f_s \left(\frac{v + v_o}{v - v_s} \right)$$

f_o = frequency heard by observer

f_s = frequency of source

v = velocity of wave transmission in the medium

v_o = component of observer velocity directed toward the source

v_s = component of source velocity directed toward the observer

$$= 50 \, \text{mph} \left(\frac{5280 \, \text{ft}}{\text{sec}} \right) \left(\frac{\text{hr}}{3600 \, \text{sec}} \right)$$

$$= 73.3 \, \text{ft/sec}$$

$$f_o = 700 \left(\frac{1087 + 0}{1087 - 73.3} \right)$$

$$= 700 \left(\frac{1087}{1013.7} \right)$$

$$= 750 \, \text{Hz}$$

Answer is (D)

PHYSICS–40

A policeman waiting at a speed trap hears a car going by honking its horn. Being an amateur musician with perfect pitch, the policeman recognizes that the pitch of the horn is $G^\sharp$ (f = 415.30 Hz), a half-step lower than its normal pitch of A (f = 440.00 Hz). The velocity of sound in the air is 1087 ft/sec. From this information, at what speed was the car travelling?

(A) 31 mph   (B) 42 mph   (C) 44 mph
          (D) 63 mph   (E) 78 mph

The change in pitch is due to the Doppler effect. The frequency of the observer, f_o, is:

$$f_o = f_s \left(\frac{v + v_o}{v - v_s} \right)$$

f_s = frequency of source

v = velocity of sound

$\quad$ = 1087 ft/sec

v_o = velocity of the observer

$\quad$ = 0 ft/sec

v_s = velocity of the source

Solve for v_s:

$$v_s = v - \left(\frac{f_s}{f_o}\right)\left(v + v_o\right)$$

$$= 1087 - \left(\frac{440.00}{415.30}\right)\left(1087 + 0\right)$$

$$= -\left(65.6\,\frac{ft}{sec}\right)\left(3600\,\frac{sec}{hr}\right)\left(\frac{mi}{5280\,ft}\right)$$

$$= -44\ mph\ (44\ mph\ \text{away from the policeman})$$

Answer is (C)

PHYSICS–41

The sun generates 1 kW/m^2 when used as a source for solar collectors. A collector with an area of 1 m^2 heats water. The flow rate is 30 l/min. What is the temperature rise in the water? The specific heat of water is 4200 J/kg·°C.

(A) 0.11 °C (B) 0.48 °C (C) 4.8 °C

(D) 28.5 °C (E) 48 °C

$$\text{power} = \left(1\,\frac{kW}{m^2}\right)\left(1\,m^2\right)$$

$$= 1000\ W$$

$$= \rho Q c_p \Delta T$$

$$1000\ W = \left(1000\,\frac{kg}{m^3}\right)\left(30\,\frac{l}{min}\right)\left(\frac{1\,min}{60\,sec}\right)\left(\frac{1\,m^3}{1000\,l}\right)\left(4200\,\frac{J}{kg\cdot°C}\right)\Delta T$$

$$= 2100\Delta T$$

$$\Delta T = \frac{1000}{2100}$$

$$= 0.48\ °C$$

Answer is (B)

PHYSICS–42

A sketch of an oven wall is shown. What is the temperature in the center of the glass wool?

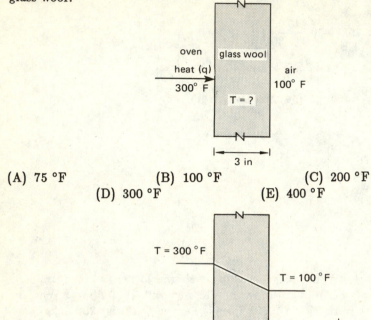

(A) 75 °F   (B) 100 °F   (C) 200 °F

   (D) 300 °F   (E) 400 °F

For steady state flow, the temperature profile is linear.

$$\text{Thus,} \quad \frac{dT}{dx} = A$$

$$T = Ax + B$$

$$A = \frac{300 - 100}{3}$$

$$= \frac{200}{3} \ {}^\circ\text{F/in}$$

$$B = 100 \,{}^\circ\text{F}$$

$$T = \frac{200x}{3} + 100$$

At the center, $x = 1.5 \, \text{in}$

$$T = \frac{(200)(1.5)}{3} + 100$$

$$= 200 \ {}^\circ\text{F}$$

Answer is (C)

PHYSICS–43

As a measure to reduce the outer wall temperature of an oven, an extra layer of insulation is added as shown. What is the new temperature of the outer wall?

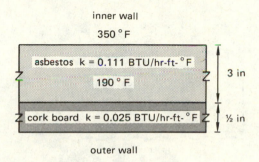

(A) 71.6 °F   (B) 118.4 °F   (C) 143.2 °F

(D) 160 °F   (E) indeterminate

$$q_{\text{(oven to asbestos)}} = q_{\text{(asbestos to cork)}}$$

$$= -kA\frac{\Delta T}{\Delta x}$$

$$-k_{\text{asbestos}}A\left(\frac{190-350}{\frac{3}{12}}\right) = -k_{\text{cork}}A\left(\frac{T_{\text{out}}-190}{\frac{1}{24}}\right)$$

$$-(0.11)\left(\frac{(-160)(12)}{3}\right) = -(0.025)\left(\frac{(T_{\text{out}}-190)(24)}{1}\right)$$

$$T_{\text{out}} = 190 + \left(\frac{0.11}{0.025}\right)\left(\frac{1}{6}\right)(-160)$$

$$= 71.6\ ^\circ\text{F}$$

Answer is (A)

PHYSICS–44

An oven has an inner wall temperature of 350 °F. The insulation is 3″ thick, and has a thermal conductivity of 0.11 BTU/hr-ft-°F. If the heat transfer coefficient of the outer wall is 3.46 BTU/hr-ft-°F, and the air is at 68 °F, what is the temperature of the outer wall?

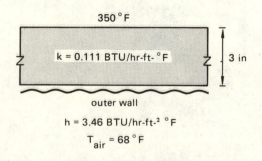

$$350\,°F$$

$$k = 0.111\ BTU/hr\text{-}ft\text{-}°F$$

$$3\ in$$

outer wall

$$h = 3.46\ BTU/hr\text{-}ft^{-2}\ °F$$

$$T_{air} = 68\,°F$$

(A) 71 °F   (B) 94 °F   (C) 100 °F

(D) 121 °F   (E) 139 °F

$$q_{(\text{oven to insulation})} = q_{(\text{outer wall to air})}$$

$$-k\frac{dT}{dx}A = hA(T_{\text{wall}} - T_{\text{air}})$$

$$-k\frac{dT}{dx} = h(T_{\text{wall}} - T_{\text{air}})$$

$$-(0.11)\left(\frac{T_{\text{wall}} - 350}{0.25}\right) = (3.46)(T_{\text{wall}} - 68)$$

$$T_{\text{wall}} = \frac{\left(\dfrac{0.44}{3.46}\right)\left(350\right) + 68}{1 + \dfrac{0.444}{3.46}}$$

$$= 100\ °F$$

Answer is (C)

PHYSICS-45

A 1 cubic foot oven is modelled as a six-sided box. The temperatures at steady state are shown. The insulating material is asbestos with a thermal conductivity of 0.111 BTU/hr-ft-°F. What is the power usage of the oven?

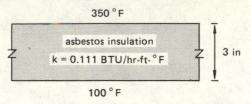

350 °F

asbestos insulation
k = 0.111 BTU/hr-ft-°F

3 in

100 °F

(A) 55.5 BTU/hr   (B) 111 BTU/hr   (C) 444 BTU/hr
        (D) 666 BTU/hr   (E) 888 BTU/hr

$$P = \text{heat transfer}$$
$$= -kA\frac{dT}{dx}$$

At steady state, the temperature profile is linear.

$$dx = \frac{3\ \text{in}}{12\dfrac{\text{in}}{\text{ft}}} = 0.25$$

Therefore, $$\frac{dT}{dx} = \frac{350 - 100}{0.25}$$
$$= -1000\ \frac{°F}{ft}$$
$$A = (6\ \text{sides})(1\ \text{ft})(1\ \text{ft})$$
$$= 6\ \text{ft}^2$$
$$P = -(0.111)(-1000)(6)$$
$$P = 666\ \text{BTU/hr}$$

Answer is (D)

PHYSICS-46

A copper sphere 1 foot in diameter radiates in space to an environment at 0 °R. Assuming that the emissivity of copper is 0.15, and that the sphere is at 140 °F, what power is radiated?

(A) 0.31 BTU/hr   (B) 2.1 BTU/hr   (C) 104.5 BTU/hr
        (D) 418.4 BTU/hr   (E) 696.7 BTU/hr

$$P = \epsilon\sigma T^4 A$$
$$= (0.15)(0.1712 \times 10^{-8})(460 + 140)^4(4\pi r^2)$$
$$= (0.15)(0.1712 \times 10^{-8})(600)^4(4\pi)(0.5)^2$$
$$= 104.5 \text{ BTU/hr}$$

Answer is (C)

PHYSICS-47

An oxidized copper sphere 3 feet in diameter with an emissivity of 0.78 is filled with water. If the water temperature is 190 °F, and the environment is assumed to be at 0 °R, what is the instantaneous rate of cooling of the water? At 190 °F, the density of water is 1.876 slug/ft³, and the specific heat is 1 BTU/lbm-°F.

(A) 3.35 °F/hr   (B) 3.82 °F/hr   (C) 5.09 °F/hr

(D) 7.89 °F/hr   (E) 9.79 °F/hr

$$\text{power out by radiation} = \epsilon\sigma T^4 A$$
$$= \epsilon\sigma T^4 4\pi r^2$$
$$\text{power lost by water} = mc_p(\dot{\Delta T})$$
$$= \frac{4\pi r^2}{3}\rho c_p(\dot{\Delta T})$$

power out by radiation = power lost by water

$$4\pi\epsilon r^2\sigma T^4 = \frac{4\pi\rho r^3 c_p(\dot{\Delta T})}{3}$$

$$\epsilon\sigma T^4 = \frac{\rho r c_p(\dot{\Delta T})}{3}$$

$$(\dot{\Delta T}) = \frac{3\epsilon\sigma T^4}{r c_p \rho}$$

$$= \frac{(3)(0.78)(0.1712 \times 10^{-8})(460 + 190)^4}{(1.5)(1.876)(32.2)(1)}$$

$$= \frac{(3)(0.78)(0.1712 \times 10^{-8})(650)^4}{(1.5)(1.876)(32.2)(1)}$$

$$= 7.89 \text{ °F/hr}$$

Answer is (D)

15 SYSTEMS MODELING

SYSTEMS–1

Which of the following matrices has an inverse?

$$\mathbf{A}_1 = \begin{pmatrix} 3 & 1 & 1 \\ 1 & 3 & 1 \\ 2 & 2 & 1 \end{pmatrix} \qquad \mathbf{A}_2 = \begin{pmatrix} 1 & 3 & 1 \\ 2 & 2 & 3 \\ 0 & 1 & 1 \end{pmatrix} \qquad \mathbf{A}_3 = \begin{pmatrix} 1 & 2 & 3 \\ 2 & 0 & 1 \\ -1 & 2 & 3 \end{pmatrix}$$

(A) $\mathbf{A}_2$ only   (B) $\mathbf{A}_1$ and $\mathbf{A}_2$ (C) $\mathbf{A}_1$ and $\mathbf{A}_3$
     (D) $\mathbf{A}_2$ and $\mathbf{A}_3$ (E) $\mathbf{A}_1$, $\mathbf{A}_2$, and $\mathbf{A}_3$

If, for matrix $\mathbf{A}$, the determinant is nonzero, the inverse matrix, $\mathbf{A}^{-1}$, exists.

$$D_1 = \begin{vmatrix} 3 & 1 & 1 \\ 1 & 3 & 1 \\ 2 & 2 & 1 \end{vmatrix} = 9 + 2 + 2 - 6 - 6 - 1 = 0$$

$\mathbf{A}_1^{-1}$ does not exist.

$$D_2 = \begin{vmatrix} 1 & 3 & 1 \\ 2 & 2 & 3 \\ 0 & 1 & 1 \end{vmatrix} = 2 + 2 - 6 - 3 = -5$$

$\mathbf{A}_2^{-1}$ exists.

$$D_3 = \begin{vmatrix} 1 & 2 & 3 \\ 2 & 0 & 1 \\ -1 & 2 & 3 \end{vmatrix} = 12 - 2 - 12 - 2 = -4$$

$\mathbf{A}_3^{-1}$ exists.

Only $\mathbf{A}_2$ and $\mathbf{A}_3$ have inverses.

Answer is (D)

SYSTEMS–2

An investor is considering a stock portfolio that costs $55. If he invests in the portfolio, there is a 0.5 probability that he will receive a total revenue of $20. If that event does not occur, he will receive a total revenue of $100. What will be the investor's expected profit if he decides to invest?

(A) $5   (B) $15   (C) $55   (D) $60   (E) $65

The expected profit is found by multiplying the expected revenues by their respective probabilities, adding them, and subtracting the initial cost.

$$\text{profit} = (0.5)(100) + (0.5)(20) - 55 = \$5$$

Answer is (A)

SYSTEMS–3

For a function of a single variable, $f(x)$, to be convex, which of the following must be true?

(A) For each pair of values of x_1 and x_2, with $0 \leq \lambda \leq 1$,

$$f\left(\lambda x_2 + (1-\lambda)x_1\right) \leq \lambda f(x_2) + (1-\lambda)f(x_1)$$

(B) For each pair of values of x_1 and x_2, with $0 \leq \lambda \leq 1$,

$$f\left(\lambda x_2 + (1-\lambda)x_1\right) \geq \lambda f(x_2) + (1-\lambda)f(x_1)$$

(C) For each pair of values of x_1 and x_2, with $0 \leq \lambda \leq 1$,

$$f\left(\lambda x_2 + (1-\lambda)x_1\right) = \lambda f(x_2) + (1-\lambda)f(x_1)$$

(D) Graphically, $f(x)$ is: (E) Graphically, $f(x)$ is:

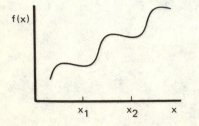

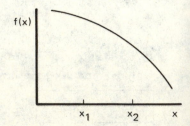

A convex function always has a minimum value.

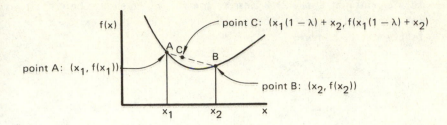

The relationship in (A) is the definition of a convex function, which implies that the function has a minimum value. For each pair of points A and B on the curve, the line segment joining these two points lies entirely above or on the graph of $f(x)$.

Answer is (A)

SYSTEMS–4

For a function of two variables, $f(x_1, x_2)$, and for all possible values of x_1 and x_2, which of the following conditions must exist so that the function is convex?

I. $\left[\dfrac{\partial^2 f(x_1, x_2)}{\partial x_1^2} \right] \left[\dfrac{\partial^2 f(x_1, x_2)}{\partial x_2^2} \right] - \left[\dfrac{\partial^2 f(x_1, x_2)}{\partial x_1 x_2} \right]^2 \geq 0$

II. $\dfrac{\partial^2 f(x_1, x_2)}{\partial x_1^2} \geq 0$

III. $\dfrac{\partial^2 f(x_1, x_2)}{\partial x_2^2} \geq 0$

(A) I only   (B) I and II   (C) I and III

(D) II and III   (E) I, II, and III

Second partial derivatives can be used to check functions of more than one variable to see if they are convex or concave. For two-variable functions, the partial derivatives in I, II, and III make up the determinant of the 2×2 Hessian matrix, which should be greater than or equal to zero. Thus, all three conditions must exist for the function to be convex.

Answer is (E)

SYSTEMS-5

If a function of n variables, $f(x_1, \ldots, x_n)$, is convex, which of the following is true about its $n \times n$ Hessian matrix?

(A) It is semidefinite.
(B) It is negative semidefinite.
(C) It is positive semidefinite.
(D) It is indefinite.
(E) It is finite.

A positive semidefinite Hessian matrix implies two conditions:

1. For all values of x, the function $f(x_1, \ldots, x_n) \geq 0$.
2. There is at least one set of nonzero values of $x_1, \ldots, x_n$ such that $f(x_1, \ldots, x_n) = 0$.

These conditions are met when the determinant of the Hessian matrix is greater than or equal to zero, which occurs if and only if the function f is convex. Thus (C) is the correct answer.

Answer is (C)

SYSTEMS–6

Which of the following statements about linear programming is not true?

(A) In mathematical notation, linear programming problems are often written in the form:

$$\text{optimize:} \quad Z = \sum_j C_j x_j$$

$$\text{subject to the constraints:} \quad \sum_i \sum_j a_{ij} x_j \leq b_i$$

($x_j \geq 0$, and a_{ij}, b_i, and C_j are constants.)

(B) Linear programming uses a mathematical model composed of a linear objective function and a set of linear constraints in the form of inequalities.

(C) The decision variables have physical significance only if they have integer values. The solution procedure yields integer values only.

(D) The simplex method is a technique used to solve linear programming problems.

(E) Linear programming deals with the problem of allocating limited resources among competing activities in an optimum manner.

By definition, $x_j \geq 0$ implies noninteger as well as integer values for the decision variable. Although it is sometimes the case that only integer values of the decision variables have physical significance, the solution procedure does not necessarily yield integer values.

Answer is (C)

SYSTEMS–7

Consider a nontrivial linear programming problem in one variable, x, with only lower- and upper-bound constraints on x. At optimum, where will x be in relation to these constraints?

(A) at its upper bound
(B) at its lower bound
(C) between its upper and lower bounds
(D) at its upper or lower bound
(E) none of the above

The constraints prevent the variable of a linear program from increasing or decreasing forever during maximization or minimization. The maximum or minimum will occur at either the upper or lower bound.

Answer is (D)

SYSTEMS-8

If all variables in a linear programming problem are restricted to be integers, which, if any, basic assumption of linear programming is violated?

(A) certainty
(B) additivity
(C) divisibility
(D) proportionality
(E) No assumption is violated.

Divisibility implies that fractional levels of the decision variables must be permissible. By restricting all variables in the problem to be integers, divisibility is not assumed.

Answer is (C)

SYSTEMS-9

If a project that has diminishing returns with scale is modeled using a linear program, which basic assumption of linear programming will be violated?

(A) certainty
(B) additivity
(C) divisibility
(D) proportionality
(E) No assumption is violated.

Proportionality assumes that a variable multiplied by a constant is equal to the contribution, regardless of scale.

Answer is (D)

SYSTEMS-10

Consider the following linear programming model:

maximize:
$$Z = 3x_1 + 5x_2$$

subject to the constraints:
$$3x_1 + 2x_2 \leq 18$$
$$x_1 \leq 4$$
$$x_2 \leq 6$$

The graphical solution is:

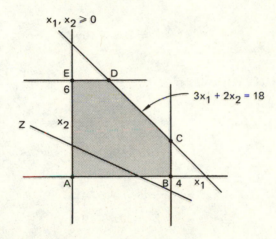

At what point does the optimum solution occur?

(A) A   (B) B   (C) C   (D) D   (E) E

If the objective function Z is plotted and moved along within the feasible region, the last point of contact will be point D. Therefore, Z will be maximized while satisfying all constraints at point D.

Answer is (D)

SYSTEMS-11

For which of the following linear programming problems can an optimum solution be found?

I.   maximize:

$$Z = 20x + 10y$$

subject to the constraints:

$$x + y \leq 4$$
$$3x + y \leq 6$$
$$x, y \geq 0$$

II.   maximize:

$$Z = 20x + 10y$$

subject to the constraints:

$$x + y \geq 4$$
$$3x + 2y \leq 6$$
$$x, y \geq 0$$

III.   maximize:

$$Z = 20x + 10y$$

subject to the constraints:

$$2x + y \leq 10$$
$$x + y \leq 6$$
$$x, y \geq 0$$

(A) I only   (B) I and II   (C) I and III

(D) II and III   (E) I, II, and III

For an optimum solution to exist, there must be a feasibility region. The graphs of the feasibility regions I, II, and III are:

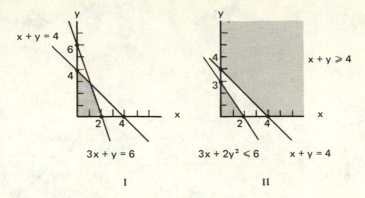

I   II

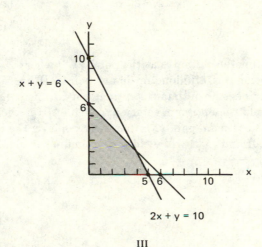

III

For II, there is no region where all four conditions are met. I and III have feasibility regions and, therefore, have optimum solutions.

Answer is (C)

SYSTEMS–12

Which of the linear programming problems in SYSTEMS–11 have multiple optimum points that yield the same optimum solution?

(A) I only   (B) II only   (C) III only
           (D) I and III   (E) none of the above

The graphs of I, II, and III are:

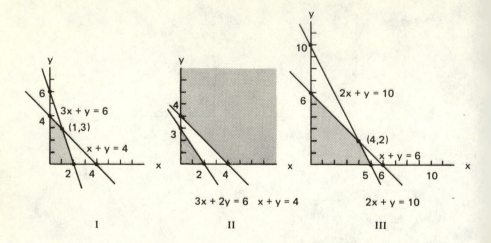

I   II   III

For I, the optimum solution is at the point $(1,3)$ where $Z = 20 + 30 = 50$; this solution has a unique optimum point. II has no feasibility region and thus has no optimum solution. For III, the points $(4,2)$ and $(5,0)$ both yield the optimum solution $Z = 100$; III is the only choice with multiple optimum points. In fact, any point on the line segment adjoining $(5,0)$ and $(4,2)$ will yield the optimum solution $Z = 100$.

Answer is (C)

SYSTEMS-13

What is the maximum value of Z for the following integer linear programming problem? (x and y are integers.)

maximize:

$$Z = 6x + 5y$$

subject to the constraints:

$$5x + 2y \leq 20$$
$$3x + 2y \leq 15$$
$$x, y \geq 0$$
$$x, y \text{ integers}$$

(A) 27   (B) 28   (C) 29   (D) 31   (E) 33

The maximum Z value is found from the extreme points in the figure:

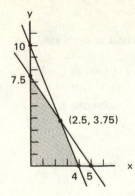

extreme point	Z
(0,0)	0
(4,0)	24
(0,7.5)	17.5
(2.5,3.75)	33.75

The largest value of Z over all real numbers is given at $(2.5, 3.75)$. Since x and y are integers though, the largest Z value will be given at either $x = 2$ and $y = 3$, or $x = 3$ and $y = 2$. The first combination gives $Z = 27$, but the second gives $Z = 28$.

Answer is (B)

SYSTEMS–14

The simplex method is extremely efficient in solving which of the following classic problems?

(A) the "transportation problem"
(B) the "assignment problem"
(C) the "transhipment problem"
(D) the "allocation problem"
(E) both (A) and (B)

Theoretically, all problem categories can be solved using the simplex method. However, the assignment, transportation, and transhipment problems are numerically inefficient when solved by the general simplex method. For allocation problems, the variables are continuous and, therefore, manageable enough in size to be solved using the simplex method.

Answer is (D)

SYSTEMS-15

Upon which of the following properties of linear programming is the simplex method based?

I.   The collection of feasible solutions constitute a convex set.
II. If a feasible solution exists, a basic solution exists where the feasible solutions correspond to the extreme points of the set of feasible solutions.
III. Only a finite number of basic feasible solutions exist.
IV. If the objective function possesses a finite maximum, at least one optimum solution is a basic feasible solution.

(A) I only
(B) IV only
(C) I and II
(D) I, III, and IV
(E) I, II, III, and IV

The simplex method is based upon all of the given linear programming properties.

Answer is (E)

SYSTEMS-16

As the simplex algorithm progresses from one solution to the next in a linear programming maximization problem, what will happen to the value of the objective function?

(A) It will increase and then decrease.
(B) It will decrease and then increase.
(C) It will increase or stay the same.
(D) It will decrease or stay the same.
(E) There is no general trend; the value varies randomly.

A characteristic of the simplex algorithm is that the value of the objective function improves (does not worsen) between iterations. For a maximization problem, only (C) can be true.

Answer is (C)

SYSTEMS-17

In the following simplex tableau, the first row of numbers represents the objective function of a maximization problem, and subsequent rows represent constraints. What are the basic variables?

x_1	x_2	x_3	x_4	x_5	right side
-10	-6	0	0	0	0
1	0	1	0	0	3
0	1	0	1	0	15
2	3	0	0	1	17

(A) x_1, x_4, and x_5
(B) x_2, x_4, and x_5
(C) x_3, x_4, and x_5
(D) x_1 and x_2
(E) x_3 and x_5

The system of equations represented by the tableau implies that $Z = 10x_1 + 6x_2 = 0$, and

$$\mathbf{A}x = \begin{pmatrix} 3 \\ 15 \\ 17 \end{pmatrix}$$

Given the negative coefficients of the objective function, Z, the values of the variables must be $x_1 = 0$, $x_2 = 0$, $x_3 = 3$, $x_4 = 15$, and $x_5 = 17$. x_3, x_4, and x_5 are "used" by the columns of the basis and are, therefore, the basic variables.

Answer is (C)

SYSTEMS–18

After several iterations, a simplex tableau is:

	Z	x_1	x_2	x_3	x_4	x_5	x_6	x_7	right
Z	1	-4	6	0	0	-3	1	0	14
x_4	0	7	2	0	1	-2	-6	0	3
x_7	0	6	2.5	0	0	4	0.8	1	12
x_3	0	3	3.5	1	0	2.5	5	0	2

Which of the following describes the solution found?

(A) It is feasible but not optimum.
(B) It is optimum but not feasible.
(C) It is optimum and feasible.
(D) It is neither optimum nor feasible.
(E) It is not in proper form.

The solution is feasible because there are no negative numbers on the right side. It is not, however, an optimum solution since there are negative numbers in the objective function.

Answer is (A)

SYSTEMS–19

In the following simplex tableau, the first row of numbers represents the objective function of a maximization problem. What are the current values of x_1 and x_2?

x_1	x_2	x_3	x_4	x_5	right side
-7.5	0	-4.5	0	0	0
1	1	0	0	0	3
0	0	1	1	0	15
2	0	3	0	1	17

(A) $x_1 = 0$, $x_2 = 0$ (B) $x_1 = 0$, $x_2 = 3$ (C) $x_1 = 2$, $x_2 = 15$

(D) $x_1 = 3$, $x_2 = 3$ (E) $x_1 = 7.5$, $x_2 = 17$

Since the coefficients of x_1 and x_2 in the objective function are negative, x_1 and x_2 must be zero. The tableau represents a system of equations with the solution: $x_1 = 0$, $x_2 = 3$, $x_3 = 0$, $x_4 = 15$, and $x_5 = 17$.

Answer is (B)

SYSTEMS-20

In the following simplex tableau, identify the current entering and exiting basic variables.

	Z	x_1	x_2	x_3	x_4	x_5	x_6	right side
Z	1	4	−6	−1	0	0	0	0
x_4	0	−3	2	−3	1	0	0	10
x_5	0	4	−8	3	0	1	0	24
x_6	0	7	3	6	0	0	1	30

(A) x_2 entering, x_4 exiting
(B) x_2 entering, x_6 exiting
(C) x_3 entering, x_5 exiting
(D) x_3 entering, x_4 exiting
(E) x_1 entering, x_4 exiting

The first row of the tableau gives:

$$Z = -4x_1 + 6x_2 + x_3$$

x_2 will cause the greatest increase in Z when it goes from 0 to 1. The entering variable is, therefore, x_2. With $x_1 = x_3 = 0$, the x_4 row gives:

$$2x_2 + x_4 = 10$$
$$x_2 = 5$$
$$x_4 = 0$$

The x_6 row gives:

$$3x_2 + x_6 = 30$$
$$x_2 = 10$$
$$x_6 = 0$$

The exiting variable is the one that goes to zero first as x_2 is increased from zero. The exiting variable is x_4.

Answer is (A)

SYSTEMS-21

In the following simplex tableau, the first row of numbers represents the objective function of a maximization problem. What will be the next variable to enter the basis?

x_1	x_2	x_3	x_4	x_5	right side
-5	-3	0	0	0	0
1	0	1	0	0	3
0	1	0	1	0	15
2	3	0	0	1	17

(A) x_1

(B) x_2

(C) x_3

(D) x_1 or x_4

(E) none of the above

If the first row represents the objective function of a maximization problem, then the tableau is already optimum, since for a maximization problem a tableau is optimum when all $C_i \leq 0$. If this had been the tableau for a minimization problem, x_1 would have been the next variable to enter the basis.

Answer is (E)

SYSTEMS–22

Find the "pivot" value in the following simplex tableau.

	x_1	x_2	u	v	
u	3	2	1	0	7
v	7	5	0	1	12
	-50	-40	0	0	

(A) 2   (B) 3   (C) 5   (D) 7   (E) 12

The first step in the method of pivot searching is to select the pivotal column by determining the column with the most negative entry in the objective row. In this problem it is the x_1 column.

Next, find the pivotal row by dividing each row's rightmost value by that row's pivotal column value. The row that has the lower quotient is the pivotal row. The quotient for row 1 is $\frac{7}{3}$, and for row 2 it is $\frac{12}{7}$. Thus, the second row is the pivotal row.

Finally, the pivot is the value that is at the intersection of the pivotal row and column. For this problem, it is 7.

> Answer is (D)

SYSTEMS–23

Find the optimum value of the slack variable x_3.

	x_1	x_2	x_3	x_4	x_5	
x_3	1	-1	1	0	0	2
x_4	2	1	0	1	0	4
x_5	-3	2	0	0	1	6
	-5	-3	0	0	0	0

(A) $\frac{7}{38}$ (B) $\frac{36}{7}$ (C) $\frac{38}{7}$ (D) $\frac{39}{7}$ (E) $\frac{40}{7}$

The simplex tableau becomes:

	x_1	x_2	x_3	x_4	x_5	
x_3	1	−1	1	0	0	2
x_4	2	1	0	1	0	4
x_5	−3	2	0	0	1	6
	−5	−3	0	0	0	0

	x_1	x_2	x_3	x_4	x_5	
x_1	1	−1	1	0	0	2
x_4	0	3	−2	1	0	0
x_5	0	−1	3	0	1	12
	0	−8	5	0	0	10

	x_1	x_2	x_3	x_4	x_5	
x_1	1	0	$\frac{1}{3}$	$\frac{1}{3}$	0	2
x_2	0	1	$-\frac{2}{3}$	$\frac{1}{3}$	0	0
x_5	0	0	$\frac{7}{3}$	$\frac{1}{3}$	1	12
	0	0	$-\frac{1}{3}$	$\frac{8}{3}$	0	10

	x_1	x_2	x_3	x_4	x_5	
x_1	1	0	0	$\frac{2}{7}$	$-\frac{1}{7}$	$\frac{2}{7}$
x_2	0	1	0	$\frac{3}{7}$	$\frac{2}{7}$	$\frac{24}{7}$
x_3	0	0	1	$\frac{1}{7}$	$\frac{3}{7}$	$\frac{36}{7}$
	0	0	0	$\frac{59}{21}$	$\frac{1}{7}$	$\frac{82}{7}$

The slack variables are, therefore, $x_3 = \frac{36}{7}$, $x_4 = 0$, and $x_5 = 0$. The optimum value is $\frac{36}{7}$.

Answer is (B)

SYSTEMS–24

What is Z equal to in the following linear programming problem?

maximize:
$$Z = 6x_1 + 5x_2 + 3x_3$$

subject to the constraints:

$$x_1 + x_2 + 2x_3 \leq 6$$
$$2x_1 + 3x_2 + x_3 \leq 8$$
$$3x_1 + 2x_2 + 2x_3 \leq 9$$
$$x_1, x_2, x_3 \geq 0$$

(A) 18   (B) $\frac{92}{5}$ (C) $\frac{94}{5}$ (D) $\frac{96}{5}$ (E) 20

Converting the problem to canonical form by adding slack variables gives:

maximize:
$$Z = 6x_1 + 5x_2 + 3x_3$$

subject to the constraints:
$$x_1 + x_2 + 2x_3 + x_4 \qquad = 6$$
$$2x_1 + 3x_2 + x_3 \qquad + x_5 \qquad = 8$$
$$3x_1 + 2x_2 + 2x_3 \qquad\qquad + x_6 = 9$$

The simplex method generates the following tableaus:

	x_1	x_2	x_3	x_4	x_5	x_6	
x_4	1	1	2	1	0	0	6
x_5	2	3	1	0	1	0	8
x_6	3	2	2	0	0	1	9
	-6	-5	-3	0	0	0	0

	x_1	x_2	x_3	x_4	x_5	x_6	
x_4	0	$\frac{1}{3}$	$\frac{4}{3}$	1	0	$-\frac{1}{3}$	3
x_5	0	$\frac{5}{3}$	$-\frac{1}{3}$	0	1	$-\frac{2}{3}$	2
x_1	1	$\frac{2}{3}$	$\frac{2}{3}$	0	0	$\frac{1}{3}$	3
	0	-1	1	0	0	2	18

	x_1	x_2	x_3	x_4	x_5	x_6	
x_4	0	0	$\frac{7}{5}$	1	$-\frac{1}{5}$	$-\frac{1}{5}$	$\frac{13}{5}$
x_2	0	1	$-\frac{1}{5}$	0	$\frac{3}{5}$	$-\frac{2}{5}$	$\frac{6}{5}$
x_1	1	0	$\frac{4}{5}$	0	$-\frac{2}{5}$	$\frac{3}{5}$	$\frac{11}{5}$
	0	0	$\frac{4}{5}$	0	$\frac{3}{5}$	$\frac{8}{5}$	$\frac{96}{5}$

The maximum value of Z is $\frac{96}{5}$.

Answer is (D)

SYSTEMS–25

Using the simplex method, in what form would one write the given objective function in order to maximize it?

$$Z = |x_1| - 3x_2$$

(A) $Z = x_1^+ + x_1^- - 3x_2$; subject to $x_1^+, x_1^- \geq 0$
(B) $Z - x_1^+ + x_1^- + 3x_2 = 0$; subject to $x_1^+, x_1^- \geq 0$
(C) $Z + x_1^+ - x_1^- - 3x_2 = 0$; subject to $x_1^+, x_1^- \geq 0$
(D) $Z - |x_1| + 3x_2 = 0$; subject to $x_1 \geq 0$
(E) $Z - |x_1| - x_1 - 3x_2 = 0$; subject to $x_1 \geq 0$

The absolute value term is written as $x_1^+ - x_1^-$, with x_1^+ and x_1^- greater than or equal to zero. The proper form for the simplex method is to have all terms on the same side of the equal sign equal to zero.

Answer is (B)

SYSTEMS–26

One constraint for a linear program is: $3x_1 - 2x_2 + 4x_3 \geq 6$. What is the proper form of this constraint for use in the simplex method?

(A) $3x_1 - 2x_2 + 4x_3 - 6 = 0$
(B) $3x_1 - 2x_2 + 4x_3 + 6 \leq 0$
(C) $3x_1 - 2x_2 + 4x_3 + \bar{x}_4 \geq -6$
(D) $3x_1 - 2x_2 + 4x_3 + x_4 = 6$
(E) $3x_1 - 2x_2 + 4x_3 - x_4 + \bar{x}_5 = 6$

The proper form for a constraint uses a slack variable to account for the inequality. For this problem, the slack variable x_4 is added.

Answer is (D)

SYSTEMS–27

How would the following problem be written if the simplex solution method is to be used?

maximize:
$$Z = 12x_1 - 4x_2$$

subject to the constraints:
$$x_1 + 3x_2 = 42$$
$$x_1, x_2 \geq 0$$

(A) $Z = 12x_1 - 4x_2;\ x_1 + 3x_2 - 42 = 0$
(B) $Z - 12x_1 - 4x_2 = 0;\ x_1 + 3x_2 + \bar{x}_3 = 42$
(C) $Z - 12x_1 - 4x_2 + Mx_3 = 0;\ x_1 + 3x_2 + \bar{x}_3 = 42;\ M$ is some large number.
(D) $Z - 12x_1 - 4x_2 = -Mx_3;\ x_1 + 3x_2 - 42 = \bar{x}_3$
(E) $Z - 12x_1 - 4x_2 - Mx_3 = 0;\ x_1 + 3x_2 - \bar{x}_3 = 42$

The slack variable x_3 is added to the objective function, multiplied by a constant, M. The restriction becomes $x_1 + 3x_2 + \bar{x}_3 = 42$. Although (D) has all the correct terms, they are not in the proper position.

Answer is (C)

SYSTEMS-28

Consider the following linear programming problem:

maximize:

$$Z = 6x_1 + 5x_2$$

subject to the constraints:

$$x_1 + x_2 \leq 4$$
$$5x_1 + 3x_2 \leq 15$$
$$x_1, x_2 \geq 0$$

If $x_1 + x_2 + x_3 = 4 + \Delta b$, find the range of Δb over which the basis remains unchanged.

(A) $-1 \leq \Delta b \leq 1$
(B) $0 \leq \Delta b \leq 1$
(C) $-2 \leq \Delta b \leq 1$
(D) $-2 \leq \Delta b \leq 2$
(E) $-1 \leq \Delta b \leq 2$

Using the slack variables x_3 and x_4, the constraints become:

$$x_1 + x_2 + x_3 \qquad = 4$$
$$5x_1 + 3x_2 \qquad + x_4 = 15$$

The simplex tableaus are:

	x_1	x_2	x_3	x_4	
x_3	1	1	1	0	4
x_4	5	3	0	1	15
	-6	-5	0	0	0

	x_1	x_2	x_3	x_4	
x_3	0	$\frac{2}{5}$	1	$-\frac{1}{5}$	1
x_1	1	$\frac{3}{5}$	0	$\frac{1}{5}$	3
	0	$-\frac{7}{5}$	0	$\frac{6}{5}$	18

	x_1	x_2	x_3	x_4	
x_2	0	1	$\frac{5}{2}$	$-\frac{1}{2}$	$\frac{5}{2}$
x_1	1	0	$-\frac{3}{2}$	$\frac{1}{2}$	$\frac{3}{2}$
	0	0	$\frac{7}{2}$	$\frac{1}{2}$	$\frac{43}{2}$

For the basis to remain unchanged, Z must also remain unchanged. Thus, for Z to remain at $\frac{43}{2}$,

$$x = \begin{pmatrix} x_1 \\ x_2 \end{pmatrix} = \begin{pmatrix} \frac{5}{2} \\ \frac{3}{2} \end{pmatrix}$$

$$x' = \begin{pmatrix} x_1 \\ x_2 \end{pmatrix} + \Delta b \begin{pmatrix} \frac{5}{2} \\ -\frac{3}{2} \end{pmatrix} \geq 0$$

This gives:

$$\frac{5}{2} + \Delta b \left(\frac{5}{2} \right) \geq 0 \qquad \frac{3}{2} - \Delta b \left(\frac{3}{2} \right) \geq 0$$

$$\Delta b \geq -1 \qquad\qquad \Delta b \leq 1$$

Therefore, $-1 \leq \Delta b \leq 1$.

Answer is (A)

SYSTEMS–29

Which of the following statements is incorrect for the primal linear programming problem in the form:

maximize:

$$Z_x = \sum_j C_j x_j$$

subject to the constraints:

$$\sum_j \sum_i a_{ij} x_j \leq b_i$$

$$x_j \geq 0$$

(A) The dual problem will be:

minimize:

$$Z_y = \sum_i b_i y_i$$

subject to the constraints:

$$\sum_j \sum_i a_{ij} y_i \geq C_j$$

$$y_i \geq 0$$

(B) The dual problem is the same as in choice (A), but with the inequality signs reversed.

(C) y_i is unrestrictive in sign if the inequality signs in the primal problem are replaced by equality signs.

(D) x_j is unrestrictive in sign if the inequality signs in the dual problem are replaced by equality signs.

(E) The coefficients in the jth constraint of the dual problem are the coefficients of x_j in the primal problem constraints and vice-versa.

By definition, the dual of a primal linear programming problem is exactly the reverse of the primal, including the reversal of the inequality signs.

Answer is (B)

SYSTEMS-30

For the following problem, what are the constraints of the dual problem?

maximize:
$$Z = 6x_1 + 3x_2 + 4x_3$$

subject to the constraints:
$$x_1 + 2x_2 + 3x_3 \leq 12$$
$$x_1 + 4x_2 + 3x_3 = 15$$
$$x_1, x_3 \geq 0$$

The dual problem statement is:

minimize:
$$Z' = 12w_1 + 15w_2$$

(A)
$$w_1 + w_2 = 6$$
$$2w_1 + 4w_2 \geq 3$$
$$3w_1 + 3w_2 \geq 4$$
$$w_1, w_2 \geq 0$$

(B)
$$w_1 + w_2 \geq 6$$
$$2w_1 + 4w_2 \geq 3$$
$$3w_1 + 3w_2 = 4$$
$$w_2 \geq 0$$

(C)
$$w_1 + w_2 \geq 6$$
$$2w_1 + 4w_2 = 3$$
$$3w_1 + 3w_2 \geq 4$$
$$w_1 \geq 0$$

(D)
$$w_1 + w_2 \geq 6$$
$$2w_1 + 4w_2 \geq 3$$
$$3w_1 + 3w_2 = 4$$
$$w_1, w_2 \geq 0$$

(E)
$$w_1 + w_2 = 6$$
$$2w_1 + 4w_2 \geq 3$$
$$3w_1 + 3w_2 \geq 4$$
$$w_1 \geq 0$$

Each of the constraints, C_i, in the primal problem corresponds to a respective variable, w_i, in the dual problem. The coefficients of the objective function in the primal problem are the constants on the right-hand side of the constraints in the dual problem. The coefficients of the ith constraint in the primal problem are the coefficients of the ith variable in the dual problem. If the jth variable in the primal problem is restricted, the jth constraint in the dual problem is an inequality. If the primal constraint is an equality, then the corresponding variable will be unrestricted in the dual problem. Thus, w_2 is unrestricted in the dual problem, eliminating (A), (B), and (D). In the dual problem, if a variable w_i is unrestricted, then the ith constraint is an equality. If the variable is restricted, the ith constraint is an inequality. Therefore, $2w_1 + 4w_2 = 3$.

Answer is (C)

SYSTEMS-31

The mathematical model for the classic transportation problem is:

Find $x_{ij}(i = 1, 2, \ldots, m; \; j = 1, 2, \ldots, n)$ in order to maximize

$$\sum_{i=1}^{m} \sum_{j=1}^{n} C_{ij} x_{ij}$$

subject to the constraints:

$$\sum_{j=1}^{n} x_{ij} = a_i \quad i = 1, \ldots, m$$

$$\sum_{i=1}^{m} x_{ij} = b_j \quad j = 1, \ldots, n$$

$$x_{ij} \geq 0$$

Which of the following is incorrect?

(A) m can be regarded as the number of factories supplying n warehouses with a certain product.

(B) a_i is the number of units produced at factory i, while b_j is the number of units required for delivery to warehouse j.

(C) C_{ij} is the shipping cost from factory i to warehouse j. x_{ij} is the decision variable, the amount shipped from factory i to warehouse j.

(D) The model has feasible solutions only if $\sum_{i=1}^{m} a_i = \sum_{j=1}^{n} b_j$

(E) x_{ij} has physical significance only for noninteger values.

A transportation problem has physical significance only when decision variables are integers.

Answer is (E)

SYSTEMS-32

Which of the following describes the optimum solution to a transportation problem?

(A) It can be determined using the simplex algorithm.

(B) It cannot be found if there are no upper-bound constraints on supplies from several sources.

(C) It can only be found if there are no upper bounds on supplies from several sources.

(D) It is trivial if the demand is unstable.

(E) Both (A) and (C) are correct.

Transportation problems are special linear programming problems that allow for the presence or lack of upper bounds on variables and assume constant demand. Only (A) is true.

Answer is (A)

SYSTEMS-33

How must the following cost and requirements table for a transportation problem be altered so that linear programming methods can be used to find an optimum solution?

source	destination				supply
	A	B	C	D	
1	C_{1A}	C_{1B}	$\cdots$		4
2	C_{2A}	C_{2B}			6
3	$\vdots$				2
	6	2	7	3	
	demand				

(A) A dummy source must be added to supply six units.

(B) A dummy destination must be added to increase the demand by six units.

(C) The sum of the costs in each row must be made equal by inclusion of a dummy cost.

(D) The sum of the costs in each column must be made equal by inclusion of a dummy cost.

(E) No alteration is necessary.

For the model to have a feasible solution, the supply and demand must be equal. A dummy source must be created which will supply an additional 6 units.

Answer is (A)

SYSTEMS–34

For the following transportation problem cost and requirements tables, how many basic variables are there?

source	destination				supply
	A	B	C	D	
1	C_{1A}	C_{1B}	$\cdots$		s_1
2	C_{2A}	C_{2B}			s_2
3	$\vdots$				s_3
	d_1	d_2	d_3	d_4	
	demand				

(A) three   (B) four   (C) six

      (D) seven   (E) twelve

The number of basic variables is equal to the number of sources plus the number of destinations minus one. Thus, there are $3 + 4 - 1 = 6$ basic variables.

Answer is (C)

SYSTEMS-35

The "Northwest Corner Rule" is to be used to find an initial solution to the following transportation simplex problem. What is the value of the fourth basic variable?

source	destination					supply
	A	B	C	D	E	
1						20
2						30
3						40
4						30
	40	20	10	30	20	
	demand					

(A) 10   (B) 20   (C) 30   (D) 40   (E) 50

The "Northwest Corner Rule" procedure for obtaining an initial basic feasible solution is:

1. Start with the cell in the upper left-hand corner.

2. Allocate the maximum feasible amount.

3. If there is supply remaining, move one cell to the right. If there is no remaining supply, move one cell down. Stop when it is impossible to do either of these. Repeat the process beginning at step 2 for the new cell. Each new cell represents a new basic variable.

Carrying out this procedure for the given problem:

source	destination					supply
	A	B	C	D	E	
1	20					20
2	20	10				30
3		10	10	20		40
4				10	20	30
	40	20	10	30	20	
	demand					

The fourth basic variable is equal to 10.

Answer is (A)

SYSTEMS-36

Four technicians, Tom, Scott, Ed, and Jeri, are each assigned a project to work on. Given the following estimated costs for each technician to complete each project, what is the optimum job assignment scheme such that all projects are completed at the minimum cost? (The order of technicians listed in the answer choices corresponds to job 1, job 2, job 3, and job 4.)

	job			
	1	2	3	4
Tom	10	14	15	12
Scott	9	13	17	10
Ed	8	12	14	11
Jeri	12	15	12	12

(A) Tom, Scott, Ed, Jeri
(B) Scott, Tom, Jeri, Ed
(C) Ed, Jeri, Scott, Tom
(D) Scott, Jeri, Ed, Tom
(E) Ed, Tom, Jeri, Scott

The cost matrix can be reduced by subtracting any constant from a row, as long as the row entries remain greater than or equal to zero. Subtract eight from each row. The matrix is then:

	job			
	1	2	3	4
Tom	2	6	7	4
Scott	1	5	9	2
Ed	0	4	6	3
Jeri	4	7	4	4

Thus, for minimum cost, Ed should do job 1, and Scott should do job 4. Jeri should do job 3, and Tom should do job 2. The correct order is Ed, Tom, Jeri, and Scott.

Answer is (E)

SYSTEMS-37

Which of the following will result if an integer programming problem is solved as a linear programming problem, and the resulting values of the decision variable are rounded off?

(A) an optimum integer solution
(B) a noninteger solution that is optimum
(C) an integer solution that may be optimum, or close to it
(D) an integer solution that is not optimum
(E) a noninteger solution that is not optimum

After rounding off, the solution will be an integer. However, it will no longer be an exact solution. Thus, it may provide an optimum solution or one that is close to optimum.

Answer is (C)

SYSTEMS-38

Which of the following is not a good application of network analysis?

(A) electrical engineering
(B) information theory
(C) cybernetics
(D) the study of transportation systems
(E) inventory theory

Inventory problems are generally not solved using network analysis. Network analysis involves maximizing the flow through a network connecting a source and a destination. Inventory theory involves the optimization of the problem of stocking goods; it is not concerned with the flow through a network.

Answer is (E)

SYSTEMS-39

For which of the following is the Program Evaluation and Review Technique (PERT) not used?

(A) construction projects
(B) computer programming assignments
(C) preparation of bids and proposals
(D) queueing problems
(E) installation of computer systems

> PERT is used to predict the completion time for large projects. All of the choices given except (D) are projects which have a finite completion time.

Answer is (D)

SYSTEMS-40

Identify the statement which is incorrect.

(A) The primary objective of PERT is to determine the probability of meeting specified deadlines.

(B) PERT identifies the activities that are most likely to "bottleneck" and the activities that are most likely to stay on schedule.

(C) PERT evaluates the effect of changes in the program.

(D) To apply PERT, one should develop a network representation of the project plan.

(E) The two basic concepts of PERT are "earliest time" and "latest time" as related to a particular event.

> Linear programming automatically performs a sensitivity analysis of the variables as a by-product of the solution process. PERT, however, cannot provide a similar sensitivity analysis to evaluate the effect of changes in the program parameters.

Answer is (C)

SYSTEMS-41

What are the basic features of dynamic programming problems?

I. The problem can be divided into stages with a policy decision required at each stage.

II. Each stage has a number of states associated with it.

III. The effect of the policy decisions at each stage is to transform the current state into a state associated with the next stage.

IV. The problem formulation is dependent on the probability distribution associated with it.

(A) I only   (B) IV only   (C) I and II

     (D) I, II, and III   (E) I, II, III, and IV

Statement IV is irrelevant. There may not be a probability distribution when deterministic problems are solved using dynamic programming. The formulation of the problem depends only on the first three statements.

Answer is (D)

SYSTEMS-42

Which of the following statements about dynamic programming is false?

(A) Dynamic programming is a mathematical technique that is often useful for making a sequence of interrelated decisions.

(B) Dynamic programming provides a systematic procedure for determining the combination of decisions that maximize overall effectiveness.

(C) Dynamic programming can be represented in standard mathematical formulation.

(D) Dynamic programming is a conceptual approach to problem solving.

(E) The "stagecoach problem" is often used to illustrate the principles of dynamic programming.

Dynamic programming is a conceptual approach to problem solving, not a mathematical one. Its formulation depends on the specifics of the problem.

Answer is (C)

SYSTEMS–43

Queueing theory provides a large number of alternative mathematical models for describing which of the following?

(A) network problems
(B) probabilistic arrivals
(C) probabilistic service facilities
(D) waiting line problems
(E) probabilistic arrivals and services

Queueing theory involves the mathematical study of waiting lines or "queues."

Answer is (D)

SYSTEMS–44

The various elements of the queueing process are depicted below:

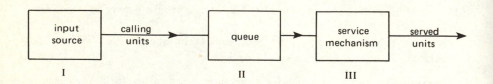

Which elements of the figure make up the queueing system?

(A) I only   (B) II only   (C) III only

          (D) I and II   (E) II and III

The queue and the service mechanism make up the actual queueing system.

Answer is (E)

SYSTEMS-45

In a queueing process of customers in a store, what statistical pattern will most likely describe the arrival of customers over time?

(A) the normal law of probability
(B) the Poisson distribution
(C) the uniform law of probability
(D) the exponential distribution
(E) the binomial law

The common assumption is that calling units in a queueing process arrive according to a Poisson distribution.

Answer is (B)

SYSTEMS-46

In a queueing process of customers in a store, what type of distribution most likely governs the time between consecutive arrivals of customers?

(A) a normal probability distribution
(B) an exponential distribution
(C) a uniform probability distribution
(D) a Poisson distribution
(E) a binomial distribution

The "inter-arrival time" is commonly assumed to be exponentially distributed.

Answer is (B)

SYSTEMS-47

The jobs to be performed by a particular machine arrive according to a Poisson input process with a mean rate of one per hour. If the machine breaks down and requires 2 hours to be repaired, what is the probability that the number of new jobs that arrive during the 2-hour period is zero?

(A) e^{-2} (B) e^{-1} (C) 1   (D) e (E) e^2

For a Poisson distribution, the probability, p, of x jobs arriving in a time, t, is given by:

$$p\{x\} = \lambda t^x \left(\frac{e^{-\lambda t}}{x!} \right)$$

λ is the mean arrival rate. Therefore,

$$p\{x\} = \left(1 \right) \left(2^0 \right) \left[\frac{e^{-(1)(2)}}{0!} \right]$$
$$= e^{-2}$$

Answer is (A)

SYSTEMS-48

In a queueing system that has an arrival rate of 5 customers per hour, the expected waiting time for any customer in the system, including service time, is 40 minutes. What is the expected number of customers in the system under steady state conditions?

(A) $\frac{5}{40}$ customers   (B) $\frac{2}{15}$ customers   (C) $\frac{10}{3}$ customers

(D) 8 customers   (E) 200 customers

Little's formula states:
$$L = \lambda W$$

L is the expected number of customers in the system, λ is the mean arrival rate of customers per hour, and W is the expected waiting time for each customer. Thus,

$$L = \lambda W = \left(5 \right) \left(\frac{2}{3} \right)$$
$$= \frac{10}{3} \text{ customers}$$

Answer is (C)

SYSTEMS–49

Consider a queueing system with three servers, such that the mean service rate for each busy server is 2 customers per hour. If the mean arrival rate of customers is 5 per hour, what is the expected fraction of total time that all servers will be busy? Assume steady state conditions.

(A) $\dfrac{2}{3}$ (B) $\dfrac{5}{6}$ (C) $\dfrac{5}{3}$

(D) $\dfrac{5}{2}$ (E) none of the above

The expected fraction of time that all servers will be busy is equal to:

$$\text{arrival rate} \times \frac{1}{\text{number of servers} \times \text{service rate}} = 5\left(\frac{1}{(3)(2)}\right)$$
$$= \frac{5}{6}$$

Answer is (B)

SYSTEMS–50

Consider the following equation:

$$F(s) = \frac{A_1}{s - s_1} + \frac{A_2}{s - s_2} + \frac{A_3}{s + s_3}$$

If s_1, s_2, and s_3 are the poles corresponding to the three terms in $F(s)$, respectively, which point on the graph may represent s_3 if s_3 is real and $s_3 > 0$?

(A) s_a (B) s_b (C) s_c

(D) s_d (E) s_c or s_d

The poles of the feedback equation fall on the left side of the vertical axis if they are negative. The only point that may represent s_3 is s_a.

Answer is (A)

SYSTEMS–51

The pole diagram is shown for the following equation:

$$F(s) = \frac{P(s)}{Q(s)} = \frac{P(s)}{(s^2 + 2\varsigma\omega_n s + \omega_n^2)(s - s_3)}$$

If s_1, s_2, and s_3 are the poles corresponding to the solution of $(s^2 + 2\varsigma\omega_n s + \omega_n^2)$ and $(s - s_3)$, respectively, what points on the diagram correspond to s_1 and s_2? (s_3 is not plotted.)

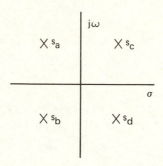

(A) s_a and s_b
(B) s_a and s_c
(C) s_a and s_d
(D) s_b and s_c
(E) s_c and s_d

The roots of $s^2 + 2\varsigma\omega_n s + \omega_n^2$ must be either s_a and s_b or s_c and s_d, since the $j\omega$ components are of the same magnitude but different in sign. The algebraic expression gives negative roots. Therefore, the solution is on the left side of the plot.

Answer is (A)

SYSTEMS–52

The following function is plotted on a pole-zero diagram:

$$F(s) = \frac{K(s - z_1)}{s(s - p_1)(s - p_2)}$$

The z_1 value and the $j\omega$ axis are not shown. The magnitude of p_1 is larger than the magnitude of p_2, and both are positive numbers. Determine which of the following statements is true.

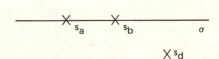

(A) $p_1 = s_a, p_2 = s_b$
(B) $p_1 = s_b, p_2 = s_a$
(C) $p_1 = s_c, p_2 = s_d$
(D) p_1 and p_2 are real and described by s_c and s_d.
(E) none of the above

Unless otherwise noted by j, a constant is a real number. Therefore, p_1 and p_2 fall on the σ-axis. p_1 is of greater magnitude than p_2 and is, therefore, to the right of p_2.

Answer is (B)

SYSTEMS–53

Which of the following best describes the function shown below?

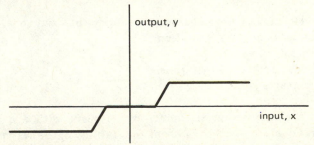

(A) It has a dead zone.
(B) It is a saturated zone system.
(C) There is both a dead zone and a saturated system zone.
(D) It is an impulse zone system.
(E) It is a ramp system.

The dead zone occurs in the region where there is no amplitude near the origin. The positive and negative saturation occurs after a short linear increase or decrease, respectively. A ramp would only have a single, continuously increasing function. An impulse is a narrow major increase, such as a single square wave pulse.

Answer is (C)

SYSTEMS-54

Which of the following is true about the function shown?

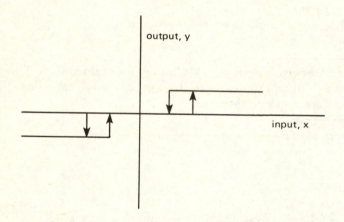

(A) It has a dead zone with linear output outside the dead zone.
(B) It has a dead zone with saturation.
(C) There is no dead zone.
(D) It has a dead zone with hysteresis.
(E) It is an unstable system with a ramp.

This is a classic hysteresis input/output curve with a dead zone. There is no information given about the stability of the system. There is no ramp present.

Answer is (D)

SYSTEMS–55

The frequency response of a system is given by:

$$\frac{M}{\alpha} = \frac{X_2(j\omega)}{X_1(j\omega)}$$

By differentiation, the peak value of M, M_{max}, and the frequency at which it occurs, ω_{max}, are expressed in terms of the damping ratio, ς, and natural frequency, ω_n:

$$M_{\text{max}} = \frac{1}{2\varsigma\sqrt{1 - \varsigma^2}} \qquad \omega_{\text{max}} = \omega_n\sqrt{1 - 2\varsigma^2}$$

Determine which curve in the figure is a correct representation for M to be the largest response.

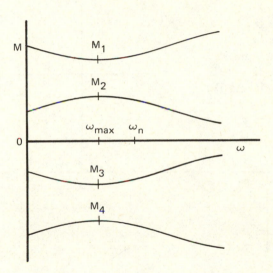

(A) $M = M_1$ (B) $M = M_2$ (C) $M = M_3$

(D) $M = M_4$ (E) $M = M_1$ or M_3

As ω goes to ω_{max}, M should increase to its peak value. Therefore, the shape of M_2 and M_4 are both correct. However, since M_{max} is always positive, M_2 is the solution.

Answer is (B)

SYSTEMS-56

The frequency response of a system is given by:

$$\frac{M}{\alpha} = \frac{X_2(j\omega)}{X_1(j\omega)}$$

By differentiation, the peak value of M, M_{max}, and the frequency at which it occurs, ω_{max}, are expressed in terms of the damping ratio, ς, and natural frequency, ω_n:

$$M_{max} = \frac{1}{2\varsigma\sqrt{1-\varsigma^2}} \qquad \omega_{max} = \omega_n\sqrt{1-2\varsigma^2}$$

For what range of ω on the curve segment shown does the peak response in amplitude occur?

(A) ω is at point II.
(B) ω is at point IV.
(C) ω is between I and II.
(D) ω is between II and III.
(E) ω is between II and IV.

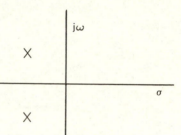

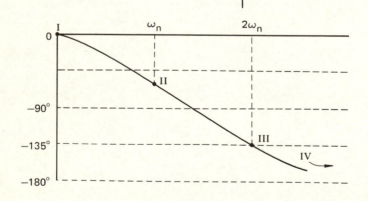

A plot of M_{max} shows that the peak falls between 0 and ω_n. The polar plot shows two conjugate poles for a simple second-order system. The damping ratio, ς, and the natural frequency are used to determine ω_{max}. The peak is at $\omega_{max} = \omega_n\sqrt{1-2\varsigma^2}$ unless $\varsigma = 0$ (completely undamped), $\omega_{max} < \omega_n$.

Answer is (C)

SYSTEMS-57

Which of these $Q(s)$ equations are unconditionally stable?

I. $4s^4 + 8s^2 + 3s + 2 = 0$
II. $4s^4 + 2s^3 + 8s^2 + 3s + 2 = 0$
III. $4s^4 + 2s^3 + 8js^2 + 5s + 2 = 0$
IV. $4s^4 + 2s^3 + 8s^2 - 3s + 2 = 0$

(A) I only   (B) II only   (C) I and IV
        (D) II and III   (E) None are stable.

According to Routh's stability criteria for unconditional stability, there can be no sign changes in the terms of the equation, nor can there be imaginary or missing exponential terms. Equation II is the only such equation.

Answer is (B)

SYSTEMS-58

The Routhian array is given below for this equation:

$$Q(s) = s^4 + 6s^3 + 13s^2 + (20 + K)s + K = 0.$$

s^4	1	13	K
s^3	6	$20 + K$	0
s^2	$\frac{58-K}{6}$	K	0
s^1	$\frac{(58-K)(20+K)-36K}{58-K}$	0	0
s^0	K	0	0

For what range of K will the system be stable?

(A) $0 < K < 33.1$
(B) $33.1 < K < 58$
(C) $58 < K < 116$
(D) $0 < K < 116$
(E) $K > 58$

For the system to be stable, there can be no sign changes in the first column of the Routhian array. Since the first two entries of that column are positive, the last three entries must also be positive. The entry in the s^2 row gives $K < 58$, while the entry in the s^0 row gives $K > 0$. The numerator in the s^1 row is equal to $-K^2 + 2K + 116$, with roots of -35.1 and 33.1, or $-35.1 < K < 33.1$. For K to satisfy all these restrictions, $0 < K < 33.1$.

Answer is (A)

SYSTEMS-59

The characteristic equation for a system is:

$$Q(s) = s^4 + 5s^3 + 10s^2 + Ks - 1 = 0$$
$$K > 0$$

The Routhian array is:

s^4	1	10	−1
s^3	5	K	0
s^2	$\frac{50-K}{5}$	−1	0
s^1	$\frac{K(50-K)+25}{5(50-K)}$	0	0
s^0	5	0	0

Which of these statements is true?

(A) The system is unstable at all points.
(B) The system is unstable for $K < 50$.
(C) The system is stable for $K < 50$.
(D) The system is stable for some point above $K = 50$.
(E) The system is stable at $K = 100$.

For the system to be stable, there can be no sign changes in the first column of the Routhian array. The $(50-K)$ term in the s^2 row means that the system should be stable for $0 < K < 50$. The term in the s^1 row requires that $-0.49 < K < 50.5$. Thus, the range of values for K for the system to be stable is $0 < K < 50$.

Answer is (C)

SYSTEMS-60

The log magnitude-angle diagram shown gives the response curves for two systems, I and II. Determine which of the following is true.

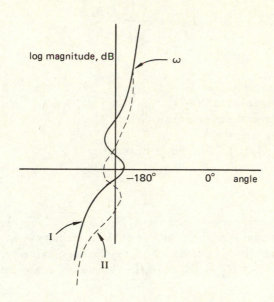

(A) System I is unstable.
(B) System II is unstable.
(C) Both systems are unstable.
(D) Both systems are stable.
(E) Both systems are conditionally stable.

The stability of a system is determined by where its response curve intersects the angle axis. System I is stable because its point of intersection is to the right of the log magnitude axis. System II is unstable because its point of intersection is to the left of the log magnitude axis.

Answer is (B)

SYSTEMS–61

Given the following transfer functions, which of these statements are true?

$$G_1(s) = \frac{K}{s\left(s + \frac{1}{B_1}\right)}$$

$$G_2(s) = \frac{K\left(s + \frac{1}{B_2}\right)}{s\left(s + \frac{1}{B_1}\right)}$$

(A) $G_2(s)$ is as stable as $G_1(s)$.
(B) $G_2(s)$ is less stable than $G_1(s)$.
(C) $G_2(s)$ has slower transients than $G_1(s)$.
(D) System $G_2(s)$ is unstable.
(E) Statements (A), (B), (C), and (D) are false.

The addition of another zero at $1/B_2$ in system $G_2(s)$ causes the plotted function to shift to the left of the imaginary axis. The result of this is that the transients will decay faster making system $G_2(s)$ more stable than $G_1(s)$. None of the choices given are true.

Answer is (E)

SYSTEMS-62

The polar plot shows a conditionally stable system. Determine which of these statements is true.

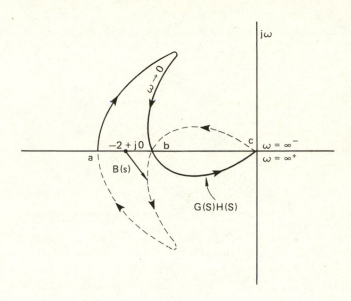

(A) The system is stable at low gain.
(B) The system is unstable at low gain.
(C) The system is unstable at high gain.
(D) The system is unstable at high and low gain.
(E) none of the above

For low values of gain, the $(-1 + j0)$ point is between points a and b. $P_r = 1$ (number of poles in the right half plane), $N = -1$ (number of net counterclockwise revolutions from $-1 + j0$) and $Z_r = 2$, which means the system is unstable. For high values of gain, the $(-1 + j0)$ point is between b and c. Here, $P_r = 1$ and $N = +1$. Therefore, $Z_r = 0$, which means the system is stable. Thus, the system is unstable at low gain.

Answer is (B)

SYSTEMS-63

Both of the curves shown represent a system response of the form:

$$G(s) = \frac{K}{s(1+s)(1+0.5s)}$$

K_1 is the K for curve I, and K_2 is the K for curve II. Which of these statements is true?

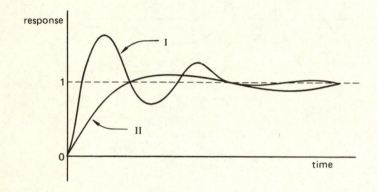

(A) K_1 is greater than K_2.
(B) K_1 is less than K_2.
(C) The size of K has no effect on the response.
(D) K is the same for both function.
(E) none of the above

Generally, the gain, K, has a direct effect on the type of response obtained. Larger values of gain give larger overshoots or longer settling times. Thus, the gain of curve I is larger than the gain of curve II.

Answer is (A)

SYSTEMS–64

A control system has a control response ratio of:

$$\frac{C(s)}{R(s)} = \frac{1}{s^2 + 0.3s + 1}$$

Given that the damping ratio is $\varsigma < 0.707$, how many peaks occur in the transient prior to reaching steady state conditions?

(A) zero
(B) one
(C) two
(D) The function gives a minimum.
(E) The function has a flat response.

The control ratio has two complex poles which are dominant, but has no zeros. These poles are $(-0.15 + j0.88)$ and $(-0.15 - j0.88)$. For a damping ratio of $\varsigma < 0.707$, a peak occurs. If the control response was in the form of:

$$\frac{C(s)}{R(s)} = \frac{1}{(s^2 + 2\varsigma\omega_n s + \omega_n^2)}$$

$$\text{Then,} \quad \omega_n^2 = 1$$
$$2\varsigma = 0.3$$
$$\varsigma = 0.15$$

Therefore, there is only one peak value given by:

$$M_{\max} = \frac{1}{2\varsigma\sqrt{1 - \varsigma^2}} = \frac{1}{2(0.15)\sqrt{1 - (0.15)^2}}$$
$$= 3.371$$

Answer is (B)

SYSTEMS–65

A function of the form:

$$\frac{s + \alpha}{(s + a)(s + b)(s + c)}$$

has the transform:

$$\frac{(\alpha - a)e^{-at}}{(b - a)(c - a)} + \frac{(\alpha - b)e^{-bt}}{(c - b)(a - b)} + \frac{(\alpha - c)e^{-ct}}{(a - c)(b - c)}$$

Which of the following is the correct transform for the following function?

$$F(s) = \frac{1}{s(s + a)(s + b)}$$

(A) $\dfrac{1}{ab}\left(1 - \dfrac{be^{-at}}{b - a} + \dfrac{ae^{-bt}}{b - a}\right)$

(B) $\dfrac{1}{ab}\left(1 + \dfrac{be^{-at}}{b - a} + \dfrac{ae^{-bt}}{b - a}\right)$

(C) $\dfrac{1}{ab}\left(1 + \dfrac{be^{-at}}{b - a} - \dfrac{ae^{-bt}}{b - a}\right)$

(D) $\dfrac{1}{ab}\left(1 - \dfrac{be^{-at}}{b - a} - \dfrac{ae^{-bt}}{b - a}\right)$

(E) $\dfrac{1}{ab}\left(1 - \dfrac{be^{-at}}{b - a} + j\dfrac{ae^{-bt}}{b - a}\right)$

The new function can almost be obtained from the old one by letting c approach zero. This makes e^{-ct} approach 1 in the transform. Thus, the functions only differ by the terms involving α. The correct sign of the terms in the transform of the new function is given in (A).

Answer is (A)

SYSTEMS–66

A circuit with inductance L and resistance R has the transfer function:

$$G(j\omega) = \frac{R}{\sqrt{R^2 + (\omega L)^2}}$$

or

$$G(j\omega T) = \frac{1}{1 + j\omega T}$$

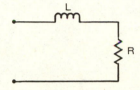

Which of these statements is true about the polar plot of the function?

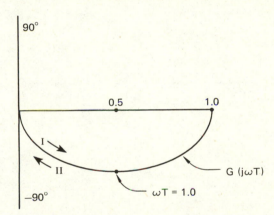

(A) I shows ω going to zero.

(B) II shows ω going to zero.

(C) If the inductor is replaced by a capacitor, the plot will be in the upper half of the phase plane.

(D) Both (A) and (C) are true.

(E) None of the above are true.

As $\omega \to \infty$, $|G(j\omega)| \to 0$. As $\omega \to 0$, $|G(j\omega)| \to 1$. I shows ω going to zero. If the inductor is a capacitor, the plot would be "reflected" on the horizontal axis into the top half of the plane.

Answer is (D)

SYSTEMS-67

A typical transfer function is as follows:

$$G(j\omega) = \frac{C(j\omega)}{E(j\omega)}$$

$$= \frac{K}{j\omega(1 + j\omega T_1)(1 + j\omega T_2)(1 + j\omega T_3)}$$

The plot is:

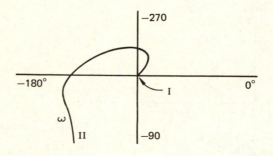

Which of the following is true?

(A) I is where $G(j\omega) = 0$.
(B) I is where $G(j\omega) \to \infty$.
(C) II is where $G(j\omega) \to \infty$.
(D) II is where $G(j\omega) \to -\infty$.
(E) none of the above

In the equation, as $\omega \to 0$, $G(j\omega) \to \infty$, and as $\omega \to \infty$, $G(j\omega) \to 0$.
Therefore, point I is where $G(j\omega) \to \infty$.

Answer is (B)

16 COMPUTER SCIENCE

COMPUTER SCIENCE-1

Where is the error in the following program?

```
670   READ L,M,N
680   DATA 33 40 60
690   PRINT L+M/N
700   PRINT L*N+M
800   END
```

(A) line 670
(B) line 680
(C) line 690
(D) line 700
(E) line 800

Line 680 should have commas separating the data points.

Answer is (B)

COMPUTER SCIENCE–2

What is the output of the following program?

```
100   DATA 'June',15,1955
200   READ N$
300   DATA 1948
400   READ C,X,X₁
500   PRINT X₁,N$,X,C
```

(A) 15,1955,June,1955
(B) June 15 1955 1948
(C) 1948 June 1955 15
(D) 1948 June 15 1955
(E) 1955 June 15 1948

June is assigned to N$, 15 is assigned to C, 1955 is assigned to X, and 1948 is assigned to X_1. The output will be: 1948 June 1955 15.

Answer is (C)

COMPUTER SCIENCE–3

In the program below, what is the value assigned to N?

```
10   N=0
15   GO TO 50
20   FOR X=-1 TO 7
25   N=6
30   FOR Y=3 TO 6
35   N=X*Y+16
40   NEXT Y
45   NEXT X
50   PRINT N
55   END
```

(A) −3678   (B) −356   (C) 0   (D) 98   (E) 1021

The statement of line 15 bypasses the loops, and N remains the value assigned to it in line 10: N=0.

Answer is (C)

COMPUTER SCIENCE-4

Determine the output of the program below.

```
10   FOR X=−1 TO 8 STEP 3
20   S=0
30   FOR Y=0 TO 4 STEP 2
40   S=S+X*Y+1
50   NEXT Y
60   PRINT S
70   NEXT X
80   END
```

(A) −3 15 33 51
(B) 0 17 23 84
(C) −5 15 18 24
(D) 72 12 21 28
(E) 81 11 23 32

S gets assigned a value three times during the first loop of X, and is then printed.

$$\left.\begin{array}{l} X = -1 \\ S = 0 \\ Y = 0 \end{array}\right\} \quad S = 0 + (-1)(0) + 1 = 1$$

$$\left.\begin{array}{l} X = -1 \\ S = 1 \\ Y = 2 \end{array}\right\} \quad S = 1 + (-1)(2) + 1 = 0$$

$$\left.\begin{array}{l} X = -1 \\ S = 0 \\ Y = 4 \end{array}\right\} \quad S = 0 + (-1)(4) + 1 = -3$$

This may be continued for the subsequent loop values of X. The corresponding S values printed are 15, 33, and 51.

Answer is (A)

COMPUTER SCIENCE–5

What are the respective values of A_1, A_2, and A_3, given the assignments below?

$A_1 = INT(5)$
$A_2 = INT(10.4)$
$A_3 = INT(-10.4)$

(A) $A_1=5$, $A_2=10$, $A_3=-10$
(B) $A_1=5$, $A_2=11$, $A_3=-10$
(C) $A_1=5$, $A_2=10$, $A_3=-11$
(D) $A_1=5$, $A_2=11$, $A_3=-11$
(E) $A_1=5$, $A_2=-11$, $A_3=10$

The INT function assigns the largest integer value not greater than the original value. 5, 10, and -11 are the respective A values.

Answer is (C)

COMPUTER SCIENCE–6

What would be the value printed as a result of the following instructions?

```
10   DEF FNA(X)=X**2+1/X
20   PRINT FNA(2)
30   END
```

(A) 3.2   (B) 4.5   (C) 6.5   (D) 7.2   (E) 8.0

$$FNA(2) = (2)^2 + \frac{1}{2} = 4.5$$

Answer is (B)

COMPUTER SCIENCE–7

What would be the output of the following program?

```
10   DEF FNA(X,Y)=X*2+X*3−X*Y
20   READ B(1),C,D
30   DATA 3,2,4
40   PRINT FNA(B(1),B(1))
50   PRINT FNA(C/D,C*D)
60   END
```

(A) 6,−1.5   (B) 8,3   (C) 9,−11
              (D) 11,−9   (E) 12,−2

From the READ and DATA statements, B(1) is 3, C is 2, and D is 4. Thus, in line 40, B(1) replaces X and Y in the function statement of line 10. The output of line 40 is:

$$(3)(2) + (3)(3) − (3)(3) = 6$$

Line 50 assigns $X = \frac{2}{4} = \frac{1}{2}$ and $Y = (2)(4) = 8$, and then prints the function output:

$$(\tfrac{1}{2})(2) + (\tfrac{1}{2})(3) − (\tfrac{1}{2})(8) = −1.5$$

Answer is (A)

COMPUTER SCIENCE–8

If table Y represents an array, and B is 2, what is the value of Y(B+2)?

I	table Y
1	90
2	−20
3	410
4	70
5	17

(A) −22   (B) −18   (C) 17   (D) 70   (E) 90

When B is 2, Y(B+2)=Y(4). The value for Y(4) is 70.

Answer is (D)

COMPUTER SCIENCE-9

Why is the following expression unacceptable as a FORTRAN integer constant?

$$237,100$$

(A) The first character must be a letter.
(B) There are more than six characters.
(C) There is no decimal point.
(D) It contains a comma.
(E) There is nothing wrong, the expression is an acceptable integer constant.

Commas are not allowed in integer constants.

Answer is (D)

COMPUTER SCIENCE-10

Which of the following is/are acceptable as integer variables?

I. NINA   II. A+10   III. INTEGER   IV. 2813   V. A160

(A) I only
(B) I and V
(C) I and III
(D) I, III, and V
(E) I, IV, and V

Variable names can be formed from up to six characters, the first character being a letter. Integer variables must begin with the letters I, J, K, L, M, or N; all other variable names represent real variables. The characters must be alphanumeric, hence the + character in II is not allowed. I is the only acceptable choice.

Answer is (A)

COMPUTER SCIENCE–11

Which of the following is not acceptable as an integer variable name?

(A) IRISH
(B) KOST
(C) MAPLE
(D) JAPAN
(E) INSERTS

INSERTS exceeds the limit of six characters.

Answer is (E)

COMPUTER SCIENCE–12

Which of the following are not acceptable as FORTRAN integer constants?

I. 5000   II. -14 III. A15   IV. 16.52   V. $+999$

(A) I and II
(B) II and III
(C) III and IV
(D) I, II, and IV
(E) II, III, and V

An integer constant cannot begin with a letter character and cannot have a decimal point. Thus, III and IV are unacceptable.

Answer is (C)

COMPUTER SCIENCE–13

Which of the following are not acceptable as real constant names?

I. APPLE   II. LAMB   III. 4ABC   IV. WATER   V. TREE

(A) II and III
(B) IV and V
(C) I, II, and III
(D) II, III, and V
(E) none of the above

LAMB is incorrect since the initial character L denotes an integer variable. 4ABC is incorrect since the first character should be a letter.

Answer is (A)

COMPUTER SCIENCE–14

What is a correct declaration form for the logical variable AZ?

(A) AZ=.TRUE.
(B) AZ=1
(C) AZ=1.
(D) AZ='TRUE'
(E) AZ=.'TRUE'.

For a logical variable, TRUE is enclosed in periods only.

Answer is (A)

COMPUTER SCIENCE–15

Which of the following is unacceptable as a FORTRAN statement?

(A) A=B
(B) A=B+C
(C) AB=C*D
(D) A*B=C*D
(E) B=A+C

The left side of an assignment statement must be a variable and not an operation. (D) is unacceptable.

Answer is (D)

COMPUTER SCIENCE–16

What value is assigned to A in the following expression?

$$A=3.4+(18*(4**2+17.2*8/6)-2/4$$

(A) 92.5   (B) 178.1   (C) 502.1
        (D) 703.7   (E) 900.1

The expressions within parentheses are evaluated first. This is followed by exponentiation, multiplication and division, and addition and subtraction. Within each level of operation, the expressions are performed left to right.

$$A = 3.4 + [(18)(16 + 22.43)] - 0.5 = 703.7$$

Answer is (D)

COMPUTER SCIENCE–17

What value would the computer assign to the variable D8=1.0E8−1.5E6?

(A) -0.5×10^2 (B) 1.5×10^2 (C) 9.85×10^7

     (D) 10.5×10^7 (E) 11.1×10^7

The letter E denotes scientific notation. Therefore,

$$D8 = (1 \times 10^8) - (1.5 \times 10^6) = 9.85 \times 10^7$$

Answer is (C)

COMPUTER SCIENCE–18

Which of the following is a valid FORTRAN statement for the algebraic expression below?

$$\frac{3x^2 - 5x}{y^2 + 1}$$

(A) 3.0*X**2.0−5.0*X/Y**2.0+1.0

(B) (3.0X**2.0−5.0X)/Y**2.0+1.0

(C) (3.0*X**2.0−5.0*X)/(Y**2.0+1.0)

(D) (3.0*X**2.0−5.0X)/(Y**2.0+1.0)

(E) (3.0*X**2.0−5.0*X)/Y**2.0+1.0

The numerator and denominator must be enclosed in parentheses. Each operation must be explicitly and unambiguously stated. 3.0X is not a substitute for 3.0*X.

> Answer is (C)

COMPUTER SCIENCE–19

Which of the following is a FORTRAN expression corresponding to

$$\frac{\ln\left(a + \sqrt{b^3}\right)}{\sin\left(c^2 + 1\right)} ?$$

(A) ALOG10(A+SQRT(B**3.0))/SIN(C**2.0+1.0)

(B) LOG(A+SQRTB**3.0)/SIN(C**2.0+1.0)

(C) ALOG(A+SQRT(B**3.0)/SIN(C**2.0+1.0))

(D) LN(A+SQRT(B**3.0))/SIN(C**2.0+1.0)

(E) ALOG(A+SQRT(B**3.0))/SIN(C**2.0+1.0)

The FORTRAN expression for the natural logarithm is ALOG, followed by the term enclosed in parentheses. The natural logarithm term in (C) includes the denominator due to a misplaced parenthesis.

> Answer is (E)

COMPUTER SCIENCE–20

How would the algebraic expression below be written in FORTRAN?

$$\left(1 + \frac{x}{y}\right)^{x-1}$$

(A) (1.0+X/Y)*(X−1.0)

(B) (1.0+X)/Y**(X−1.0)

(C) (1.0+X/Y)**(X−1.0)

(D) (1.0+X/Y)**X−1.0

(E) (1.0+X)/Y**X−1.0

The exponent must be enclosed in parentheses in this case. (A) has a multiplication sign instead of an exponent sign, and the first term in (B) is incorrect. Only (C) is correct.

Answer is (C)

COMPUTER SCIENCE-21

How would the following expression be written in FORTRAN?

$$2 - \frac{x}{\sqrt{x^2 - \dfrac{1}{y}}}$$

(A) (2.0−X)/SQRT(X**2.0−1.0/Y)

(B) 2.0−(X/SQRTX**2.0−1.0/Y)

(C) 2.0−X/SQRTX**2.0−1.0/Y

(D) 2.0−X/SQRT(X**2.0−1.0/Y)

(E) 2.0−X/SQRT((X**2.0−1.0)/Y)

The parentheses in (A), (B), and (E) are wrong, and (C) is missing parentheses around the SQRT term.

Answer is (D)

COMPUTER SCIENCE-22

Which of the following is the mathematical equivalent of the following FOR-TRAN expression?

SIN(2.0PI)**ABS(Y−1.0)

(A) $(\sin 2\pi)^{y-1}$ (B) $(\sin 2\pi)^{|y-1|}$ (C) $\sin 2\pi * * |y-1|$

(D) $\sin(2\pi)^{|y-1|}$ (E) none of the above

The exponent in (A) should have absolute value signs, the (**) character in (C) is meaningless in math, and (sin 2π) should be enclosed in parentheses in (D).

Answer is (B)

COMPUTER SCIENCE–23

What is the output of the following statement? ($\sqcup$ denotes a blank space.)

```
       F=50.0*4.6
       WRITE(6,100)F
  100  FORMAT(' ',T6,E9.3)
```

(A) 0.230E$\sqcup$03
(B) $\sqcup$0.230E$\sqcup$03
(C) $\sqcup\sqcup\sqcup\sqcup\sqcup\sqcup$2.30E$\sqcup$02
(D) $\sqcup\sqcup\sqcup\sqcup\sqcup$2.30$\sqcup$E$\sqcup$02
(E) $\sqcup\sqcup\sqcup\sqcup\sqcup$0.230E$\sqcup$03

The T format in line 100 specifies the column location where the output starts. Thus, there are 5 blank spaces before the output. E9.3 means there are to be nine character positions reserved for the output, with three digits printed to the right of the decimal point in the scientific notation.

Answer is (E)

COMPUTER SCIENCE–24

Which of these format specifications reads the input -567610 as -5676.10?

(A) F6.2   (B) F7.2   (C) F7.3
         (D) F8.2   (E) F8.3

There are a total of eight character positions used, including the negative sign and the decimal point. Since there are two digits to the right of the decimal, the correct format specification is F8.2.

Answer is (D)

COMPUTER SCIENCE-25

Determine the output of the following program segment:

```
10   A=25.357
20   B=0.25
30   L=15
40   WRITE(6,50)A,B,C
50   FORMAT(T2,'A=',F5.2,1X,'B=',F4.2,1X,'L=',I2)
```

(In the choices, ⊔ denotes a blank space.)

(A) A=25.357⊔B=0.25⊔L=15
(B) ⊔A=25.357⊔B=0.25⊔L=15
(C) ⊔A=25.36⊔B=0.25⊔L=15
(D) ⊔A=25.36⊔B=025⊔L=15
(E) ⊔A=253.57⊔B=0.25⊔L=15

> There should be one blank space before the output begins, thus (A)
> is wrong. The format for A specifies that there are two digits to the
> right of the decimal so A becomes 25.36. (D) is missing the decimal
> point for B.

> Answer is (C)

COMPUTER SCIENCE-26

Which of the following is the corresponding FORMAT statement for this output?
(⊔ denotes a blank space.)

⊔⊔⊔⊔⊔SUM=⊔2230.

(A) 10 FORMAT(5X,'SUM=',F5.1)

(B) 10 FORMAT(5X,'SUM=',1X,F5.0)

(C) 10 FORMAT(5X,'SUM=',F6.1)

(D) 10 FORMAT(5X,'SUM=',1X,F5)

(E) 10 FORMAT(5X,SUM=,1X,F5.0)

> For the output, there are five blanks followed by text, then one blank,
> and finally a five-character number with no digits to the right of the
> decimal point. (B) is the correct choice.

> Answer is (B)

COMPUTER SCIENCE–27

What should be the READ and FORMAT statements for this data output?

315.66

(A)   READ(5,20)I
    20 FORMAT(I6)

(B)   READ(5,20)S
    20 FORMAT(F6.2)

(C)   READ(5,20)S
        FORMAT(F6.2)

(D)   READ(5,30)S
    20 FORMAT(F6.2)

(E)   READ(5,20)S
    20 FORMAT(I6)

> Since the output is a real number, neither an integer variable I nor an integer field I6 can be used, thus (A) and (E) are wrong. (C) is missing its FORMAT statement number, and the FORMAT statement number in (D) does not match the READ statement number.

Answer is (B)

COMPUTER SCIENCE–28

Which of the following is the corresponding output of this WRITE statement? (⊔ denotes a blank space.)

    KX=130
    SUM=125.3869
    WRITE(6,3)KX,SUM
  3 FORMAT(I6,2X,F8.2)

(A) 130⊔⊔⊔⊔⊔⊔⊔125.39
(B) ⊔⊔130⊔125.39⊔⊔⊔⊔
(C) ⊔⊔⊔130⊔⊔⊔⊔125.39
(D) ⊔⊔⊔130125.39⊔⊔⊔⊔
(E) ⊔⊔⊔13012539⊔⊔⊔⊔⊔

The integer 130 will be written first, with six character positions allotted to it. There will be three blank spaces preceding the numerals, since numbers are always flush with the right side of the specified column block. After 130, two blanks are output due to the 2X in the FORMAT statement. Finally, the real number SUM is written, with two digits to the right of the decimal and eight character positions overall. The number will be flush right in its block.

Answer is (C)

COMPUTER SCIENCE-29

Which of the following are correct WRITE and FORMAT statements for the output A=$_{\sqcup}$0.023E$_{\sqcup}$05? ($_{\sqcup}$ denotes a blank space.)

(A)   WRITE(6,2)A
    2   FORMAT(T2,'A=',1X,I9)

(B)   WRITE(6,2)A
        FORMAT(T2,'A=',T4,E9.3)

(C)   WRITE(6,2)A
    2   FORMAT(T2,'A=',1X,F9.3)

(D)   WRITE(6,2)A
    2   FORMAT(T2,'A=',1X,E9.3)

(E)   WRITE(6,2)A
    2   FORMAT(T2,'A=',T3,E9.3)

For an output in scientific notation, the FORMAT statement cannot use an I or F-mask as in (A) and (C). There is no FORMAT statement number in (B), and the FORMAT statement in (E) will not print out a blank space after the equal sign.

Answer is (D)

COMPUTER SCIENCE–30

Which of the following is the output of this WRITE statement?

        WRITE(6,10)
    10   FORMAT(1X,'THE NUMBER=',2X,'33.15')

($\sqcup$ denotes a blank space.)

(A) $\sqcup$THE$\sqcup$NUMBER$\sqcup$=$\sqcup\sqcup$33.15
(B) $\sqcup$THE$\sqcup$NUMBER=$\sqcup\sqcup$33.15
(C) $\sqcup$THE$\sqcup$NUMBER=$\sqcup\sqcup\sqcup$33.15
(D) THE$\sqcup$NUMBER=$\sqcup\sqcup\sqcup\sqcup$33.15
(E) $\sqcup$THE$\sqcup$NUMBER=$\sqcup\sqcup$331.5

There is a blank space before "THE NUMBER=," then two more blanks, and then 33.15. There is no space between the letter R and the equal sign.

Answer is (B)

COMPUTER SCIENCE–31

A data input is 22.2$\sqcup$55184.5$\sqcup$750. ($\sqcup$ denotes a blank space.) What numbers will the computer read using the following FORMAT statement?

        READ(5,8) X1, X2, X3, X4
    8   FORMAT(4F4.2)

(A) 22.2, 5.51, 8.45, 7.50
(B) 2.22, 5.51, 84.5, 7.50
(C) 22.2, 55.10, 84.5, 7.50
(D) 22.2, 5.51, 84.5, 75.0
(E) 22.2, 5.51, 84.5, 7.50

The FORMAT statement specifies that the input is to be read four character spaces at a time from left to right, four times successively. The number read in will consist of four characters, with a maximum of two decimal places. If a decimal point was one of the initial four characters, it will be kept. Thus, the input is read as: 22.2, $\sqcup$551, 84.5, and $\sqcup$750, which become evaluated as 22.2, 5.51, 84.5, and 7.50.

Answer is (E)

COMPUTER SCIENCE-32

Which of the following is the correct WRITE statement to print the even numbers from 2 to 40, assuming conventional output device reference numbers?

(A)   WRITE(6,8) (I=1,40,2)
   8   FORMAT(20I2)

(B)   WRITE(5,8) (I=2,40,2)
   8   FORMAT(20I2)

(C)   WRITE(6,8) (I=2,2,40)
   8   FORMAT(20I2)

(D)   WRITE(6,8) (I=2,40,2)
   8   FORMAT(20I2)

(E)   WRITE(6,8) (I=2,40,2)
   6   FORMAT(20I2)

The first number in the WRITE line should be 6, the conventional reference number for a printer; the WRITE line in (B) refers to a card reader instead of a printer. The second number is the FORMAT line number which is to be used; (E) does not have the correct number in this case. Both (A) and (C) have errors in their loop specifications: the first two numbers specify the beginning and end of the loop, and the third number specifies the increment within the loop. The correct loop specification should be (I=2,40,2) as in (D).

> Answer is (D)

COMPUTER SCIENCE-33

A data line is as follows: 21.459214.5307421557.82134524. Which of the following is an incorrect storage of this data using the following READ statement?

    READ(5,98)A,B,I,J,C,K
 98   FORMAT(2F7.4,2I2,F6.2,I3)

(A) A=21.4592   (B) B=14.5307   (C) J=15
      (D) C=57.82   (E) K=134

The FORMAT statement specifies that the data is to be read from left to right in two blocks of seven character spaces, two blocks of two character spaces, one block of six, and one last block of three spaces. Since each block corresponds to a variable, K=345 and not 134.

Answer is (E)

COMPUTER SCIENCE–34

Which of the following is a correct IF statement?

(A) IF.(I.GE.5) GO TO 7
(B) IF (I,GE,5) GO TO 7
(C) IF (I.GE.5)
(D) IF (I.GE.5) GO TO 7
(E) IF I.GE.7, GO TO 7

The logical IF statement has the form:

IF [le] [statement]

[le] is a logical expression and [statement] is any executable statement except DO or IF. The operator within the logical expression should be of the form .[op]., and there should be no periods or commas between the components of the statement. There are no commas as in (B).

Answer is (D)

COMPUTER SCIENCE–35

What value is assigned to Q in this program?

```
     R=18.0
     S=6.0
     T=2.0
     Q=R/S**T−T
     IF Q 10,20,30
10   Q=10
15   GO TO 40
20   Q=100
25   GO TO 40
30   Q=1000
40   END
```

(A) −1.5   (B) 7   (C) 10   (D) 100   (E) 1000

Before the numbered lines, Q gets the value:

$$Q = \frac{R}{S^T} - T$$
$$= \frac{18}{(6)^2} - 2 = 0.5 - 2$$
$$= -1.5$$

Since Q is less than zero, the IF statement passes control of the program to line 10, where Q is reassigned the value of 10. The program then skips to the END line.

Answer is (C)

COMPUTER SCIENCE–36

Which of the following DO statements counts J from 1 to 11 in increments of 2?

(A) DO 5 J=1,2,11
(B) DO 5, J=1,11,2
(C) DO 5 J=1.11.2
(D) DO 5 J,1,11,2
(E) DO 5 J=1,11,2

The general form of the DO statement is DO s $i = j, k, l$. s is a statement number, i is the integer loop variable, j is the initial value assigned to i, k is an inclusive upper bound on i, which must exceed j, and l is the increment for i. Thus, the correct form of the statement is DO 5 J $= 1, 11, 2$.

Answer is (E)

COMPUTER SCIENCE–37

What is wrong with this program?

```
      DO 20 I=1,5
      A(I,J)=5.8
      I=5
  20  CONTINUE
```

(A) A cannot be a real number.
(B) I cannot be an integer.
(C) There is an excess comma in the DO statement.
(D) The value of I cannot be changed in the DO loop.
(E) none of the above

The integer loop variable I is made a constant, creating an error in the DO loop.

Answer is (D)

COMPUTER SCIENCE–38

Which of these expressions for a one-dimensional array, A, is acceptable? Assume B, H, I, J, K, and L are greater than zero.

(A) A(I,K)   (B) A(H+J)   (C) A(2*L−1)
         (D) A(−3)   (E) A(B)

(C) is the only acceptable one-dimensional array. Since the array variable or element must be an integer variable or a positive integer, (B), (D), and (E) are wrong. (A) is an acceptable form for a two-dimensional array.

Answer is (C)

COMPUTER SCIENCE–39

Which is the correct DIMENSION statement for a 9×9 matrix, A, and a vector, B, with 9 elements?

(A) DIMENSION A(9.9),B(9)
(B) DIM A(9,9),B(9)
(C) DIMENSION A(9,9),B(9)
(D) DIMENSION, A(9,9),B(9)
(E) DIMENSION. A(9.9),B(9)

The correct choice is (C). The DIMENSION statement does not use periods, thus (A) and (E) are wrong. (B) does not have DIMENSION written completely, and (D) has an extra comma.

Answer is (C)

COMPUTER SCIENCE–40

A 2×2 matrix, Z, is loaded from a DATA statement in the following order: 15.0, 10.0, 7.0, 20.0. List the matrix elements of Z in order of descending magnitude.

(A) $Z(1,1), Z(1,2), Z(2,1), Z(2,2)$

(B) $Z(2,2), Z(1,1), Z(1,2), Z(2,1)$

(C) $Z(1,1), Z(2,1), Z(1,2), Z(2,2)$

(D) $Z(2,2), Z(1,1), Z(2,1), Z(1,2)$

(E) $Z(2,2), Z(1,2), Z(2,1), Z(1,1)$

The matrix elements are assigned in the order $Z(1,1), Z(2,1), Z(1,2)$, and $Z(2,2)$. These elements get the respective values of 15.0, 10.0, 7.0, and 20.0. Thus, in order of descending magnitude, the matrix elements are: $Z(2,2), Z(1,1), Z(2,1), Z(1,2)$.

Answer is (D)

COMPUTER SCIENCE–41

Find the values of X and Y at the end of this logic statement:

$$X=2$$
$$Y=4$$

$$Z=X$$
$$X=Y$$
$$Y=Z$$

(A) X=2, Y=4
(B) X=4, Y=4
(C) X=4, Y=2
(D) X=2, Y=2
(E) X=3, Y=3

The logic sequence is:

$$Z=X=2$$
$$X=Y=4$$
$$Y=Z=2$$

The final X and Y values are 4 and 2, respectively.

Answer is (C)

COMPUTER SCIENCE–42

What is the value of A after this program segment is performed?

```
    A=0
    DO 77 I=1,4
77  A=A+I
```

(A) 4   (B) 5   (C) 8   (D) 10   (E) 12

A table of the value of A after each loop is:

A	I	A=A+I
0	1	1
1	2	3
3	3	6
6	4	10

The final value of A is 10.

Answer is (D)

COMPUTER SCIENCE–43

What is the final value of SUM in this program?

```
    SUM=0
    DO 77 I=1,3,2
    SUM=SUM+1.0/I
77  CONTINUE
```

(A) 1   (B) $\frac{4}{3}$ (C) $\frac{5}{3}$ (D) 2   (E) $\frac{7}{3}$

Since I goes from 1 to 3 in an increment of 2, SUM goes through two iterations. At the end of the first iteration,

$$SUM = 0 + \frac{1}{1} = 1$$

After the second iteration,

$$SUM = 1 + \frac{1}{3} = \frac{4}{3}$$

Answer is (B)

COMPUTER SCIENCE–44

Find the value of J in the following program:

```
     I=1
     J=0
15   J=J**2+I**2
     IF (I.EQ.3) GO TO 16
     I=I+1
     GO TO 15
16   CONTINUE
```

(A) 9   (B) 15   (C) 34   (D) 40   (E) 42

The value of J after each loop in the program is:

I	$J=J^2+I^2$
1	1
2	5
3	34

Answer is (C)

COMPUTER SCIENCE–45

Which of the following sets of statements gives a value for ISUM different than the others?

(A)
```
        ISUM=0
        DO 10 I=1,5
        ISUM=ISUM+I*2-1
  10    CONTINUE
```

(B)
```
        ISUM=0
        ISIG=1
        DO 20 I=1,9
        IF (ISIG.EQ.-1) GO TO 10
        ISUM=ISUM+I
  10    ISIG=-ISIG
  20    CONTINUE
```

(C)
```
        ISUM=0
        DO 10 I=2,10,2
        ISUM=ISUM+I
  10    CONTINUE
        ISUM=ISUM-10
```

(D)
```
        ISUM=0
        DO 10 I=1,9,2
        ISUM=ISUM+I
  10    CONTINUE
```

(E)
```
        ISUM=0
        DO 10 I=1,5
        ISUM=ISUM+5
  10    CONTINUE
```

Choice (C) gives ISUM=20. All other choices give ISUM=25.

Answer is (C)

COMPUTER SCIENCE–46

What is the object of the following program?

```
      DIMENSION KK(100)
 10   KK(I)=I
      DO 60 J=2,7
      DO 30 K=J,50
      N=J*K
 30   KK(N)=0
 60   CONTINUE
      DO 80 I=1,100
      IF (KK(I).EQ.0) GO TO 80
      WRITE(5,101) KK(I)
101   FORMAT(3X,I5)
 80   CONTINUE
```

(A) to find a sum of the even numbers between 1 and 100
(B) to find a sum of the odd numbers between 1 and 100
(C) to find the prime numbers between 1 and 100
(D) to sort the numbers of KK(100) in descending order
(E) to sort the numbers of KK(100) in ascending order

This program finds the prime numbers between 1 and 100. It sets all nonprime values of the array KK equal to zero, and then writes out what remains.

Answer is (C)

COMPUTER SCIENCE–47

Given the following statements, what is the value of R in the MAIN program?

EXTERNAL EXP   **main program**
A=0.0
R=FUNCT(A,EXP)+20.0

FUNCTION FUNCT(X,FX)   **function definition**
FUNCT=FX(X)
RETURN
END

(A) 0.0   (B) 1.0   (C) 20.0   (D) 21.0   (E) 101.0

The main program reads a value of 0 into the function FUNCT. This function is the natural exponential function e^x. Therefore, the value $e^0 = 1$ is returned to the main program, where it is summed with the value 20.0. The final value of R is 21.0.

Answer is (D)

COMPUTER SCIENCE–48

Of the following, which is the only acceptable FORTRAN statement?

(A) DIMENSION Z(TEN)
(B) WRITE(6,101) (MATRIX(J,J=1,8)
(C) READ(5,99) (TIME(X),X=1,10)
(D) WRITE(6,100) XDAT YDAT
(E) READ(5,99) (TOTAL(N),N=1,20)

The only correct statement is (E). The others are not allowed in FORTRAN.

Answer is (E)

COMPUTER SCIENCE–49

If A and B are false, and C is true, which of the following logical expressions will be true?

(A) .NOT.C.AND.A
(B) B.AND.C.OR.A
(C) .NOT.(A.AND.B)
(D) .NOT.C
(E) .NOT.A.OR..NOT.C

Only (C) is true. Since (A.AND.B) is false, NOT(A.AND.B) is true.

Answer is (C)

COMPUTER SCIENCE–50

What will be the final value for ISUM in the following statement?

```
    ISUM=0
    DO 10 I=1,20,3
    ISUM=ISUM+I
10  CONTINUE
```

(A) 3   (B) 30   (C) 50   (D) 70   (E) 150

ISUM will be the sum of the sequence of numbers generated by the DO loop:

$$ISUM = 1 + 4 + 7 + 10 + 13 + 16 + 19 = 70$$

Answer is (D)

COMPUTER SCIENCE–51

What are the values assigned to A and B in this subroutine?

COMMON ALPHA,BETA   **main program**
ALPHA=4.0
BETA=3.0

COMMON B,A   **subroutine**

(A) A=3.0, B=3.0
(B) A=4.0, B=3.0
(C) A=3.0, B=4.0
(D) A=4.0, B=4.0
(E) none of the above

It is the order of the common variables which determines their value. In the subroutine, B corresponds to ALPHA=4.0 in the main program while A corresponds to BETA=3.0.

Answer is (C)

COMPUTER SCIENCE–52

Which of the following is/are true statements regarding user-defined functions?

I.   The function is more versatile than a subroutine as it is not limited to mathematical calculations.
II.   The function is defined as a variable in the main program.
III. The function may be used repeatedly and anywhere in the program.

(A) I only   (B) I and II   (C) I and III

              (D) II and III   (E) none of the above

I is false, since the subroutine is more versatile. II and III are true.

Answer is (D)

COMPUTER SCIENCE–53

Which of the following would have to be added to the main program as a result of the variable XINCH being changed to INCH? Before the change, the function is:

```
FUNCTION FOOT(XINCH)
FOOT=XINCH/12.0
RETURN
END
```

(A) a COMMON statement
(B) a CALL statement
(C) a DIMENSION specification
(D) a REAL variable declaration
(E) none of the above

Without a REAL variable declaration, the variable INCH would be read by the computer as an integer variable, and FOOT would have an error.

Answer is (D)

COMPUTER SCIENCE–54

What will be the output of the following program? (ᵤ denotes a space in the output.)

```
        REAL K
        DATA K,X/1.3,14.5
        READ(5,100) K,X
100     FORMAT(F5.1,3X,F5.1)
        F=K*X
        WRITE(6,200)K,X,F
200     FORMAT(1X,'K=',E9.2,3X,'X=',E10.3,3X,'F=',E9.2)
        STOP
999     END
```

(A) ᵤK=ᵤ0.1Eᵤ01ᵤᵤᵤX=ᵤ0.1Eᵤ02ᵤᵤᵤF=ᵤ0.2E02

(B) K=10ᵤᵤᵤX=100ᵤᵤᵤF=20

(C) ᵤK=ᵤ0.13Eᵤ01ᵤᵤᵤX=ᵤ0.145Eᵤ02ᵤᵤᵤF=ᵤ0.19Eᵤ02

(D) ᵤK=0.13E01ᵤᵤᵤX=0.145E02ᵤᵤᵤF=0.19E02

(E) none of the above

The key line in the program is line 200. Line 100 is a FORMAT line for the data read into the program, but line 200 is the FORMAT line for the output.

> Answer is (C)

COMPUTER SCIENCE–55

Of the following statements, which is false?

(A) The FORMAT statement may or may not be executed by the program.

(B) More than one STOP statement is permitted in a program.

(C) The arithmetic IF can only be used to evaluate an arithmetic expression.

(D) No FORMAT statement is required with unformatted READ or WRITE statements.

(E) When using nested implied DO's, the inner implied DO is always completed first.

The FORMAT statement must be executed by the program. Therefore, (A) is false.

Answer is (A)

COMPUTER SCIENCE-56

What is the output of the program below? ($\sqcup$ denotes a space, and $\sqsubset$ designates a skipped line.)

```
      DO 30 I=1,3
      WRITE(6,10) X(I)
 10   FORMAT(1X,E8.2)
      WRITE(6,20)
 20   FORMAT(2X,/)
 30   CONTINUE
```

(A) $\sqcup$0.41E$\sqcup$04$\sqcup\sqcup\sqcup$0.68E$\sqcup$04$\sqcup\sqcup\sqcup$0.58E$\sqcup$04

(B) $\sqcup$0.41E$\sqcup$04
     $\sqcup$0.68E$\sqcup$04
     $\sqcup$0.58E$\sqcup$04

(C) $\sqcup$0.41E$\sqcup$04
     $\sqsubset$
     $\sqcup$0.68E$\sqcup$04
     $\sqsubset$
     $\sqcup$0.58E$\sqcup$04

(D) $\sqcup$0.41E$\sqcup$04
     $\sqsubset$
     $\sqsubset$
     $\sqcup$0.68E$\sqcup$04
     $\sqsubset$
     $\sqsubset$
     $\sqcup$0.58E$\sqcup$04

(E) none of the above

The "/" symbol designates a skipped line. The 2X that precedes the line skip symbol in line 20 results in two skipped lines between output lines. (D) is the correct choice.

Answer is (D)

17 ATOMIC THEORY

ATOMIC THEORY–1

According to the Bohr model of the hydrogen atom, which of the following statements are true?

I. As the electron orbits the proton, it constantly radiates light with a frequency equal to its frequency of revolution.

II. The electron orbits the proton in certain orbits that can be found by assuming that its angular momentum is quantized.

III. Because of the quantization of angular momentum, calculations using the Bohr model and those based on classical physics can never give the same results.

IV. When an electron orbiting a proton changes states to a lower energy level, the frequency of the radiation given off is proportional to the change in energy. This accounts for the hydrogen spectrum.

(A) I and II   (B) II and IV   (C) I, II, and III

        (D) II, III, and IV   (E) III and IV

The Bohr postulates are:

1. The electron moves in a certain orbit without radiating.

2. The frequency of the emitted photon is proportional to the change in energy of the electron.

3. The correspondence principle states that, in the limit as energies and orbits become large, quantum calculations must agree with classical calculations.

Thus, only II and IV are true.

Answer is (B)

ATOMIC THEORY–2

What is the deBroglie wavelength of an electron $(m = 9.1 \times 10^{-31}$ kg) travelling at a speed of 4×10^6 m/s?

(A) 0.182 nm   (B) 0.697 nm   (C) 1.320 nm

(D) 1.744 nm   (E) 2.106 nm

The deBroglie wavelength (λ_B) of a particle is given as follows:

$$\lambda_B = \frac{h}{mv}$$
$$= \frac{6.626 \times 10^{-34} \text{ J} \cdot \text{s}}{(9.1 \times 10^{-31})(4 \times 10^6)}$$
$$= 1.82 \times 10^{-10} \text{ m}$$
$$= 0.182 \text{ nm}$$

Answer is (A)

ATOMIC THEORY-3

An electron orbiting a proton at $r = 4.76\,\text{Å}$ drops into the first Bohr radius $(r_0 = 0.529\,\text{Å})$, and emits a photon. What is the wavelength of the photon emitted?

(A) $2.74 \times 10^{-6}\,\text{Å}$ (B) $3.48 \times 10^{-3}\,\text{Å}$ (C) $2.38\,\text{Å}$
         (D) $1030\,\text{Å}$ (E) $3990\,\text{Å}$

The frequency of the light emitted is:

$$f = \frac{E_i - E_f}{h}$$

where $E_i = $ initial energy

$E_f = $ final energy

$h = $ Planck's constant

$\quad\quad = 4.136 \times 10^{-15}\ \text{eV}\cdot\text{s}$

$$\frac{1}{\lambda} = \frac{f}{c}$$

$$= \frac{E_i - E_f}{hc}$$

$$\lambda = \frac{hc}{E_i - E_f}$$

The energy at a particular radius is:

$$E_n = -z^2\frac{E_0}{n^2}$$

where $E_0 = $ energy at first Bohr radius

$z = $ number of protons $= 1$

$n = $ quantum number corresponding to the radius

According to the Bohr model of the atom, the following holds true:

$$n^2 = \frac{r}{r_0}$$

$$= \frac{4.76}{0.529}$$

$$= 9$$

$$n = 3$$

$$E_i = E_3$$

$$= -1\left(\frac{13.6\,\text{eV}}{9}\right)$$

$$= -1.511\,\text{eV}$$

$$E_f = -E_0$$

$$= -13.6\,\text{eV}$$

$$\lambda = \frac{(4.136 \times 10^{-15})(3.0 \times 10^8)}{-1.511 - 13.6}$$

$$= 1.03 \times 10^{-7}\text{m}$$

$$= 1030\,\text{Å}$$

Answer is (D)

ATOMIC THEORY–4

Consider the binding energy of an electron in the K-shell (E_K) and L-shell (E_L) of the same metal. Which of the following statements is true?

(A) $E_K = \dfrac{E_L}{4}$ (B) $E_K = \dfrac{E_L}{2}$ (C) $E_K = E_L$

(D) $E_K = 2E_L$ (E) $E_K = 4E_L$

The binding energy for an electron in a shell is defined as the energy required to remove the electron from that shell.

$$E = -\frac{m}{2\hbar}\left(\frac{e^2}{4\pi\epsilon_0}\right)\frac{1}{n^2}$$

$$\propto \frac{1}{n^2}$$

$$n_K = 1 \text{ for the } K\text{-shell}$$

$$n_L = 2 \text{ for the } L\text{-shell}$$

$$\frac{E_K}{E_L} = \frac{n_L^2}{n_K^2}$$

$$E_K = 4E_L$$

Answer is (E)

ATOMIC THEORY–5

Determine the wavelength of light emitted when an electron jumps from the fourth Bohr orbital to the second Bohr orbital if the orbital energies are as follows: $E_2 = -5.43 \times 10^{-19}$ J, $E_4 = -1.36 \times 10^{-19}$ J, and $h = 6.626 \times 10^{-34}$ J·s.

(A) 212 nm   (B) 396 nm   (C) 488 nm

        (D) 613 nm   (E) 754 nm

$$\Delta E = E_4 - E_2$$
$$= hf$$
$$= \frac{hc}{\lambda}$$
$$\lambda = \frac{hc}{E_4 - E_2}$$
$$= \frac{(6.626 \times 10^{-34})(3 \times 10^8)}{(-1.36 \times 10^{-19}) - (5.43 \times 10^{-19})}$$
$$= 4.88 \times 10^{-7} \text{m}$$
$$= 488 \, \text{nm}$$

Answer is (C)

ATOMIC THEORY–6

The Balmer series is a series of energy transitions to the energy level corresponding to the quantum number $n = 2$. Which of the following are the first three wavelengths of light in the series?

I.	4300 Å	IV.	5200 Å
II.	4600 Å	V.	6600 Å
III.	4900 Å	VI.	7400 Å

(A) I, III, and V   (B) II, III, and V   (C) II, V, and VI

        (D) II, III, and IV   (E) III, V, and VI

In the Balmer series, $n_f = 2$, $n_i = 3, 4, 5 \ldots$

$$E_n = -\frac{13.6}{n^2}\,\text{eV}$$

$$\frac{1}{\lambda} = \frac{f}{c}$$

$$= \frac{E_i - E_f}{hc}$$

$$E_i - E_f = -13.6\left(\frac{1}{n_i^2} - \frac{1}{n_f^2}\right)$$

$$\frac{1}{\lambda} = \frac{13.6\left(\frac{1}{n_i^2} - \frac{1}{n_f^2}\right)}{hc}$$

For the Balmer series,

$$\frac{1}{\lambda} = -\frac{(13.6)\left(\frac{1}{n^2} - \frac{1}{2^2}\right)}{hc}$$

The three lowest transitions are for $n = 3, 4$, and 5. For $n_i = 3$,

$$\frac{1}{\lambda} = -\frac{(13.6)\left(\frac{1}{3^2} - \frac{1}{2^2}\right)}{(4.13 \times 10^{-15})(3 \times 10^8)}$$

$$= 1.52 \times 10^6\,\text{m}^{-1}$$

$$\lambda = 6600\,\text{Å}$$

For $n_i = 4$,

$$\frac{1}{\lambda} = -\frac{(13.6)\left(\frac{1}{4^2} - \frac{1}{2^2}\right)}{(4.13 \times 10^{-15})(3 \times 10^8)}$$

$$= 2.06 \times 10^6\,\text{m}^{-1}$$

$$\lambda = 4900\,\text{Å}$$

For $n_i = 5$,

$$\frac{1}{\lambda} = -\frac{(13.6)\left(\frac{1}{5^2} - \frac{1}{2^2}\right)}{(4.13 \times 10^{-15})(3 \times 10^8)}$$

$$= 2.31 \times 10^6\,\text{m}^{-1}$$

$$\lambda = 4300\,\text{Å}$$

Thus I, III, and V give the correct wavelengths.

Answer is (A)

ATOMIC THEORY-7

What is the frequency of the photon emitted when neon undergoes a transition between the 20.66 eV level and the 18.70 eV level?

(A) 4.74×10^{14} Hz (B) 5.00×10^{14} Hz (C) 8.10×10^{14} Hz
(D) 3.00×10^{15} Hz (E) 3.75×10^{15} Hz

$$\Delta E = hf$$
$$= 20.66 - 18.70$$
$$hf = 1.96 \, \text{eV}$$

$$f = \frac{\left(1.96 \, \text{eV}\right)\left(1.66 \times 10^{-19} \, \dfrac{\text{J}}{\text{eV}}\right)}{6.626 \times 10{-34} \, \text{J} \cdot \text{s}}$$

$$= 4.74 \times 10^{14} \, \text{Hz}$$

Answer is (A)

ATOMIC THEORY-8

In a three-level laser, electrons in atoms are excited into an energy state E_3, then decay spontaneously to an energy E_2, which is a metastable state. The atoms are struck by photons of a specific frequency, and make stimulated emissions to the ground state, E_1. If the photons cause all of the emissions to be of the same frequency, the light will be amplified. What frequency must the photons be for this to occur?

(A) $f = \dfrac{E_2 - E_1}{h}$

(B) $f = \dfrac{E_3 - E_1}{h}$

(C) $f = \dfrac{E_3 - E_1}{h} + \dfrac{E_2 - E_1}{h}$

(D) $f = \dfrac{E_3 - E_1}{h} - \dfrac{E_2 - E_1}{h}$

(E) $f = \dfrac{E_3 E_1 - E_2 E_1}{h}$

The transition that must be amplified is the E_2 to E_1 transition. Thus, the frequency of the radiated light is $(E_2 - E_1)/h$. Photons of this frequency will be more likely to cause this transition. Therefore, if photons of this frequency are used, more transitions will take place, and the light will be amplified.

Answer is (A)

ATOMIC THEORY-9

A photoelectric surface has a work function of 8.0×10^{-19} joules. What is the maximum velocity of the photoelectrons emitted by light of frequency 4×10^{15} Hertz? (Planck's constant $= 6.626 \times 10^{-34}$ J $\cdot$ s, and the mass of the electron $= 9.11 \times 10^{-31}$ kg)

(A) 3.1×10^4 m/s

(B) 6.5×10^5 m/s

(C) 2.0×10^6 m/s

(D) 4.2×10^7 m/s

(E) 1.8×10^8 m/s

The energy of the photon is equal to the work function of the material plus the maximum kinetic energy of the photoelectron that is emitted.

$$hf = \phi + \mathrm{KE}$$

where hf = energy of the photon

ϕ = work function

E_k = energy of emitted electron

$$= \frac{1}{2}mv^2$$

$$\frac{1}{2}mv^2 = hf - \phi$$

$$v = \left[\frac{2(hf - \phi)}{m}\right]^{1/2}$$

$$= \left\{\frac{2[(6.626 \times 10^{-34})(4 \times 10^{15}) - 8.0 \times 10^{-19}]}{9.11 \times 10^{-31}}\right\}^{1/2}$$

$$= 2.02 \times 10^6 \text{ m/s}$$

Answer is (C)

ATOMIC THEORY-10

Assume the threshold wavelength for a material is 6200 Å. What is the maximum energy of the photoelectrons emitted by light of wavelength 4000 Å?

(A) 1.1 eV   (B) 1.3 eV   (C) 1.5 eV

          (D) 2.0 eV   (E) 3.1 eV

The maximum energy of the photoelectrons is given as follows:

$$E_{max} = E - \phi$$

where $\quad E =$ energy of the photons

$$= \frac{hc}{\lambda}$$

$$= \frac{(4.136 \times 10^{-15})(3 \times 10^8)}{4.0 \times 10^{-7}}$$

$$= 3.1\,eV$$

$\phi =$ work function of the material

$$= \frac{hc}{\lambda}$$

$$= \frac{(4.136 \times 10^{-15})(3 \times 10^8)}{6.2 \times 10^{-7}}$$

$$= 2\,eV$$

$$E_{max} = 3.1 - 2$$

$$= 1.1\,eV$$

Answer is (A)

ATOMIC THEORY–11

Which of the following statements concerning quantum theory and the shell model of the atom are false?

I.   The energy of an atom is dependent upon all four of the quantum numbers $(n, l, m,$ and $m_s)$ associated with it, except for hydrogen, which only depends on n.

II.   According to the Pauli exclusion principle, two electrons in an atom may have the same values for quantum numbers n, l, and m only.

III. The three quantum numbers n, l, and m arise from boundary conditions in the solution of Schrodinger's equation.

IV.   There is a fourth quantum number associated with the spin of the electron, m_s, which can only have values $+\frac{1}{2}$ and $-\frac{1}{2}$.

V.   If an atom in the ground state has its last electron in the M-shell $(n = 3)$, there must be at least ten electrons in that atom.

(A) I

(B) IV

(C) II and III

(D) II and V

(E) IV and V

The Pauli exclusion principle states that no electrons may have the same values for the four quantum number n, l, m, and m_s. Therefore, II is false.

An atom with its last electron in the M-shell must have at least 11 electrons. Since it is in the ground state, all of the previous shells must be full. There are two electrons in the K-shell, eight in the L-shell, and at least one in the M-shell. This makes a total of at least 11 electrons. Therefore, V is also false.

Answer is (D)

ATOMIC THEORY–12

In an atom such as sodium, there is one electron in the outermost shell (in this case, $n = 3$). Which of the following statements is true regarding the energy required to excite an electron in the $n = 1$ shell compared to that required to excite an electron in the $n = 2$ shell?

(A) It is greater because the electron is closer to the proton, and thus the Coulomb attractive force is much stronger.

(B) It is greater because the shell next to it is full. Thus, by the Pauli exclusion principle, it must jump to the first shell which is not full, in this case, the $n = 3$ shell.

(C) It is greater because an electron must first jump to the $n = 3$ shell from the $n = 2$ shell. Then the electron from the $n = 1$ shell can jump to the $n = 2$ shell.

(D) It is equal to the energy required to excite an electron in the $n = 2$ shell because in both cases the electron makes a jump to the next shell.

(E) It is less. Because of the crowding of electrons close to the nucleus, the Coulomb repulsion pushes the closer electrons out more easily.

In an atom, an excited electron will jump to the next highest unfilled shell (in this case, the $n = 3$ shell). So the electrons in both the $n = 1$ shell and the $n = 2$ shell will jump to the $n = 3$ shell. However, the energy difference between the $n = 1$ and the $n = 3$ shell is greater than the energy difference between the $n = 2$ shell and the $n = 3$ shell. Thus, (B) is correct.

Answer is (B)

ATOMIC THEORY–13

A particle is in a one-dimensional box of length $2L$. What is the energy level of the ground state?

(A) $\dfrac{h^2}{32mL^2}$ (B) $\dfrac{h^2}{16mL^2}$ (C) $\dfrac{h^2}{8mL^2}$

(D) $\dfrac{h^2}{4mL^2}$ (E) $\dfrac{h^2}{2mL^2}$

Use Schrodinger's equation:

$$-\frac{\hbar^2}{2m}\Psi'' = E\Psi$$

$$\Psi'' = -\frac{2mE}{\hbar^2}\Psi$$

$$k^2 = \frac{2mE}{\hbar^2}$$

$$\Psi'' = -k^2\Psi$$

The solution of the differential equation is:

$$\Psi = A\exp(ikx) + B\exp(-ikx)$$

Apply the boundary conditions that $\Psi = 0$ when $x = 0, 2L$.

$$0 = A + B$$
$$A = -B$$
$$0 = A\exp(2ikL) + B\exp(-2ikL)$$
$$0 = -B\exp(2ikL) + B\exp(-2ikL)$$
$$\exp(2ikL) = \exp(-2ikL)$$

This can only be satisfied when $k = \frac{n\pi}{2L}$ where $n = 1, 2, 3 \ldots$

$$E_n = \frac{\hbar^2 k^2}{2m}$$

$$= \left(\frac{\hbar^2}{2m}\right)\left(\frac{n^2\pi^2}{4L^2}\right)$$

$$= \frac{n^2 h^2}{32mL^2}$$

$$n = 1 \quad \text{for the ground state}$$

Therefore, $\quad E_1 = \dfrac{h^2}{32mL^2}$

$$\boxed{\text{Answer is (A)}}$$

ATOMIC THEORY-14

Electrons are in a one-dimensional box of length $2L$. What is the relationship between E_F, the Fermi energy, and $\overline{E}$, the average energy?

(A) $\overline{E} = \dfrac{E_F}{4}$ 　　　 (B) $\overline{E} = \dfrac{E_F}{3}$ 　　　 (C) $\overline{E} = \dfrac{E_F}{2}$

　　　 (D) $\overline{E} = E_F$ 　　　 (E) $\overline{E} = 2E_F$

For a particle in a one-dimensional box of length $2L$, the energy of the nth level is given by $E_n = n^2 h^2 / 32mL^2$. There are two electrons per level. The Fermi energy corresponds to $n = N/2$, where N is the total number of electrons. At this level, all levels where $n < N/2$ are occupied, and all levels with $n > N/2$ are unoccupied. Therefore,

$$E_F = \left(\frac{N}{2}\right)^2 \frac{h^2}{32mL^2}$$

The definition of average energy is the total energy divided by the total number of particles.

$$\text{Thus,} \quad \overline{E} = \frac{1}{N} \sum_{n=1}^{\frac{N}{2}} \frac{2n^2 h^2}{32mL^2}$$

$$\text{For } N \gg 1, \quad \sum_{n=1}^{\frac{N}{2}} n^2 \approx \int_0^{\frac{N}{2}} n^2 dn$$

$$= \frac{n^3}{3} \Big|_0^{\frac{N}{2}}$$

$$= \left(\frac{1}{3}\right)\left(\frac{N}{2}\right)^3$$

$$\overline{E} = \left(\frac{1}{3}\right)\left(\frac{N}{2}\right)^2 \frac{h^2}{32mL^2}$$

$$= \tfrac{1}{3} E_F$$

Answer is (B)

ATOMIC THEORY–15

Consider the allowed energy states for a particle in a three-dimensional cubic box. When $n^2 = 41$ in the energy eigenvalue E_n, how many degenerate levels are there?

(A) four   (B) six   (C) eight   (D) nine   (E) twelve

A three-dimensional box can be treated exactly as a two-dimensional box. Therefore, $E_n = n^2 h^2 / 32mL^2$, where n is a positive integer. However, there is one quantum number for each set of boundary conditions. Thus, there is one quantum number for each dimension of the box. This means that $n^2 = n_x^2 + n_y^2 + n_z^2 = 41$. The combinations of n_x, n_y, and n_z that will give a value of $n^2 = 41$ are as follows:

n_x	n_y	n_z
1	2	6
1	6	2
2	1	6
2	6	1
6	1	2
6	2	1
3	4	4
4	4	3
4	3	4

Since there are nine combinations of n_x, n_y, and n_z that give the same value of $n = 41$, there are nine degenerate levels.

Answer is (D)

ATOMIC THEORY-16

Consider an atom in the $l = 2$, $s = \frac{1}{2}$ state. If m_j are the eigenvalues of J_z, how many values of m_j are there?

(A) four   (B) five   (C) six   (D) eight   (E) ten

$$J_z = L_z + S_z$$

In terms of eigenvalues,

$$m_j = m_l + m_s$$
$$l = 2$$

Therefore, $m_l = +2, +1, 0, -1, -2$

$$s = \frac{1}{2}$$

Therefore, $m_s = \frac{1}{2}, -\frac{1}{2}$

$$m_j = 2.5, 1.5, 0.5, -0.5, -1.5, -2.5,$$

Thus, there are six possible values for m_j.

Answer is (C)

ATOMIC THEORY–17

n the normal Zeeman effect, how many possible transitions consistent with the election rules for levels $l = 2$ and $l = 1$ are there?

A) three   (B) six   (C) nine   (D) twelve   (E) fifteen

$$\text{for} \quad l = 2$$
$$m_2 = 2, 1, 0, -1, -2$$
$$\text{for} \quad l = 1$$
$$m_1 = 1, 0, -1$$

The selection rule for the normal Zeeman effect is $\Delta m_l = 0, \pm 1$. Between the $l = 2$ and the $l = 1$ levels, there are three transition for $\Delta m_l = 1$, three transitions for $\Delta m_l = 0$, and three transitions for $\Delta m_l = -1$. These transitions are shown on the following figure.

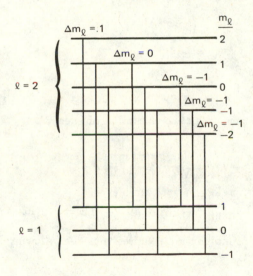

Thus, there are $3 + 3 + 3 = 9$ transitions.

Answer is (C)

ATOMIC THEORY–18

In the Compton effect, a photon scatters off a free electron as shown. What does the frequency ν' (after collision) depend on?

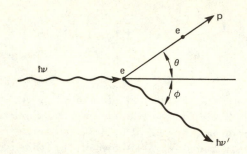

$$
\begin{aligned}
h\nu &= \text{incident photon energy} \\
\text{p} &= \text{electron momentum after collision} \\
h\nu' &= \text{photon energy after collision}
\end{aligned}
$$

(A) θ only
(B) ϕ only
(C) θ and ϕ
(D) neither θ nor ϕ
(E) the sum of θ and ϕ

First, use the principle of conservation of energy.

$$h\nu + m_0 c^2 = h\nu' + \sqrt{m_0^2 c^4 + \text{p}^2 c^2}$$

$$m_0 c^2 = \text{electron rest mass energy}$$
$$\text{p}^2 c^2 = (h\nu)^2 - 2(h\nu)(h\nu') + (h\nu')^2 + 2(h\nu - h\nu')m_0 c^2 \qquad (1)$$

Conservation of momentum requires that the following holds true.

$$\frac{h\nu}{c} = \text{p}\cos\theta + \frac{h\nu'}{c}\cos\phi \qquad (2)$$

$$0 = \text{p}\sin\theta - \frac{h\nu'}{c}\sin\phi \qquad (3)$$

$$\cos\theta = \frac{\sqrt{\text{p}^2 - \left(\dfrac{h\nu'}{c}\right)^2 \sin^2\phi}}{\text{p}} \qquad (4)$$

Substitute the value of $\cos \theta$ from equation 4 into equation 2.

$$h\nu - h\nu' \cos \phi = \sqrt{p^2 c^2 - (h\nu')^2 \sin^2 \phi}$$

$$(h\nu)^2 - 2(h\nu)(h\nu') \cos \phi + (h\nu')^2 = p^2 c^2 \qquad (5)$$

Substitute the value of $p^2 c^2$ from equation 1 into equation 5 and simplify.

$$2(h\nu)(h\nu')(1 - \cos \phi) - 2(h\nu - h\nu')m_0 c^2 = 0$$

$$\nu - \nu' = \left(\frac{h}{m_0 c^2}\right)(\nu\nu')(1 - \cos \phi)$$

$$\frac{1}{\nu'} - \frac{1}{\nu} = \frac{h}{m_0 c^2}(1 - \cos \phi)$$

Thus, ν' depends on ϕ only.

Answer is (B)

ATOMIC THEORY-19

A state of energy E_1 with a lifetime of τ_1 decays into the state of energy E_2. The state of E_2 then decays with a lifetime of τ_2 into the state of E_3. It is known that $\tau_1 = 2\tau_2$. Initially, all of the atoms (quantity N_0) are in the E_1 state. Calculate the number of atoms N_2 which are in the state E_2 at any instant in time, t.

(A) $N_0 \left[e^{(-t/\tau_1)} - e^{(-2t/\tau_1)}\right]$

(B) $N_0 \left[e^{(-t/\tau_1)} + e^{(-2t/\tau_1)}\right]$

(C) $2N_0 \left[e^{(-t/\tau_1)} - e^{(-2t/\tau_1)}\right]$

(D) $2N_0 \left[e^{(-t/\tau_1)} + e^{(-2t/\tau_1)}\right]$

(E) $N_0\, e^{(-3t/\tau_1)}$

The number of atoms that decay from E_1 to E_2 is:

$$N_1 = N_0\, e^{(-t/\tau_1)}$$

The number of atoms in state E_2 equals the number of atoms coming from E_1 to E_2 minus the number of atoms decaying from E_2 to E_3. Thus, the number of atoms in state E_2 at a given time is:

$$N_2 = N_0\, e^{(-t/\tau_1)} - N_{2,0}\, e^{(-t/\tau_2)}$$
$$= N_0\, e^{(-t/\tau_1)} - N_{2,0}\, e^{(-2t/\tau_1)}$$

Since there are no atoms in state E_2 at $t = 0$, the initial conditions are:

$$N_2(0) = 0$$
$$0 = N_0 e^0 - N_{2,0} e^0$$
$$N_0 = N_{2,0}$$
$$N_2 = N_0 \left[e^{(-t/\tau_1)} - e^{(-2t/\tau_1)} \right]$$

Answer is (A)

ATOMIC THEORY–20

A source of radiation has a mean nucleus life of $\tau = 35.8$ sec. There are initially $N_0 = 5.37 \times 10^{10}$ nuclei in the source. Which of the following statements are true?

I.   The decay constant is $\lambda = 0.0279\, \text{sec}^{-1}$.

II.   The half-life is $t_{\frac{1}{2}} = 24.8$ sec.

III.   If a rate counter with an 80% efficiency is placed near the source, it will show a rate of 4.22×10^7 after 2 minutes.

IV.   The sample will essentially have all decayed (0.01% left) in 5.5 minutes.

(A) I and III   (B) II and III   (C) I, III, and IV

(D) I, II, and IV   (E) all are true

First, find the decay constant:

$$\lambda = \frac{1}{\tau}$$
$$= \frac{1}{35.8}$$
$$= 0.0279 \, \text{sec}^{-1}$$

Therefore, statement I is true.

Next, find the half-life:

$$\frac{1}{2} = e^{(t_{\frac{1}{2}}/\tau)}$$
$$t_{\frac{1}{2}} = \tau \ln 2$$
$$= (35.8) \ln 2$$
$$= 24.8 \, \text{sec}$$

Thus, statement II is true.

Next, find the count rate after 2 minutes. The count rate at time t is:

$$R = R_0 e^{-\lambda t}$$
$$\text{where} \quad R_0 = \text{initial count rate}$$
$$= \lambda N_0$$
$$t = \text{time in seconds}$$

Since the detector is only 80% efficient, the rate shown by the detector after 2 minutes, R_d, is:

$$R_d = (0.80)R$$
$$= (0.80)(1.50 \times 10^9)e^{-(0.0279)(120)}$$
$$= 4.22 \times 10^7 \, \text{decays/sec}$$

Therefore, statement III is true.

The time that it takes for the sample to decay to an amount N is:

$$t = \frac{-1}{\lambda} \ln \frac{N}{N_0}$$
$$N = 1 \times 10^{-4} N_0$$
$$\frac{N}{N_0} = 1 \times 10^{-4}$$
$$t = \frac{-1}{0.0279} \ln 1 \times 10^{-4}$$
$$= 330 \,\text{sec}$$
$$= 5.5 \,\text{min}$$

Thus, IV is also true. Statements I, II, III, and IV are all true.

$$\boxed{\text{Answer is (E)}}$$

ATOMIC THEORY–21

A fossil fern containing 0.110 lbm of carbon is carbon dated to determine its age. The decay rate of C^{14} in the fossil is 191 decays/min. How old is the fern? (The half-life of C^{14} is 5730 years, and the rate of decay of C^{14} in a living organism per lbm of carbon is 6.80×10^3 decays/min-lbm C.)

(A) 7290 years   (B) 11,300 years   (C) 14,100 years

(D) 23,800 years   (E) 126,000 years

The rate of decay of C^{14} in a dead organism is given by:

$$R = R_0 e^{-\lambda t}$$
where λ = decay constant
R_0 = initial decay rate
t = time elapsed

R_0 is simply the decay rate of carbon−14 in a living organism, because up to the point it dies, it replenishes its carbon. Thus, until the organism's death, the decay rate is fairly constant.

$$R_0 = \left(6800 \, \frac{\text{decays}}{\text{min-lbm C}}\right) \left(0.110 \, \text{lbm C}\right)$$

$$= 748 \text{ decays/min}$$

$$\lambda = \frac{\ln 2}{t_{\frac{1}{2}}}$$

$$= \frac{\ln 2}{5730}$$

$$= 1.21 \times 10^{-4} \, \text{yr}^{-1}$$

$$t = -\frac{1}{\lambda} \ln \frac{R}{R_0}$$

$$= -\frac{1}{1.21 \times 10^{-4}} \ln \left(\frac{191}{798}\right)$$

$$= 11,300 \text{ years}$$

Answer is (B)

ATOMIC THEORY–22

For a fission reactor to remain in control, it must have a reproduction factor K very close to unity. If the average fission generation (i.e., the time for a neutron emitted in one fission to cause another fission) is 0.010 seconds, and the reproduction factor rises to $K = 1.0001$, how long will it be before the rate doubles?

(A) 1.8 sec   (B) 24.2 sec   (C) 45.9 sec
         (D) 69.3 sec   (E) 123.7 sec

The reaction rate rises as K^N where N is the number of fission generations. For the reaction rate to be doubled, $K^N = 2$. Solve for N.

$$N = \frac{\log 2}{\log K}$$

$$= \frac{\log 2}{\log 1.0001}$$

$$= 6930 \text{ generations}$$

Since each generation takes 0.01 seconds, the time for 6930 generations is:

$$t = (6930)(0.01)$$

$$= 69.3 \text{ sec}$$

Answer is (D)

ATOMIC THEORY–23

Which of the following statements about nuclear fusion is false?

(A) A typical fusion reaction releases less energy than a typical fission reaction.

(B) Because of the energy required to produce nuclear fusion, it cannot yet be used by man.

(C) A typical fusion reaction releases more energy per unit mass than a typical fission reaction.

(D) Fusion has advantages over fission because of a greater availability of fuel and a lack of some of the dangers of fission.

(E) The largest problems in using fusion as a power source arise in trying to achieve the huge temperatures and pressures needed for a sufficient time. It is not yet technically possible.

Nuclear fusion is used by man. Its current use is in thermonuclear weapons. Thus, (B) is false.

Answer is (B)

ATOMIC THEORY–24

A body has a rest mass of 1 slug. What is the body's mass when it is travelling with a speed $v = c/2$? (c = speed of light)

(A) 1 slug   (B) 1.15 slug   (C) 1.20 slug
            (D) 1.25 slug   (E) 1.30 slug

$$m = \frac{m_0}{\sqrt{1 - \left(\frac{v}{c}\right)^2}}$$

$$= \frac{1}{\sqrt{1 - \frac{c^2}{4c^2}}}$$

$$= \frac{1}{\sqrt{0.75}}$$

$$= 1.15 \text{ slug}$$

Answer is (B)

ATOMIC THEORY-25

A body has a rest mass of 10 slugs. What is the body's relativistic kinetic energy when it is travelling at a speed $v = 0.5c$? (c = speed of light)

(A) $1.55c^2$ (B) $1.80c^2$ (C) $1.90c^2$

           (D) $1.95c^2$ (E) $2.00c^2$

$$m = \frac{m_0}{\sqrt{1 - \left(\frac{v}{c}\right)^2}}$$

$$= \frac{10}{\sqrt{1 - \frac{0.25c^2}{c^2}}}$$

$$= \frac{10}{\sqrt{0.75}}$$

$$= 11.55 \text{ slugs}$$

$$E_k = \text{relativistic kinetic energy}$$

$$= c^2(m - m_0)$$

$$= c^2(11.55 - 10)$$

$$= 1.55c^2$$

Answer is (A)

ATOMIC THEORY-26

What is the total relativistic energy of a particle if its mass is equal to 1 slug when it is travelling with a certain speed, v?

(A) 1   (B) $c^2 - 1$ (C) c^2

           (D) $c^2 + 1$ (E) $5c^2$

Regardless of speed, the total relativistic energy is:

$$E_{\text{total}} = mc^2$$

$$= (1)c^2$$

$$= c^2$$

Answer is (C)

Send me more information!

Please send me descriptions and prices of all available books in the Engineering Review Series. I understand there will be no obligation on my part.

Name _____

Address _____

City _____

State _____ Zip _____

A friend of mine is also taking the exam. Send additional literature to:

Name _____

Address _____

City _____

State _____ Zip _____

I have a comment!

Your suggestions are highly valued, as they frequently lead to improvement in the contents of our publications. If you think you have discovered an error, or if you feel that some pertinent information is missing from this book, please let us know.

Title of book: _____

Edition: _____

_____

_____

_____

Contributed by (optional):

Name _____

Address _____

City _____

State _____ Zip _____